This study has been made with the cooperation of the
Secretariat of the Institute of Pacific Relations
and constitutes a report in its
International Research Series

PIONEER SETTLEMENT IN THE ASIATIC TROPICS

Studies in Land Utilization and Agricultural Colonization in Southeastern Asia

KARL J. PELZER
The Johns Hopkins University

GREENWOOD PRESS, PUBLISHERS
WESTPORT, CONNECTICUT

Library of Congress Cataloging in Publication Data

Pelzer, Karl J. (Karl Josef), 1909-
Pioneer settlement in the Asiatic tropics.

Reprint. Originally published: New York, N.Y. : American Geographical Society, 1945. (Special publication / American Geographical Society ; no. 29) (International research series / Institute of Pacific Relations)
Bibliography: p.
Includes index.
1. Land settlement--Asia, Southeastern--History--20th century. 2. Agricultural colonies--Asia, Southeastern--History--20th century. 3. Land tenure--Asia, Southeastern--History--20th century. I. Title. II. Series: Special publication (American Geographical Society of New York) ; no. 29. III. Series: International relations series.
HD880.8.P44 1983 333.76'13'0959 83-1484
ISBN 0-313-23853-7 (lib. bdg.)

Originally published in 1945 by the American Geographical Society

Reprinted in 1983 by Greenwood Press
A division of Congressional Information Service, Inc.
88 Post Road West, Westport, Connecticut 06881

Printed in the United States of America

10 9 8 7 6 5 4 3 2 1

To

Betty

Kristin and Anno

CONTENTS

LIST OF ILLUSTRATIONS

LIST OF TABLES

FOREWORD

Victory over Japan will establish a wider range of American interest in the Western Pacific and Southeast Asia. To deal with that interest we shall need a larger fund of knowledge than we have been able to draw on in the past; and the fund will have to be continuously replenished by workers in the field. The provision of factual knowledge alone will require a larger number of Americans than have ever before worked on the problems of tropical and subtropical areas and of colonial and dependent peoples.

We shall need, however, more than additional facts and additional investigators. The differences between the world before Pearl Harbor and the world after victory over Japan will demand of us a new approach to the problems of peoples and politics, based on a new integration of the technical study of soil, climate, land utilization, extraction of minerals, and creation of industries, and the social study of the structure and functioning of human communities—families, tribes, villages, peoples, and nations—and of cultural complexes which extend over or into more than one kind of community, and of the relations between different kinds of units.

A fresh integration is demanded of us because of the historical juncture. Of all the phases of history, the phase that immediately follows a great crisis is the one that demands the most subtle analysis of the way in which change works, and the most delicate handling of the process of change. In such a phase new problems appear that make an appeal to innovation, but at the same time old problems reappear that make an appeal to precedent. Most intricate of all is the old problem that reappears in a new guise, defying simple classification.

An analogy drawn from Dr. Pelzer's data may be used to illustrate the complex problem as a whole, though solution of the problem, or any part of it, will require more than analogy. Dr. Pelzer has shown what great effect the actions even of primitive men can sometimes have in changing the geographical environment. Doubtless the intention of primitive man is merely to *use* the environment, over and over again in the same way, rather than to *change* it. Nevertheless, in using it he may also change it. When the primitive shifting cultivator burns a tract of forest in order to grow a crop, it is because "things have always been done that way." As far as his intention reaches, the forest may recover that tract when he has abandoned it and moved on to burn a new tract.

But, as Dr. Pelzer shows, it is not always the forest that recovers the burned area. Grass may take over, delaying the appearance of second-growth forest and the regeneration of the original soil fertility. The shifting cultivators may be limited to a smaller range of territory by the spread of a sedentary people whose higher agricultural technique enables them to cultivate permanent fields. If so, the burned fields of the shifting cultivators lie fallow for a shorter period, because they have to be burned over more often. The change of environment is then accelerated; the character of wide landscapes is altered; some kinds of trees and other vegetation prove to have a greater survival value than others, or erosion and impoverishment of the soil may become very rapid.

In an analogous way we may study changes of the very greatest importance taking place in societies whose rulers direct their intentions toward maintaining continuity rather than toward initiating change. In the very maintenance of continuity changes are inadvertently encouraged which in time develop to such importance that they affect the old methods of administration—and administration may be called the "farming" of a society. Thus Dr. Pelzer describes the land policy of the Dutch in Indonesia as directed toward the maintenance of the old landholding systems, including communal village landownership; in particular, the "grain values" of the old agrarian society were to be protected against the "money values" of the Dutch themselves by an ordinance forbidding the purchase of agricultural land by any but Indonesians. Yet the pressure of money values could not be prevented from working; through such indirect channels as cash loans, it affected the lives of people who had the legal status of landowners in such a way that in practice they became tenants working for outsiders who had lent money to them. To continue the analogy, one may say that here and there in the "forest" of the pre-Dutch Indonesian society there began to appear gaps of "grassland" resulting from the Dutch methods of "cultivating" the society. Comparable changes may be studied in the Philippines; comparable, but different, partly because of differences between the pre-Dutch economies and societies of the Indies and the pre-Spanish and pre-American economies and societies of the Philippines, and partly because of differences in the Dutch impact, the Spanish impact, and the American impact.

Problems of this kind, problems of transformation, demand recognition when we come to study the alternative systems of trusteeship now being proposed for subject and dependent peoples. Change will be induced by any system that we impose. The question is whether change is to be planned and directed in each area and under each administrative authority, and coordinated as between areas and authorities, or whether it is to be allowed to overtake us as an unplanned consequence of undirected, uncoordinated policy.

The chief gap is between the American standard of planning for freedom and the colonial standard. Under the American standard freedom is planned as the outcome of trustee administration, standards of self-government are planned adequate to cope with the problems that confront a newly independent nation, and the problem as a whole is coordinated by setting in advance a date that synchronizes the *termination* of trusteeship and the *completion* of preparation for self-government. Under the colonial standard demands for freedom arise spontaneously and sporadically, and have to be deferred, fenced with, and temporized with, because they are inadvertent consequences of social changes that are not the aim of the colonial system but are inevitable by-products of it.

Dr. Pelzer has provided invaluable materials for the study of all these interacting problems. It is encouraging that his book should be the result of farsighted support for fact gathering in the field and directly in the tradition of the American interest in wanting to find out how things work. We need more of such studies, and we should look upon the need as a challenge.

Owen Lattimore

AUTHOR'S PREFACE

For a number of years I have been deeply interested in problems of pioneering, especially of pioneering in the tropics. My earlier studies were concerned with the possibilities of settlement in the tropics by peoples of other latitudes, both Westerners and Asiatics. In the course of these studies I became more and more interested in the fact that in many parts of the tropics, above all in the Asiatic tropics, there has been a great need for alleviation of the pressure of population. Occasionally this has led to spontaneous pioneering by small groups of settlers, but more frequently to the encouragement and support of pioneering by governments.

It was my original plan to visit all of the countries around the South China Sea and to study government-sponsored settlement activities in the Philippines, the Netherlands Indies, Malaya, and Indochina at first hand. The war, however, prevented me from visiting the mainland of Southeastern Asia; instead, I had to limit myself to the island world of the Philippines and the Netherlands Indies. From May 1940 to May 1941 I traveled throughout the islands of Luzon, Negros, Cebu, Mindanao, Java, Sumatra, and Bali. The war has also made impossible a rechecking of many of the references.

In the field I enjoyed the generous help of a large number of government officials as well as of private persons. In the Philippines I turned frequently for advice and assistance to Mr. Arthur F. Fischer, then President Quezon's adviser, who, as former Director of the Bureau of Forestry, has an unrivaled knowledge of the Philippine Islands. I am especially grateful to the staff of the National Land Settlement Administration.

In the Netherlands Indies I found the officials of the departments of Economic Affairs and of the Interior most kind and helpful. I wish to express thanks especially to Dr. H. J. van Mook, then Director of the Department of Economic Affairs, and to Messrs. A. Luytjes, J. J. Ochse, E. de Vries, A.M.M.P. Scheltema, A. K. Pringgodigdo, Th. J. Odenkirchen, P. Creutzberg, J. C. L. Götz van der Vet, J.F. Dekker, H. A. A. M. Wirtz, L. Gonggrijp, J. F. Kools, W. F. H. van Ameron, L. H. Vreeling, H. Loos, and W. E. K. Baron van Lynden, all of the Department of Economic Affairs. I am also greatly indebted to Governors C. O. van der Plas and L. Adam and to the following members of the Department of the Interior: Messrs. C. C. J. Maassen, J. M. Pieters, H. J. Friedericy, A. Jonkers, J. van der Zwaal, A. Deelemen, G. Ijsselstijn and G. de Lassequere.

To my wife, who accompanied me, I am deeply indebted for inspiration and encouragement as well as for keen criticism of the manuscript. Dr. Robert Bowman read the first part of the book with painstaking care; his criticism and suggestions were of great value to me. The late G.H.C. Hart read and criticized the chapters on the Netherlands Indies. I wish also to express my gratitude to Professor Leo Waibel, Dr. George F. Carter, and Messrs. Bruno Lasker and Philip Lilienthal for their helpful suggestions and to the editors for their work. The guidance of the U.S. Board on Geographical Names has been used whenever possible in the spelling of place names.

My field work was done under the auspices of the International Secretariat of the Institute of Pacific Relations and The Johns Hopkins University. I cannot express thanks in terms that completely convey my feelings of gratitude to Mr. W. L. Holland, formerly Research Secretary of the International Secretariat, and to Dr. Isaiah Bowman, President of The Johns Hopkins University.

Finally, I should like to acknowledge my indebtedness to the American Geographical Society for opening the Special Publications Series to this study. It should be pointed out that although the book is being published under the auspices of the International Secretariat of the Institute of Pacific Relations and of the American Geographical Society, the Secretariat, the Institute's National Councils, and the American Geographical Society assume no responsibility for statements of fact or opinions. The author alone is responsible for these.

KARL J. PELZER

Washington, D.C.
May 1, 1945

Part I

LAND USE

Chapter I

FROM HUNTING AND GATHERING TO AGRICULTURE

Hunting and Gathering

In Southeastern Asia today there are peoples of different cultural levels carrying on some of the major economies that man has developed in the course of time. The first people to invade the tropical forest regions around the South China Sea were alithic, or prepalaeolithic, food gatherers,[1] who obtained their food by collecting edible wild plants and by hunting and fishing. Today geographically separated peoples are still living in a prepalaeolithic or palaeolithic food-gathering and hunting stage.[2] In Southeastern Asia they are represented by the Semang of the Malay Peninsula, the Andaman Islanders, and the Aeta of the Philippines—all of whom are Negritos, or dwarfed Negroids—the Kubu of Sumatra, the Sakai of the Malay Peninsula, the Toala of Celebes,[3] and the Punan of Borneo.[4]

The degree to which collecting, hunting, or fishing contributes to the food supply varies with the season; it also varies with different peoples, depending upon the resources of the region, the type of equipment the people may possess, and the skills they may have acquired. Food-gathering and hunting peoples have a division of labor under which the men hunt while the women and children gather vegetable matter and catch insects and small animals. The men may lend a helping hand, however, in digging up some of the very deep-growing yams. Men will also participate in the picking of fruits, especially when it is necessary to climb tall trees. The food gatherers and hunters of Southeastern Asia are nomadic in the sense that they do not remain long in one place. However, they do not wander aimlessly through the jungle. Each group has its definite territory—among

---

Note: The titles of the works here cited in footnotes have been abbreviated, whenever possible. The full titles may be found in the Bibliography, p. 269 ff.

1 Nippold (Rassen- und Kulturgeschichte der Negrito-Völker, 1936, pp. 253–255 and 398–400) considers the alithic wood culture of the Negritos of Southeastern Asia to be older than palaeolithic culture. These people use wood rather than stone as raw material for their tools. Because of the perishable nature of wood, archaeologists and prehistorians have not yet been able, so far as I know, to go farther back than the palaeolithic age and therefore have not been able to confirm the hypothesis of an alithic age. Menghin (Weltgeschichte der Steinzeit, 1931, p. 91) believes that, from an archaeological point of view, an alithic wood culture may be regarded only as a possibility, but Kern (Die Anfänge der Weltgeschichte, 1933, p. 35), who compared pygmy culture with palaeolithic cultures, comes to the conclusion that alithic wood cultures probably preceded palaeolithic culture.

2 By divesting the word "palaeolithic" of its specific chronological connotation the term may be used for present-day peoples who have a food-gathering and hunting culture.

3 The racial affinities of the Kubu, the Sakai, and the Toala are not clear; by some they are called Veddoids, by others, Australoids.

4 They are considered to be Proto-Malays.

the Semang, for example, each band roams about in an area of from 15 to 20 square miles,[5] and its territorial claims are recognized by its neighbors. The habitations of these semi-nomads are usually crude, temporary shelters—sometimes an overhanging rock or cave.

These primitive food gatherers have acquired an amazing knowledge of the flora of their habitat[6] and have learned to make use of the leaves, flowers, seeds, nuts, berries, fruits, bark, tubers, and roots. Some have not only found ways of making poisonous vegetable substances edible but have also developed elaborate techniques of food preservation. For instance, at certain times of the year the Semang live largely on the poisonous seeds of the *Piah* tree (*Elateriospermum tapos* Blume), which contain hydrocyanic acid.[7] If the seeds are needed immediately they are boiled or roasted; otherwise they are pounded in water or rasped, packed into bark cloth or bamboo tubes, buried in wet earth for a month or more, and allowed to ferment. This treatment not only removes the poison but also preserves the food for several weeks.[8]

Primitive and Advanced Agriculture

There are practices among contemporary gatherers that can be regarded as contributing the link between their economy and primitive cultivation. The Semang, who are especially fond of durians, will camp under, or near, a durian tree in order to prevent monkeys or wild pigs from getting the ripe fruit. It is reported that they may even cut down a ring of forest around these trees during the fruiting season to eliminate the competition of other forest vegetation. This procedure represents a form of husbandry. The Semang will also put the tops of wild yams back into the ground after they have cut off the tubers, so that the plants may continue to grow and produce new tubers. Such practices are but one step short of primitive cultivation.

---

5 By way of contrast, the territory of an Australian group may include as much as 8,000 or 10,000 square miles of arid country.

6 Milsum (Fruit Culture in Malaya, 1919) lists as many as 30 wild trees of which the fruits are collected and eaten by food gatherers and others.

7 Burkill: A Dictionary of the Economic Products of the Malay Peninsula, 1935, Vol. 1, p. 905.

8 This technique of burying a paste, or dough, and allowing it to ferment seems to be one of the oldest and most primitive ways of preserving a surplus supply of food. The method described above is applied to the seeds of *Pangium edule* Reinw. The boiled seeds are buried with ashes and allowed to ferment, a process which requires about 40 days. It is possible to shorten the process by digging up the buried mass after 15 days. It is then boiled again, soaked in running water, and left for another 4 days of fermentation. When needed immediately the seeds are crushed, boiled, and put into running water for a day, after which they are boiled once more and are then ready for consumption.

The Polynesians pack breadfruit into pits and cover them with earth. Fermentation will turn the fruit into a soft cheese-like mass from which cakes are baked. This practice enables the Polynesians to keep a supply of breadfruit after the fruiting season is over. See Burkill: *op. cit.*, Vol. 2, pp. 1652–1653, and Vol. 1, p. 250.

Food-gathering people in semi-arid and arid regions preserve food by drying it in the sun, but this technique cannot be used in the tropical rain forests of Southeastern Asia.

During untold centuries in which they collected plant matter, women acquired an intimate knowledge of vegetable life. Eduard Hahn and others have therefore reasoned that women were the inventors of the cultivation of plants and hence were the first food producers. Hahn has given the name *Hackbau* (hoe culture or hoe tillage) to primitive cultivation, and maintains that the hoe is woman's implement; he draws a sharp distinction between Hackbau and more advanced cultivation, which he calls *Pflugbau* (plow culture or plow tillage) and ascribes to men.[9] It is indeed true that men generally farm with plows and that many examples may be cited of women using hoes; nevertheless, there are so many instances of male primitive cultivators employing hoes or similar instruments that Hahn's classification should be modified. While there seems to be no reason for abandoning his theory of the invention of plant cultivation by women, the classification of cultivation by tools, such as the hoe and the plow, ascribed to one sex or the other, is unsatisfactory. For, in spite of the fact that the collection of vegetable matter is almost exclusively the woman's task in the gathering economy, among primitive cultivators men either assist women in the cultivation of plants or entirely take over this activity whenever hunting or fishing become less important than soil tillage. Furthermore, many primitive cultivators do not use the hoe, but continue to employ the digging stick, characteristically used by the food-gathering women. From Formosa southward throughout the Malay Archipelago, as well as on the mainland of Southeastern Asia, the digging stick, also known as the dibble stick or planting stick, is frequently used by the primitive cultivator. On the other hand, there are advanced cultivators who till their land intensively with hoes; they use manure, rotate their crops, terrace fields that lie on slopes, and practice irrigation, yet they do not use plows drawn by draft animals (however, in some instances they have livestock). In other words, advanced agriculture does not depend upon the animal-drawn plow nor does primitive agriculture depend upon the hoe.

A more fundamental distinction between primitive and advanced agriculture may be made in terms of land use rather than of tools. Primitive cultivators, with the possible exception of those living in arid regions, raise one or two crops on a given plot of land, then abandon the plot and make a new field, preferably in the forest, so that the old plot lies fallow sometimes for many years and may itself revert to forest; in short, primitive cultivators shift the place of their agricultural activities with regularity. The crop land consists of small patches scattered throughout the forest; these patches are extremely ragged in appearance and do not look at all like fields. A primitive cultivator would never think of removing stumps and fallen trees; instead, he deliberately raises his crop among them. This kind of cultivation is best described as shifting-field, or shift-

[9] Hahn: Das Alter der wirtschaftlichen Kultur, 1905; idem: Die Entstehung der Pflugkultur, 1909; idem: Von der Hacke zum Pflug, 1914; Kolff: "Hahn's hypothese," 1926, pp. 884–907.

ing, cultivation. In contrast to this method, advanced agriculture is practiced year after year on the same well tended plot, the fertility of which is maintained by manuring, crop rotation, intensive cultivation, terracing, and often the regular addition of mineral matter in the form of silt deposited on the land by irrigation water. Only occasionally may the land be allowed a short fallow period. This type of tillage may be called permament-field, or sedentary, cultivation.[10]

Shifting cultivation and sedentary cultivation are generic terms used to describe fundamental differences in land use, differences that find their outward expression in the cultural landscape created by those who maintain these economies.[11] Can there be a greater contrast than that between the shifting cultivators' small, irregular, temporary clearings with their chaotic jumble of fallen and half-fallen tree trunks and stumps, on the one hand, and, on the other, an irrigated plain, subdivided into small fields, each surrounded by a dyke—or a steep slope where a succession of beautifully built terraces leads up for several hundreds or even thousands of feet? These kinds of landscape are physical expressions of totally different economies, or modes of life, with different cultural patterns, concepts of land tenure, attitudes toward the soil, and material equipment.

Historically, shifting cultivation without doubt antedates sedentary cultivation, which requires among other things more elaborate techniques and deeper insight into the relationships between soil, tillage, and yields.

Shifting and sedentary cultivation may each be further classified in terms of tools, as we have seen, or in terms of crops, systems of rotation, and the like.[12] For instance, in terms of crops, we may distinguish between grain and grainless culture. Of the two, grainless culture is generally conceded to be the older. One argument for this thesis is that fruits and many of the root and leaf crops may be eaten raw or simply prepared, whereas cereals require more elaborate preparation. The fact that grainless agriculture is largely confined to the fringes of Southeastern Asia also supports the thesis that, at least in this region, grainless agriculture preceded the cultivation of cereals. Still further substantiation is found in Polynesia, where the inhabitants cultivated no cereals when they were first visited

---

10 Although sedentary cultivation is commonly practiced by people who use plows drawn by animals, there are large areas where it is practiced with little or no use of the plow. Such areas are found in parts of the Philippines, Java, Sumatra, and southern China, where steep hillsides have been terraced and where the nature of the terrain makes the use of work animals impossible. Sometimes individual farms or patches of land are so small that they must be tilled with hoes, even though plows may be used on adjacent farms. In places custom and religious tradition forbid the use of plows and draft animals.

11 Waibel: Probleme der Landwirtschaftsgeographie, 1933, p. 11.

12 For example, we may distinguish: (a) shifting dibble cultivation, shifting hoe cultivation, and shifting plow cultivation; (b) grainless shifting cultivation and shifting grain cultivation; (c) shifting field-forest cultivation and shifting field-grass cultivation; (d) permanent plow cultivation and permanent planting-stick cultivation; (e) permanent four-field cultivation, and so on. Not all of these are found in Southeastern Asia.

by Westerners. This may be explained by the assumption that the ancestors of the Polynesians migrated from Southeastern Asia to the islands of the Pacific before cereals had replaced root crops as dominant plants.[13]

In Southeastern Asia we find both shifting and sedentary cultivators who depend exclusively upon root crops as a source of starch. The Nias, Mentawei, and Engano islanders off the western coast of Sumatra grew no grains until recently. Beside such root crops as taro (*Colocasia esculenta* Schott), yams (*Dioscorea* sp.), and *bira* (*Alocasia macrorhiza* (L.) Schott), they cultivated such other foods as bananas (*Musa sapientum* L.), sugar cane (*Saccharum officinarum* L.), and coconuts (*Cocos nucifera* L.). In the eastern parts of the Malay Archipelago and New Guinea the inhabitants depend chiefly on the starch of the sago palm (*Metroxylon sagu* Rottb.) which they either plant or collect in the wild state. Here, again, taro, yams, and bira are other sources of starch, as are such imported American plants as cassava (*Manihot utilissima* Pohl) and sweet potatoes (*Ipomoea batatas* Poir.). The folklore and religious practices of many tribes in Borneo and Celebes who today grow cereals recall the time when these tribes had only root crops.

In many parts of the Pacific taro is grown by sedentary cultivators on terraced, flooded fields.[14] We do not know whether the techniques of terracing and irrigation developed independently in Oceania and on the mainland, or whether the ancestors of the Polynesians brought their techniques with them from Southeastern Asia. If the latter be true, they must have left before the cultivation of rice was known to them; otherwise rice would have been among their crops. That it was not among them suggests that

13 Forde: Habitat, Economy and Society, 1934, pp. 428–430. In remote centuries the Polynesians lived in the Malay Archipelago, where they cultivated such root crops as taro, yams, "arrowroot" (*Canna orientalis* Rosc.), and turmeric (*Curcuma* sp.), in addition to bananas and plantains (*Musa paradisiaca* L.), breadfruit (*Artocarpus communis* Forst.) and coconuts, all of which, according to the botanists, are natives of the forest regions of Southeastern Asia, whether the mainland or the Malayan Archipelago. All these plants were probably carried to Polynesia, either via a northern route through Micronesia or by a more southerly course through New Guinea and Melanesia. See Buck: Vikings of the Sunrise, 1938, pp. 305–309. The names of these plants were also carried into the islands of the Pacific. Variants of the names "taro" and "bira," to mention two examples, range from the Malay Archipelago far into Oceania. *Colocasia* is *talo* in Nias, *talos* in Sundanese, *tales* in Javanese, *dalo* in Fiji, *talo* in Samoan, *kalo* in Hawaiian, *taro* in Maori. *Alocasia* is known as *birah* or *bira* in Batak, *bira* in Madurese, *biga* in the Philippines, *piga* in Guam, and *via* in Fiji. Later, the Polynesians, following the example of other peoples of the Pacific, were quick to adopt such New World crops as cassava and sweet potatoes as well as American fruit trees into their economy because these are closely related to their customary crops, but they still remain most reluctant to change from roots to grains as staple, possibly because "the cultivation of roots and fruit trees has little technical relation to the raising of grains, for they are not grown from seed but from eyes and cuttings." See Forde: *op. cit.*, p. 380. Another illustration of the fact that a plant, the introduction of which involves a change in methods of cultivation, is accepted only under pressure, is found in the history of the introduction of the potato into Europe.

14 Krämer: "Der Taro und die Nasskultur," 1928, pp. 165–172; Rivers: "Irrigation and the Cultivation of Taro," 1926, pp. 262–287.

taro had been cultivated on permanent flooded fields earlier than had rice. Another possibility would be that taro, together with the technique of irrigation, was introduced into Oceania later than were the yams and alocasia, and the Polynesians' more important fruit trees.

A folktale of the Bontocs in the mountains of northern Luzon furnishes us with another clue. According to Bontoc lore, their hero, Lumauig, taught them agriculture in a tiny irrigated garden, which is still tended from generation to generation by a special priest. The crop of this garden is taro and not rice.[15] This would seem to lend support not only to the thesis that the cultivation of root crops preceded that of grain but also to the thesis that terracing and irrigation were already being practiced by cultivators of root crops.

A beautiful example of sedentary cultivation with root crops as a staple has recently been discovered in the remote Grand Valley of the Balim River in the highlands of Netherlands New Guinea. On the valley floor of the Balim and its tributaries neolithic people practice intensive cultivation, including such techniques as the use of green and animal manure, deep tillage with stone adzes, the digging of ditches to improve the run-off and drainage of swampy land, and the construction of terraces.[16] On the slopes they practice shifting cultivation. In both their shifting and their sedentary agriculture the staple crop is the sweet potato, while taro is a secondary crop. Also subsidiary are bananas, sugar cane, and pandanus.

In most parts of Southeastern Asia, however, root crops have been replaced by such cereals as millet, Job's tears, upland and lowland rice,[17] and maize. There seem to be good reasons to believe that the first cereals cultivated in Southeastern Asia under the system of shifting cultivation were foxtail, also known as Italian millet (*Setaria italica* Beauv.), and Job's tears (*Coix lacryma-jobi* L.). Before their domestication these grasses, like the root crops and fruit trees of Southeastern Asia, had been used by alithic and palaeolithic hunting and gathering people. Shifting cultivators succeeded in domesticating these plants by selecting over the course of years the species most valuable to them. Although foxtail and Job's tears have largely been replaced by the more valuable grains, rice and maize, there are still shifting cultivators in India, Thailand, Indochina, Malaya, the Indies, and the Philippines who raise the former cereals. Burkill says of foxtail,

> Such a crop as this probably came into cultivation before man had settled enough to grow the better cereals. Its grain ripens in even as little as two months and a half from sowing, and the wandering tribes of the Malay forests are able to remain in one neighbourhood long enough to catch a crop: but the usual duration of the crop seems to be four months.[18]

---

15 Keesing and Keesing: Taming Philippine Headhunters, 1934, p. 56.

16 Brass: "Stone Age Agriculture in New Guinea," 1941, pp. 555–569.

17 Upland rice is also known as hill, mountain, dry-land, or dry, rice; lowland rice is also known as irrigated, wet-land, or wet, rice.

18 Burkill: *op. cit.*, Vol. 2, p. 1999.

Probably the grain known as Job's tears was originally domesticated somewhere in the area between India and southern China, and it is still of importance in this area, although maize has competed with it since the latter's introduction into Southeastern Asia in the sixteenth century.[19]

Rice, The Basis of the Agricultural Economy of Southeastern Asia

By far the most important grain of both shifting and permanent cultivators in Southeastern Asia is rice, which occurs in the wild state in Africa as well as in various parts of southern Asia. The cultivated species (*Oryza sativa* L.) is probably descended principally from two Asiatic wild rices, *Oryza fatua* Koen. and *Oryza minuta* Presl. *Oryza fatua* is thought to be the dominant ancestral strain. These two ancestors are found in moist places from the Himalaya Mountains to Ceylon and from southern China through Indochina, Thailand, Burma, and Malaya to Java, Borneo, and the Philippines.[20] Rice was probably first domesticated in northeastern India, somewhere north of the Bay of Bengal, whence the domesticated varieties have spread in all directions: southeastward through Burma into the Malay-inhabited areas, eastward and northward into Farther India, China, Japan, Formosa, and the Philippines, and westward into the Near East and beyond.

The remarkably large number of distinct varieties of cultivated rice[21] has been interpreted as evidence of the great antiquity of its domestication. In the course of thousands of years of selection, varieties have been developed that are adapted to a wide range of natural conditions and especially to varying amounts of water. These varieties may be divided into two main classes: upland, or dry, rice and lowland, or wet, rice. The division, based on the two main types of agronomic practice, is not a very sharp one. Some upland varieties can be grown in flooded fields, and some lowland varieties can be grown in dry fields; however, it is said to be easier to grow dry rices in water than to grow wet rices on dry land.[22] In general, the areas suitable for the cultivation of upland rice are much more limited than those suitable for lowland rice, which can be grown outside the regions of assured rains, provided that water is available for irrigation; whereas upland rice must be grown in a humid atmosphere with an assured rainfall over a three- to four-month period.[23]

Upland rice is grown by the shifting cultivator, into whose economy it

---

19 *Ibid.*, Vol. 1, pp. 629–631.

20 *Ibid.*, Vol. 2, pp. 1593–1595.

21 According to Copeland (Rice, 1924, p. 131) some 3500 varietal names are known in the Philippines; these may be presumed to describe more than 2000 different varieties; in India, where conditions seem to favor mass mutation, some 8000 varietal names have been recorded, but we do not know how many distinct varieties these represent.

22 Burkill: *op. cit.*, Vol. 2, p. 1595.

23 Lacking direct evidence, students of rice and its cultivation have frequently speculated on the comparative age of the two types. Some argue that the cultivation of wet rice preceded that of dry rice. They maintain that rice is, by nature, an aquatic plant. There are, however, grounds for thinking that the cultivation of dry rice is the older form. For one thing, there are wild varieties of upland rice.

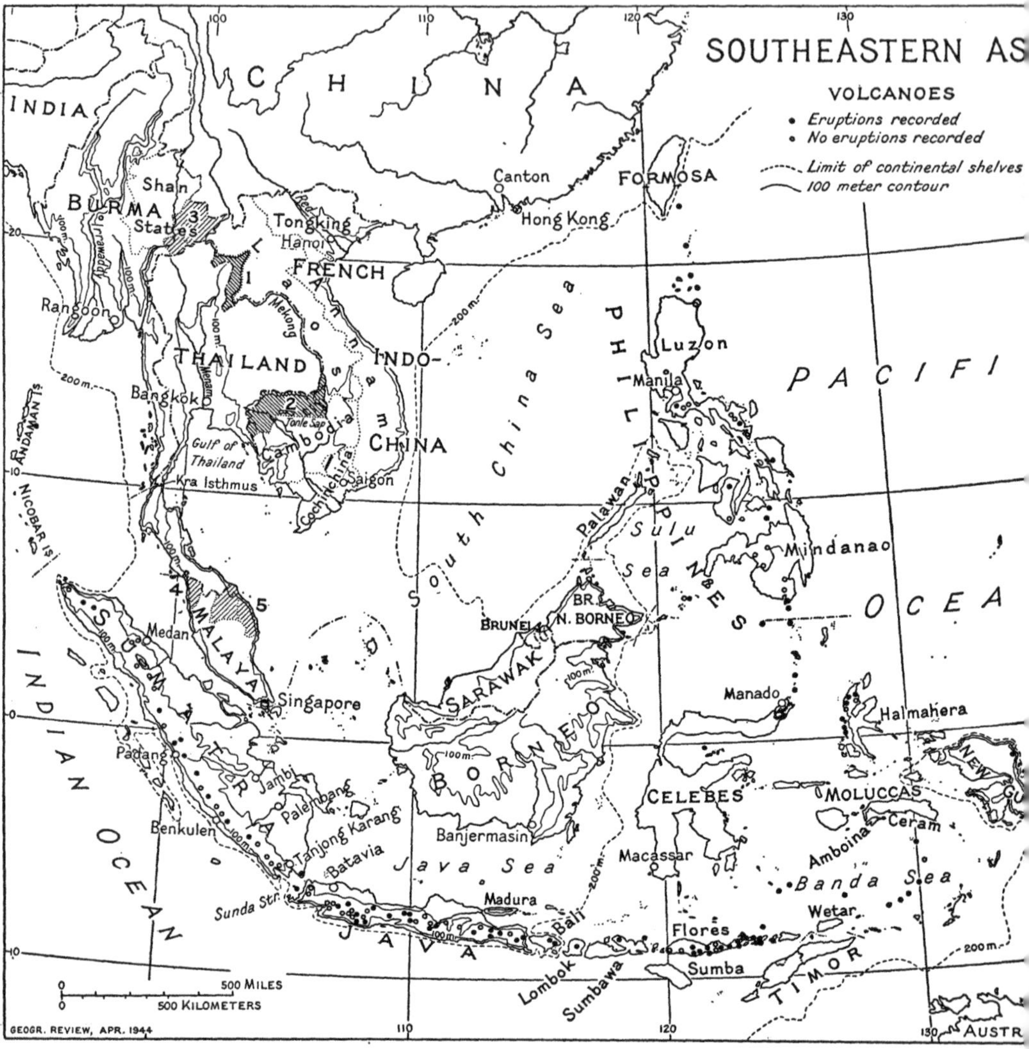

FIG. 1A—Orientation map of Southeastern Asia. The 100-meter contour line (omitted for the smaller islands) brings o small extent of lowland areas in Southeastern Asia. The 200-meter bathymetric line shows the extent of the continenta Notice especially the distribution of volcanoes. Recent territorial changes, allotting parts of Indochina, Burma, and M to Thailand, are indicated by shaded areas numbered 1 to 5. The spelling of place names on the original map does not the recommendations of the U. S. Board on Geographical Names, as the map was drawn before the publication of the B most recent recommendations. The spelling of additional place names has been kept consistent with that on the origina (Reproduced through the courtesy of the *Geographical Review*).

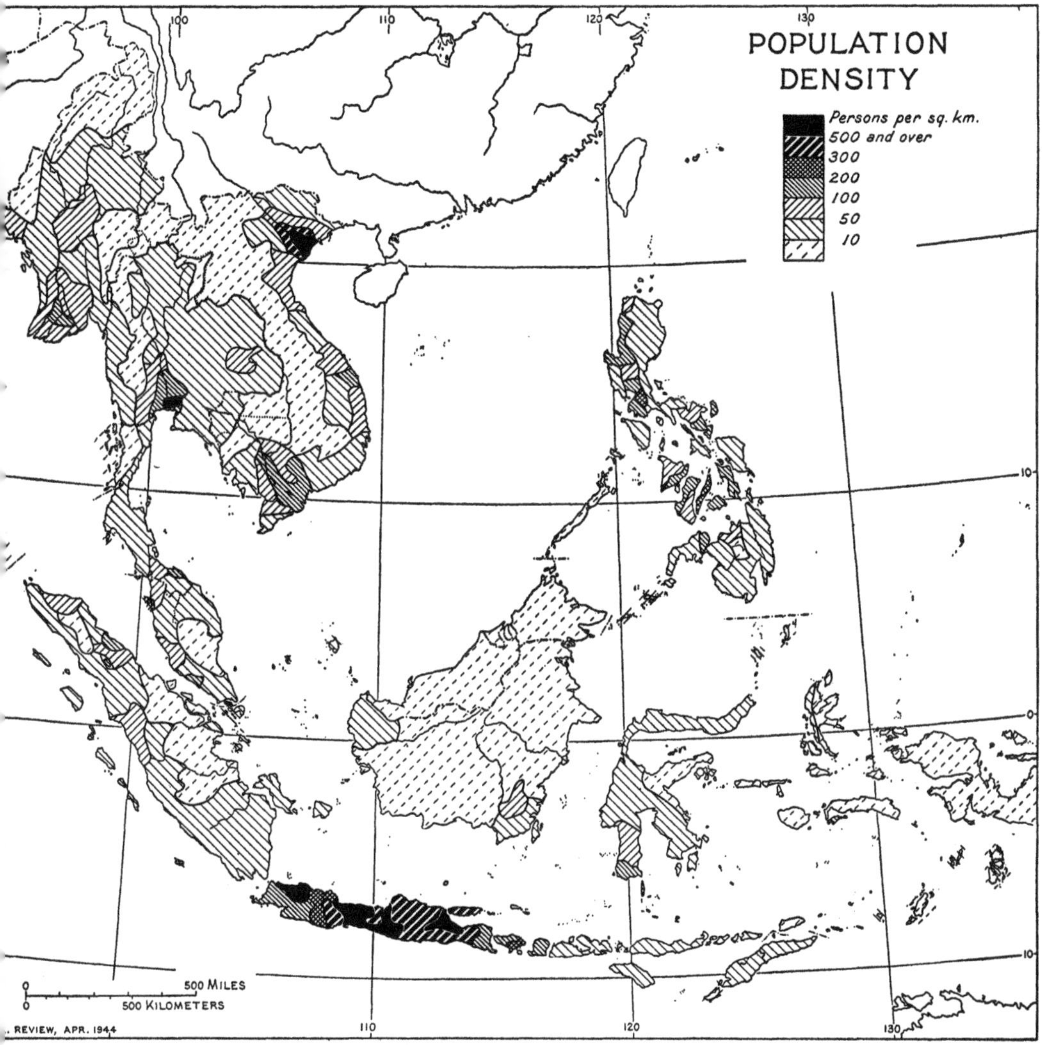

FIG. 1B—Population Density in Southeastern Asia. This map, like the one opposite, was originally prepared for Professor O. M. Broek's article, "Diversity and Unity in Southeast Asia." He has described his method of compiling the data densities as follows: "In order to present a comparable picture for the whole of Southeast Asia, the year 1940 or 1941 s chosen, and densities were calculated for administrative districts. This use of administrative districts for the sake of formity explains, for instance, the rather low density of the Annam coastal region, where the districts include large as of sparsely settled uplands adjacent to the crowded coastal strip. The scale of the map required, of course, some neralizations where there were many small districts with differing densities (Tongking, Cochinchina, Java). Cities, ere they formed separate census units, were included in the adjacent districts." A list of the sources used and further es about the compilation are given by Professor Broek in his article (p. 180). (Reproduced by courtesy of the *Geo-phical Review*).

fits perfectly, for it does not require a permanent field but can be grown anywhere, on flat land or steep slopes, provided that the rainfall is sufficiently heavy and the temperature and soil are suitable. In contrast, the cultivation of wet rice requires a sophisticated technique. To adapt to agriculture in Southeastern Asia Owen Lattimore's remark on the place of irrigation in Chinese agriculture,[24] we may say that the cultivation of wet rice lies at the core of agriculture in Southeastern Asia but not at the very roots.[25]

The sedentary cultivators of Southeastern Asia have developed not only a notable range of varieties of wet rice adapted to varying natural conditions but also many differing techniques of rice culture, which embody the experience of generations of cultivators. Every important phase of the cultivation of wet rice has many variants. The reasons for these are often hidden to the casual observer, but Western agricultural specialists after painstaking research and experimentation have been forced to concede their value.

The ordinary varieties of wet rice differ greatly in respect to the time required for maturing and the amount of water needed. Selection has led to such remarkable achievements as the development of floating rice and of salt-loving varieties. The floating rice of Thailand and Cochinchina can be grown on land subject to prolonged high floods, a condition that ordinary varieties of wet rice do not tolerate. The growing plants are able to keep up with the rising floodwater; in extreme cases, the stems may reach a length of as much as 12 or 15 feet.[26] Salt-loving rices, which can be grown on saline soils or in brackish water, are cultivated along coasts or estuaries.

The raising of seedlings in seedbeds, after which they are transplanted to the field, is the main characteristic of most Oriental wet-rice culture. This practice has given Oriental rice relatively high yields. The preparation of the seedbed itself varies considerably.[27] The commonest method of raising rice seedlings is to make the seedbed on thoroughly soaked and puddled ground. Frequently the seed is soaked and allowed to germinate before it is sown in the bed. In the northern Philippines the seedlings are raised in seedbeds that are not flooded and puddled but are left dry. When the seedlings are pulled up from either a wet or a dry seedbed their roots

---

24 Lattimore: Inner Asian Frontiers of China, 1940, p. 312. Elsewhere (p. 29) he says, "To assume the working out of such a technique [irrigated rice-growing] at the earliest stages of cultural and social development implies the ability to take a second major step forward at a time when the first tentative step had barely become possible."

25 In Java, for example, the cultivation of wet rice has spread only relatively recently into Sundanese territory and did not reach northern Bantam until the sixteenth century. Even today the Badoei of West Java have religious scruples against the planting of wet rice as well as against the use of the hoe or of the plow and carabao. Notwithstanding strenuous efforts on the part of the government to induce the Badoei to adopt a more intensive form of agriculture they continue to practice shifting cultivation and in the main to plant upland rice, using the dibble stick. See Kools: Hoema's, 1935, pp. 22–23.

26 Tran-van-Huu: "Note sur la culture du riz flottant en Cochinchine," 1920, pp. 46–54.

27 Copeland: *op. cit.*, pp. 226–229.

suffer a certain amount of damage. The seedlings must therefore be given time to grow strong enough to endure the temporary uprooting. They usually need from 30 to 40 days before they are ready for transplanting. Whenever early transplanting is necessary a technique is employed which is known in the Philippines as *dapog*. The puddled seedbed is completely covered with banana leaves, which are pressed slightly into the mud so that a thin layer of mud spreads over the leaves. On top of this is placed a layer, an inch thick, of rice hulls or finely cut straw. The whole is thoroughly soaked and sown with seeds. The banana leaves prevent the roots of the seedlings from growing into the ground and suffering the usual damage when transplanted, but the seedlings must be transplanted within two weeks after the sowing of the seed.

In the riparian marshes of Sumatra and Borneo—the *lebak* areas—all the land is flooded at the time when the rice seed should be put into the ground. The inhabitants of these districts, however, have found an ingenious solution to their problem: the floating seedbed. They build small rafts of bamboo, reed, banana stems, or the leaf-ribs of the sago palm, on which they place a layer of rush, or some other aquatic plant, and then a thick layer of fine mud. As the water level rises and falls the seedlings have a chance to develop on the float unharmed.

Numerous also are the techniques employed to supply wet-rice fields with the necessary water (see pp. 47–51). These reflect the nature of the terrain, the social organization, and the stage of technical development of the inhabitants. For it is obvious that the cultivation of wet rice on the plains of the Red River in Tongking would require very different techniques from those required on the steep mountain slopes of northern Luzon.

Culture, Types of Land Use, and Population Pressure

One of the most striking characteristics of Southeastern Asia is the great variation in the density of the population (Map, Fig. 1B). This variation reflects differences in culture as well as differences in physical environment. The culture and social organization of a people are intimately linked to their technique and the intensity of their land use, which in turn guide the people in their selection of new land when it becomes necessary for them to seek new homes. The Cambodians, for example, when living in an environment similar to that of the Annamese, do not create the large rural settlements characteristic of areas inhabited by the latter.[28] Likewise, the Madurese prefer to settle in small hamlets or in isolated homesteads (*Einzelhof*) and to till unirrigated land, whereas the Javanese live in large villages and are by preference wet-field cultivators. Again, while the Bontocs and Ifugaos of northern Luzon build terraces and cultivate wet rice as their primary crop, the neighboring Apayaos, most of the Kalingas, the Tinggians, and the Gaddangs have so far resisted all attempts on the

[28] Gourou: L'utilisation du sol, 1940, p. 94.

part of the government and of missionaries to introduce terracing and wet-rice cultivation but, instead, continue to practice shifting cultivation and to depend upon dry rice as their staple food.

Wherever we find shifting cultivation the density of population is low; on the other hand, permanent cultivation provides the basis for higher densities. The higher the density, the greater is the care bestowed upon the cultivation of the soil. The development of permanent-field cultivation, the construction of terraces on mountainsides and hillsides, the dyking and draining of plains, and other complex techniques associated with wet-rice cultivation are but expressions of a population density that demands an intensive utilization of the soil and a maximum production of food crops.

The areas where wet rice was first cultivated were probably hollows and small valleys and the well watered slopes of low hills rather than the great plains and wide valleys that are today the rice granaries of the world. The utilization of extensive plains and broad river valleys for the cultivation of wet rice requires a social organization unknown to the peoples who first began wet-rice culture.

It is thought that the cultivation of wet rice originated in days when the hoe and the dibble stick were the only agricultural tools, that is, before the animal-drawn plow reached Southeastern Asia. The broadcasting of seeds is held to be characteristic of plow culture, whereas the cultivators of wet rice in Monsoon Asia generally plant their rice seedlings in the manner of shifting cultivators,[29] a circumstance that would seem to support the hypothesis that the carabao and the plow were not a part of the original technique but were adopted later. Even today wet rice is raised in some parts of Southeastern Asia without the use of the plow. For example, peoples in central Celebes prepare their fields for cultivation by driving carabaos back and forth until the fields are thoroughly puddled.[30] In other areas where wet-rice fields are built on such steep slopes that it is impossible to get carabaos to the terraces, farmers are frequently obliged to till their land entirely by hand—employing only such tools as wooden spades and sticks, which may be shod with iron. The Ifugaos of northern Luzon, for example, have terrace walls reaching, in some cases, heights of 60 feet or more that can be scaled only by means of rocks jutting out from the faces of the walls.

Although most sedentary cultivators in Southeastern Asia make use of

29 Although rice growers in other parts of the world generally use the broadcast method of sowing, one finds it in Southeastern Asia only as a rare exception. According to Copeland (*op. cit.*, p. 224) the *sabog*, or broadcasting method, is used in the Philippines over a rather limited area. This method produces lower yields but its advantages are "some saving of time, saving the labour of transplanting, and the possibility of maturing a crop with less water; rice grown directly in its place can get along with less water than transplanted rice." According to Kruyt (" De To Rongkong in Midden-Celebes," 1920, p. 380), the To Rongkong of central Celebes transplant their wet rice only in exceptional cases but usually broadcast it.

30 Kruyt: "De To Seko in Midden-Celebes," 1920, p. 409.

the animal-drawn plow, in the densely populated regions the parcelling of land has proceeded so far that many peasants, whose fathers, or ancestors, owned draft animals (carabaos or oxen) and used plows, have farms so diminished in size that they cannot afford to keep draft animals and therefore must till their fields entirely with hand tools. In spite of the fact that they often make very intensive use of their fields, still they are not able to produce enough to meet the modest needs of their families.

In nearly all instances, however, where the governments concerned have attempted to solve this problem by opening up land suitable for sedentary cultivation to settlers coming from problem areas, the governments have been confronted by the fact that these districts, which might offer outlets for the congested areas, are inhabited by shifting cultivators or in some cases by food gatherers. The danger is then that the proposed migration may infringe upon the rights of the indigenous peoples, whose economy and mode of life, as we have noticed, require considerably more land than those of the incoming settlers, who are accustomed to permanent-field cultivation. It is, therefore, necessary for us to understand the various forms of land use in Southeastern Asia and their peculiar exigencies in order rationally to approach the problems of pioneering in the countries around the South China Sea.

Chapter II

THE SHIFTING CULTIVATOR

The field of the shifting cultivator is known by various local terms. It is called *caiñgin* in the Philippines, *humah* in Java, *ladang* in the Indies as a whole and in Malaya, *taungya* in Burma, *tamrai* in Thailand, and *rây* in Indochina.[1] Practically without exception, it is a clearing in either primeval or second-growth forest. The use of grassland is so unusual that we shall limit the following discussion to shifting cultivation on forest land.

As his name indicates, the shifting cultivator does not use the same piece of land every year; instead, he kills or cuts down at regular intervals—every year, every other year, or every third year—the trees of a small forest patch. He burns as much of the woody growth as is possible or convenient. Among the charred stumps and half-burned tree trunks he plants his favorite crops, be they cereals such as millet, upland rice, or maize; or tubers such as yams or sweet potatoes (Figs. 4–10). The latter are planted extensively by the *caiñgero*, the shifting cultivator in the Philippines. In planting cereals it is usually the men who take the planting stick to make little holes and the women who drop in a few seeds, using their feet to cover the seed with soil (Figs. 11–13). As forest land is generally free from weeds or grasses and the soil is usually rich in humus and well supplied with the ash of burned plant matter after clearing, it produces a very good, or even excellent, first harvest; the second harvest begins to show a decline in yield, and thereafter the returns diminish rapidly.[2] Grasses and weeds invade the clearing. Rather than battle these, the peasant abandons his old ladang and cuts and burns a new patch of forest (Fig. 15). The old plot reverts, under favorable conditions, to second-growth forest, the so-called *blukar* of the Malays, and is not cleared again for a period, the length of which may vary from 8 to 15 years, or more.

Shifting cultivation is characterized by a lack of tillage; the soil is hardly disturbed, as no plow turns it. There is less labor involved than in other methods of cultivation, and the work is more evenly spread over the year—an advantage where labor is relatively scarce. The tasks involved are cutting, burning, and planting; sometimes simple fencing of the field against animals, guarding—especially before the harvest and usually by women and children—and, finally, harvesting. In addition, there may be some weeding in the first year and certainly during the second.

---

1 In British India *jhum*, *kumri*, *dhya*, *bewar*, or *dullee;* in Ceylon *chena;* in Japan, Korea, and Formosa *karen* (a Japanese term); in Central America *milpa;* in parts of Africa *chitemene*.

2 Danhof reports for parts of the Lampoeng Districts that a ladang will produce from 15 to 25 quintals of rice in the first year and approximately 5 in the second year. In areas where the second yield shows such a serious decline the peasants usually prefer to make a new ladang—Danhof: "Bijdrage," 1940, p. 199.

Shifting cultivation, the most primitive type of agriculture, may be briefly defined as an economy of which the main characteristics are rotation of fields rather than of crops; clearing by means of fire; absence of draft animals and of manuring; use of human labor only; employment of the dibble stick or hoe; short periods of soil occupancy alternating with long fallow periods. Other terms for shifting cultivation are field-forest economy and field-forest rotation.

Shifting cultivators prefer flat land, but in Formosa, the Philippines, Indochina, and Thailand they have been forced back into the hills and mountains and must often be content with fields on steep slopes (Figs. 16, 17). This has led some observers to believe that shifting cultivation is exclusively practiced in hilly or mountainous regions, or that such regions are preferred in this economy. However, in the Outer Islands[3] of the Netherlands Indies, one frequently finds ladangs in flat country (Fig. 18). If the ladang is very steep it is rarely used for more than one crop, so that new clearings must be made every year.

Many shifting cultivators no longer change their dwellings when they turn to new land; they have become sedentary at least as far as their houses are concerned. They may build a hut in the ladang if it is too far away from their village and may occasionally live in such a shelter, especially at the time when the ripening crop has to be guarded against animals, but they will return with the harvest to their permanent houses in the village (Figs. 10, 19–21).

Ladang culture, as is frequently true of any form of agriculture, is intricately interwoven with religion. Magic practices and sacrificial offerings must accompany every step in the culture, even if the people are no longer pagans but, for example, Mohammedans. Friedericy has made a careful study of these magical-religious customs in the Ponre mountains at the head of the Gulf of Bone in southern Celebes.[4] In this region the people live in small permanent villages. Whenever they are in need of new ladangs—every second year—the heads of the families meet to choose new locations. At this stage they try to find out from the spirits if these agree to the clearing of the land that was selected. For this purpose the village headman (*dukun*), or somebody else who is familiar with magic, is sent out to the part of the forest that has been chosen. Here the delegate places betel leaves (*Piper betle* L.), each with a small slab of lime and a piece of areca nut (*Areca catechu* L.), under a tree and expresses the desire of his community to cut the trees and to use the soil. He ends with the wish that the betel leaves and the areca nuts be scattered should the choice

[3] I have used the term "Outer Islands" to denote all the islands of the Netherlands Indies not comprised in the jurisdiction of Java and Madura (Madoera). The Dutch administrative term is *Buitengewesten*.

[4] Friedericy: "Ponre," 1932, pp. 1–34; idem: "De taak van het bestuur," 1937, pp. 488–501. For an excellent description of the magical-religious rites that are associated with the agricultural efforts of the Bemba of Northern Rhodesia, see Richards: Land, Labour and Diet in Northern Rhodesia, 1939, pp. 351–380.

of the new ladang be a poor one. After three days the same person returns, and if he finds that the offerings have been scattered a new location is chosen and a new attempt is made to get a favorable answer from the spirits. In case the betel leaves have been found undisturbed, all male members of the village have a meal on the chosen land. Then the area is divided among the villagers by the headman. Each individual appeals to the spirits by placing a betel leaf with lime and areca nut on the ladang that has been turned over to him. If the answer is unfavorable he tries again. The betel leaf may be placed three times. If the offering is disturbed each time, the headman turns over a part of his share, which is considerably larger than any of the others, to the less fortunate villager. Throughout the following agricultural year with its stages of clearing, planting, and harvesting, the peasant repeatedly requires the services of the dukun, or *shaman*, as well as the betel leaf.

In other regions details differ, but the essential fact remains that the social-religious life and thinking of the shifting cultivator is bound up with his characteristic form of cultivation.

Shifting cultivators have no concept of private land ownership. The land belongs to the community, the tribe, or the village, and each member has the right to use as much land as he needs and can cultivate. The location of the land that he is to use may be decided upon by the village council or the headman. The person who clears the land and plants the crop gets the produce. Nobody has the right to plant a plot that has been cleared by someone else. As soon as the land is abandoned all rights are forfeited, except the right of the community. However, if trees have been planted, the ownership of the trees and their produce remains with the person who planted them, and he is given preference when the land has had a sufficient fallow period and can again be cultivated.

The Effect of Shifting Cultivation on the Land

As has been stated above, ladangs decline in fertility after the first crop, and obnoxious grasses and weeds appear among the stooling tree stumps. As long as there is an abundance of forest land at his disposal the cultivator prefers to abandon the ladang. Under favorable conditions the grasses will, in turn, give way to second-growth forest. In the course of centuries this second-growth forest may develop into a climax forest, that is, a forest having the same composition as the primeval forest. In any case, when tall forest growth again occurs, the land will have had time to recover its original fertility.

Unfortunately, however, conditions are often not sufficiently favorable for a regeneration of soil fertility and the development of a good second-growth forest. As a matter of fact, the practice of shifting cultivation has altered the character of vegetation beyond recognition over wide areas in Southeastern Asia. Instead of a dense primeval rain-forest, we now find extensive open grasslands resulting from the most important tool of the shifting cultivator, fire.

Generally, the fallow period allows time only for the evolution of mediocre second-growth forest, for the shifting cultivator usually returns to an old ladang before even a small percentage of those tree species that made up the original stand have had a chance to come back. Repeated cultivations further impoverish the forest and reduce the number of species. Certain light-loving trees that played a minor role in the climax forest will spread rapidly and dominate, together with those species that show a resistance to fire and those that are capable of sending new sprouts out of the old stumps.[5]

If fire repeatedly invades an abandoned ladang, the tree flora is reduced until finally only grasses and herbs survive. These persist because their root systems, tubers, or bulbs send out new shoots after the fire. The more important grasses and herbs that finally triumph over forest vegetation are *Imperata cylindrica* Beauv., *Imperata constricta* Hitchc., *Saccharum spontaneum* L., *Themeda gigantea* Hack, and *Eupatorium odoratum* L. These burn very rapidly and with a hot flame, even when still green. After they are well established they catch fire or are set afire every year, almost without exception; this prevents natural reforestation. Furthermore, the fire will also invade the neighboring forest and force it gradually back, so that the grass area steadily increases. This is especially true in hilly or mountainous country. In the Philippines about 18 per cent of the total area is covered with grass vegetation as a result of shifting cultivation and fires. These grass areas are known locally as *cogonales*. In the Indies, where they are called *alang-alang*,[6] they are also extensive, as in southern Sumatra. While I was traveling through southern Sumatra by rail during the dry season I counted twenty grass fires started within the space of an hour by sparks from our locomotive. In Thailand there is as yet very little cogonal land, according to Pendleton.[7] In Indochina culturally induced grasslands, either pure or with scattered trees—the latter called by the French *forêts claires* or *savanes-parcs*—are very extensive[8] (Figs. 23, 24).

Once the vegetation of an area has been changed from forest to grass, the land is of no more agricultural use to the shifting cultivator. The only benefit that he then gets out of the land is for hunting. For this, too, he often utilizes fire—either to drive wild animals out of hiding or to get a growth of young grass to attract his prey.

---

[5] In Malaya and Sumatra the spread of such bushes and trees as *Lantana camara* L., *Schima bancana* Miq., *Peronema canescens* Jack, *Vitex pubescens* Vahl, *Melastoma malabathricum* L., and *Lagerstroemia speciosa* Pers. is due to the cutting of clearings.

[6] *Imperata* (both *constricta* and *cylindrica*) is known in the Philippines by the Spanish term *cogon*, in the East Indies as *alang-alang*, and in Malaya as *lalang*. The last two are Malay words.

[7] Pendleton: "Some Interrelations between Agriculture and Forestry," 1939, p. 41.

[8] Gourou: L'utilisation du sol, 1940, pp. 368–372.

When fires no longer ravage the grasslands, however, trees return and finally drive out the grasses completely, for the latter cannot stand shade.[9]

The spread of grass is especially rapid in regions with a long, pronounced dry season, whereas reforestation is more likely to take place in areas with a well distributed, high annual rainfall. This may explain why the Philippines and Indochina have a larger percentage of open grasslands than do Malaya and the Netherlands Indies.

A relatively high density of population is an important factor in preventing successful reforestation and encouraging grasses. In a thinly settled forest area there is sufficient land to allow for a long rest period, but an increasing density of population results in the shortening of the fallow period to such a degree that the soil is given insufficient time to recover its fertility. The shorter the fallow period the poorer the succeeding forest —if any—and the poorer the following crop yields, until finally the land can no longer satisfy the needs of the shifting cultivator.

Pendleton believes that this cycle may have contributed materially to the downfall of the Khmers in Cambodia (Cambodge). He states that although the Khmers used the plow on very limited areas of wet riceland, they had largely to depend on the planting stick and shifting cultivation for their food supply. As the demands for more food arose from the rapidly increasing population concentrated in the city of Angkor, the fallow period granted to the uplands had to be progressively shortened. At the same time fires became so frequent that culturally induced grasslands spread while the cultivable area decreased, thus further diminishing the food supply available to the people of Angkor. "While the actual fall of Angkor is ascribed to the superior military power of invaders, the resistance of the inhabitants was doubtless much less than it would have been had the lands been productive and they had had adequate supplies of food."[10]

Declining crop yields and the spread of grasses have been advanced as important reasons for the breakdown of the Mayan civilization,[11] and Pendleton further raises the question "whether the development of cogonals might not have been an important contributing factor in the dying out of some of the other civilizations of the tropics, notably that of Anuradapura, in Ceylon."[12]

Today the ruins of the Mayan cities, as well as those of Angkor, are surrounded by relatively luxuriant forest vegetation, which has had time to displace the pernicious grasses since man and his tool, fire, have abandoned the inhospitable grasslands.

Not only does shifting cultivation alter the character of the vegetation,

---

9 When an old ladang remains unburned for several years the following trees spring up in addition to those already mentioned in footnote 5: *Erechthites hieracifolia* Rafin., *Blumea balsamifera* DC., *Peltophorum dasyrachis* Kurz, *Macaranga tanarius* Muell. Arg., and *Piper aduncum* L.

10 Pendleton: *op. cit.*, pp. 46 and 47.

11 Cook: "Milpa Agriculture," 1921, pp. 307–326.

12 Pendleton: *op. cit.*, p. 47.

but also it may induce soil erosion, especially when it is practiced on slopes. Forest vegetation keeps the soil from being washed or blown away, but at the end of the dry season, when the trees have been cut and burned, the soil lies exposed. Consequently, during the early days of the rainy season large quantities of topsoil are washed downhill and carried by rivers to the plains. Erosion, by lowering the moisture-holding capacity of the soil as well as the rate of absorption, may thus cause further acceleration of run-off and lead finally to destructive floods on the plains. Such floods are frequent in the Philippines and parts of the East Indies. Pendleton points out that in Thailand the main rivers receive their great load of silt from the more steeply sloping mountains of northern Thailand where shifting cultivators have started the forces of accelerated erosion.[13]

The large quantities of valuable timber destroyed year after year by the shifting cultivators in virgin areas and the prevention of reforestation by the fires that spread through the abandoned fields (Fig. 22) cause the forester to regard the shifting cultivator as his enemy. In the Philippines the forester has for many years been pleading for strictly enforced laws and severe penalties against shifting cultivators.

> The greatest enemy of our forest is caiñgin making. It is to the persistent practice of this crude form of agriculture from time immemorial that the existence of vast areas of denuded waste lands is attributable. Much of our valuable timber has been and is still being wasted by this illegal form of forest clearing. There are actually over five and a half million hectares of open lands that are practically idle. Due to continued caiñgins and the recurrence of fires on these open lands, the forest area is decreasing year after year. . . . Destructive floods in various parts of the Islands and the yearly appearance . . . of locust pest, the extermination of which is costing the Government hundreds of thousands of pesos yearly, are chiefly the direct results of this obnoxious practice.[14]

Conflicting Views on the Merits of Shifting Cultivation

Judgments on the merits of shifting cultivation have been rather contradictory since the practice first came to the attention of scientific observers. During the past century the Dutch in Java spoke of it as *Raubbau* (robber economy), and one still finds the expression used. Sauer protests against this term. According to him, shifting cultivation is a "primitive economy, in which the per capita area of land needed for a community is rather large and in which the forest growth is to be considered as a part of a long term rotation of fields." He is of the opinion that it is "an excellent device for maintaining permanent productivity."[15]

Lord Hailey calls shifting cultivation "less a device of barbarism than a concession to the character of a soil which needs long periods for recovery and regeneration; and the trekking habit, whether of Dutch farmers

[13] Pendleton: "Soil Erosion in the Tropics," 1940, p. 757.

[14] Letter of the Director of the Bureau of Forestry to the Secretary of Justice, September 10, 1934.

[15] Sauer: "Destructive Exploitation," 1938, p. 496.

[in South Africa] or African tribes, has been due to the requirements of that 'extensive' system of occupation which the character of the soil has seemed to demand."[16] But he knows from his African experiences that shifting cultivation has there reached the point where it has become the source of problems that tax the ability of the colonial administrator. "The problems which arise today are mainly due to the changes forced on this system by the restriction of land, the stabilization of populations . . . and the extension of cultivation in order to produce crops for sale."[17] He considers shifting cultivation, or *chitemene* as it is called in parts of Africa, "most wasteful, but, as experience has taught the African that burnt forest land is particularly suited to the staple crop of small millets, it is practised throughout the forest areas of Africa, and in some areas all high forest has already disappeared."[18] Worthington thinks that "shifting cultivation is admirably adapted to the needs of primitive peoples, provided there is sufficient land available"; but he also states that "after a long period of shifting cultivation the whole character of the vegetation may be altered beyond recognition. In countries with an equatorial climate, without a dry season, shifting cultivation usually changes virgin forest into secondary forest and thence to a kind of bush vegetation."[19]

The growth of population and the change from a subsistence to a money economy have increased the land requirements of the African hoe-culture peasant to such an extent that shifting cultivation has led to large-scale deforestation and has brought soil erosion, which today is recognized as one of the main problems of agriculture, in its wake. Deforestation and soil erosion have forced the colonial administrators of Africa to study shifting cultivation carefully in an attempt to find a solution to the problem. The Council of the Royal African Society passed the following resolution on October 20, 1937:

> That this Council views with the gravest concern the widespread destruction of the African soil by erosion consequent on wasteful methods of husbandry [shifting cultivation] which strike at the basis of rural economy and Native welfare, and is of opinion that immediate steps should be taken for the adoption of a common policy and energetic measures throughout British Africa in order to put an effective check upon this growing menace to the fertility of the land and to the health of its inhabitants.[20]

Shifting cultivation has been practiced on hilly land for countless centuries in India, where examples are found in Bengal and Assam, and in the northern parts of Burma; but because of disease and feuds, there was

16 Hailey: An African Survey, 1938, p. 1. We refer especially to shifting cultivation in Africa because there it has attracted much more attention than in Southeastern Asia. In the greater part of Africa it is the only form of agriculture, whereas in Southeastern Asia it takes a minor position.

17 *Ibid.*, p. 879.

18 *Ibid.*, pp. 879–880.

19 Worthington: Science in Africa, 1938, pp. 376–377.

20 Land Usage and Soil Erosion in Africa, 1938, p. 3.

always sufficient land for the population without undue pressure on the forests. A result of British rule has been to increase the population and correspondingly to reduce the rest period of the soil. The consequences have been deforestation and soil erosion. For that reason the authorities have attempted to stop this mode of land use.

In Java shifting cultivation attracted the attention of the government as early as the beginning of the nineteenth century. It was practiced especially in the mountains. The fact that a shifting cultivator required more land than a peasant who cultivated the same piece of land year after year made shifting cultivation extremely undesirable to a colonial administration that wanted to use all available land for the production of cash crops—either under the "forced culture system" of 1830 to 1870[21] or under the private plantation system that has existed since 1870. This circumstance, together with an alarming increase in population, prompted the government to do everything possible to stop the practice of shifting cultivation in Java, and today it is found only in the quite thinly settled residency of Bantam in West Java.[22] In the Outer Islands, with their sparse population and large forest areas, it is widely practiced. But, in contrast to their predecessors, Dutch officials of recent times rarely use the term *Raubbau* for shifting cultivation. Instead they consider this form of land use justifiable if the abandoned fields revert to forest and have the possibility of regaining their former fertility. In the past fifteen years, however, warning voices have pointed out that in large parts of the Outer Islands ladang culture has become a source of destruction, for the forest is no longer being given an opportunity to recuperate.[23] Both agricultural and forestry officials have consequently become concerned over shifting cultivation. In 1930 Hagreis presented a paper on ladang culture before the congress of officials of the Agricultural Advisory Service in the Netherlands Indies.[24] At that time it was decided to appoint a special commission to study ladang culture.[25] The foresters, also, devoted a good deal of attention to the problem during their congress in 1937. In 1940 Van Beukering of the Department of Economic Affairs summarized the result of the ladang *enquête*.[26] He came to the conclusion that shifting cultivation is a logical and economically justifiable form of land use wherever there is a large amount of land available and a primitive economy exists. From a social point of view no objection can be raised if, after each occupation, the fertility of the soil is regained. There is no reason to insist upon more intensive forms of land use so long as the population density does not increase beyond 50 persons per square kilometer, or 130 persons per square mile. If at this density we

21 From 1830 to 1870 the peasants of Java were forced, under the notorious "culture system," to devote part of their land and labor to producing crops for the government.

22 Kools: Hoema's, 1935.

23 Gonggrijp: "Soil Management," 1938, pp. 397–404.

24 Hagreis: "Ladangbouw," 1930, pp. 43–75 and 629–642.

25 "Enquête betreffende het ladangvraagstuk," 1931, pp. 1022–1033.

26 Van Beukering: Het ladangvraagstuk. Unpublished typescript, 1940.

allow for five persons per family and assume that the family shifts its cultivation plot annually but returns to the same plot after an average period of 10 years, each family could occupy annually one hectare, or 2.47 acres. This is the amount of land that an average shifting cultivator in the Indies needs in order to cover the food requirements of his family.

This calculation does not make any allowance for permanent forests. Actually, however, the land requirements are not quite so high, because very often the same land is used for two years instead of one, which would either allow for a longer period of fallow or keep a part of the virgin forest intact. The latter is improbable because of the shifting cultivator's preference for virgin forest.

To summarize these opinions, shifting cultivation from the agricultural point of view is not considered objectionable so long as vegetation and the fertility of the soil are restored by nature. The opinion of those who regard it as a fatal agricultural system under all circumstances is widely felt to be unjustified. From the silvicultural point of view, shifting cultivation is held to be objectionable if it destroys timber that would otherwise be exploited by the forest service or by private concessionaries.

Supplementing Shifting Cultivation With Cash Crops in the Netherlands Indies

The products of the typical shifting cultivator never reach a distant market. There is seldom any surplus, for the shifting cultivator is a subsistence farmer who lives from hand to mouth. In the Outer Islands of the Netherlands Indies, however, the shifting cultivator has been drawn into modern economy and has modified this age-old system. In parts of Sumatra and Borneo he has begun to diversify his economy by planting tree crops on the ladang among his rice or maize. After one or two harvests the fields are abandoned and left to the trees that have been planted. The trees then furnish an additional source of income as soon as they reach the bearing stage. Pepper has been raised as a ladang crop for many centuries, especially in the Lampoeng Districts. Coffee was introduced in the days of the Dutch East India Company and adopted as a commercial crop by the peasants. Benzoin (*Styrax benzoin* Dryander), also, has long been grown in this manner. More recently coconuts and, since about 1905, rubber have become ladang crops (Figs. 25–33).

A shifting cultivator may, over a period of years, convert his old ladangs into rubber gardens, for example, until he is assured of a sufficient income from his permanent tree crop to buy his food from the outside and no longer requires a ladang for the production of food. This process has led to the creation of the large native rubber areas in Borneo and Sumatra, where some shifting cultivators have become dependent upon the market price of rubber and upon the policy of the government in regard to taxation and export control. In times of low prices and export restrictions they may suffer severe economic setbacks and even hardship. In Bengkalis in the East Coast of Sumatra a few years ago,[27] rubber gardens occupied so large an area that there was actually a scarcity of land suitable for the making of new ladangs for food crops, and rubber trees had to be cut down. So long as the Indonesians[28] had their rice ladangs the market quotations of New York and London meant nothing to them; but as cash-crop producers they began to feel the effect of every fluctuation on the world market.

Although this modification of the old economy has led to permanent gardens of rubber, coconut, and benzoin trees and thus to a partial abandonment of shifting cultivation, where pepper and coffee supplement the traditional crops of the shifting cultivator no fundamental change takes place, because these gardens are not permanent. The cultivation of coffee, especially, on ladangs without shade trees or other protective measures, such as terraces, causes erosion and exhausts the soil very rapidly. The lifetime of coffee bushes on ladangs is only from 3 to 5 years once they have started to yield berries—much shorter than that of those on well-managed European estates.[29] The lifetime of a pepper garden is considerably longer, perhaps 15 years. On the slopes of Mt. Dempo and in other parts of the residency of Palembang, shifting cultivators plant coffee bushes 1½ to 2 meters apart in the midst of upland rice fields during the first year that they occupy a ladang. In the following year the bushes are still small enough to permit the growing of a rice crop among them. In the third year a new ladang is made and planted with rice and coffee, while a coffee harvest is gathered from the first ladang. In the fourth year the first ladang produces an excellent crop of coffee. In the fifth year the

TABLE I—SCHEME OF OPERATION OF SHIFTING CULTIVATORS RAISING COFFEE AS A CASH CROP IN LADANGS, IN TERMS OF HARVESTS

YEAR	LADANGS				
	FIRST	SECOND	THIRD	FOURTH	FIFTH
1st	rice				
2nd	rice				
3rd	coffee	rice			
4th	coffee	rice			
5th	coffee	coffee	rice		
6th	coffee	coffee	rice		
7th	coffee	coffee	coffee	rice	
8th		coffee	coffee	rice	
9th		coffee	coffee	coffee	rice
10th			coffee	coffee	rice
11th			coffee	coffee	coffee
12th				coffee	coffee
13th				coffee	coffee
14th					coffee
15th					coffee

27 Van Suchtelen: "Parallelen," 1938, pp. 575–584.
28 Native inhabitants of the East Indies.
29 Vonk: "Systematisch beheer van het ladangareaal," 1937, p. 621.

first ladang yields its third coffee harvest and the second its first coffee harvest, while rice and coffee are planted in a third ladang. In the seventh year the cultivator may have as many as four ladangs, the first producing

FIG. 2—Sakai ladangs in the midst of the forest in the Tasek (Lake) Bera region, Pahang, Malaya. (Reproduced through the courtesy of the *Geographical Review*.)

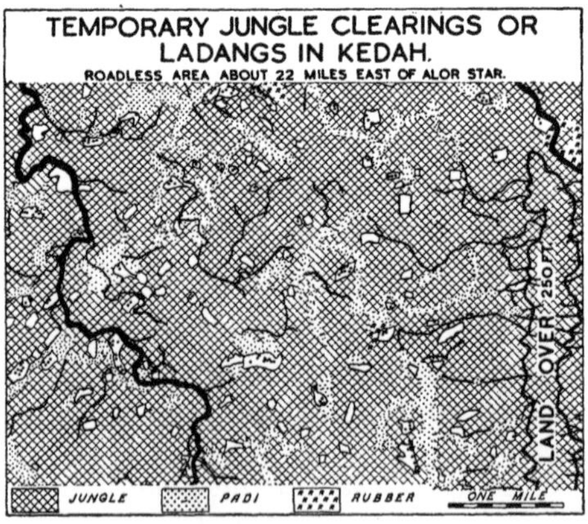

FIG. 3—Small ladangs in the forest, Kedah, Malaya. (Reproduced through the courtesy of the *Geographical Review*).

its last coffee crop before it is abandoned because of declining yields, the second yielding its third coffee harvest, the third just entering the bearing stage, while the fourth ladang supplies the shifting cultivator with rice grown among young coffee bushes.

The Areal Extent of Shifting Cultivation in Southeastern Asia

It is difficult to form an idea of the areal extent of shifting cultivation because of its migratory nature and the fact that it is often practiced in remote regions. The only country of Southeastern Asia in which an attempt has been made to estimate the area used for shifting cultivation is the Netherlands Indies, but even there it was not possible to include all parts of the Outer Islands in the estimate. Van Beukering concluded that one-third of all food consumed in the Outer Islands was grown on ladangs; this would require a total of 1,100,000 hectares of land. If we assume that all ladangs are occupied for 2 years and lie fallow for 12 years, the shifting cultivation system would need seven times 1,100,000, or 7,700,000, hectares. This estimate does not include the ladang area under coffee and pepper cultivation. (Rubber, coconut, and benzoin cannot be considered because they are, for practical purposes, permanent crops.)

Table II—Philippine Islands: Illegal Caiñgins Detected, 1914-1940

	Illegal Caiñgins	
Year	Number	Area in Hectares
1914	895	....
1915	1 198	....
1916	845	....
1917	665	....
1918	776	....
1919	603	....
1920	721	....
1921	1 602	1 429.48
1922	3 110	2 803.55
1923	1 759	1 516.33
1924	1 753	1 396.90
1925	1 762	1 391.29
1926	1 829	1 476.61
1927	2 002	1 743.92
1928	1 774	1 602.95
1929	1 312	1 370.41
1930	3 227	3 279.77
1931	2 534	3 345.74
1932	2 547	5 586.27
1933	2 427	2 831.38
1934	3 207	3 215.46
1935	2 631	2 455.73
1936	2 314	2 510.59
1937	1 967	2 308.24
1938	1 766	2 029.68
1939*	906	1 331.12
1940†	2 646	4 416.56
Total	48 778	48 071.98

Source: Data supplied by the Bureau of Forestry, Manila.
* Six months only, ending June 30, 1939, because of change of fiscal year.
† From July 1, 1939, to June 30, 1940.

Van Beukering allowed an additional 200,000 hectares for coffee and pepper ladangs; so that the total area amounts to about 8,000,000 hectares. If we assume that the population of the Outer Islands increases at the rate of 1½ to 2½ per cent each year, it would double in from 29 to 47 years. This would mean that after about 30 to 50 years 16,000,000 hectares would come within the scope of shifting cultivation.

In Malaya ladang cultivation is illegal, but it is practiced in the hinterland well away from roads and railways, in eastern sections of the peninsula. Dobby recently described it in Pahang and northeastern Kedah, near the Thai border.[30] The Sakai of Pahang regularly abandon their primitive huts as well as the clearings when they shift to a new field (Map, Fig. 2). The Malays of northeastern Kedah, on the other hand, as a rule live permanently in one place and change only their fields (Map, Fig. 3). There are no statistics available on the areal extent to which shifting cultivation is carried on in Malaya, but according to Dobby the system gives way to permanent culture "as the standard of cultivation rises and as the effect of law becomes stronger."[31]

In Thailand shifting cultivation is extensively practiced by the tribes of the mountains of the north and west, on the eastern part of the Khorat Plateau, and on the peninsula. Probably about 1,000,000 people depend upon it. According to the same kind of calculation that Van Beukering made for the Indies, a total of 1,400,000 hectares would be required, or 200,000 hectares annually, if we allow five persons per household and consider that one hectare covers the needs of a family for 2 years.[32]

In Indochina only the tribes of the mountains of Tongking, Laos, and Annam now practice rây, or shifting, cultivation. The people of the plains, Annamese and Cambodians, who represent 84 per cent of the population, practice a more intensive form of agriculture. Gourou has estimated the areal extent of shifting cultivation in Indochina. He states that a rây is used for 2 years and has to lie fallow for at least 20 years, which seems a rather long period as compared with other countries. In addition to its extensive plains, Indochina has a mountainous or hilly area of 580,000 square kilometers. Only one per cent of this land is permanently cultivated. Gourou estimates that about 40 per cent of the mountainous country is suitable for rây cultivation and he assumes that one-tenth of this, or 4 per cent of the total mountain land, is under rây cultivation in any one year.[33] If this calculation is correct it would mean that 23,200 square kilometers, or 2,320,000 hectares, are occupied annually by shifting cultivators. The numerical strength of the shifting tribes shows, however, that this estimate cannot be correct. Their number is not much greater than

30 Dobby: "Settlement Patterns in Malaya," 1942, p. 222.

31 *Ibid.*

32 For an excellent description of the life of the shifting cultivators in Thailand, see Credner: Siam, 1935, pp. 148–173.

33 Gourou: L'utilisation du sol, 1940, pp. 177–182, especially pp. 180 and 182.

1,500,000, or at the most 2,000,000, which would mean about 300,000 to 400,000 households. This number of households could hardly occupy such a large rây area as Gourou assumes.

In Formosa only the Malay tribes of the mountains in the eastern half of the island practice shifting cultivation. They number, according to estimates, about 150,000. From this we may assume that not more than 30,000 to 50,000 hectares are occupied annually by shifting cultivators.

In the Philippines it is unlawful to make caiñgins on public land without the permission of the Bureau of Forestry.[34] Permission will be given if the forest on the land involved is of no commercial value and if the caiñgero accepts the obligation to reforest his patch with seeds supplied by the forest service. Several hundred caiñgeros are brought annually before the justices of the peace, who in the past, because their sympathy was with the caiñgeros, were usually rather reluctant to fine the culprits (Table II). Nobody knows, of course, how many illegal caiñgins escape detection, but it is certain that they greatly outnumber those that are made with the permission of the foresters added to those that have been brought to the attention of the courts. Special consideration is given to the "non-Christian" tribes, for whom tracts of land have been set aside within which they continue their age-old system of cultivation. In Mindanao probably three-fourths of the population, from the primitive Manobos to the more advanced Moros, practice the caiñgin system of cultivation. Other regions where this system is practiced extensively are Mindoro, Cebu, southern Negros, and parts of Luzon—the central and eastern cordillera and the Camarines. About 10 per cent of the population of the Philippines obtain a part or all of their food from caiñgins, most of which are made either on private land or on land that belongs to non-Christian tribes who know no other form of agriculture.

Methods of Eliminating the Undesirable Aspects of Shifting Cultivation

In regions where the density of population exceeds 50 persons per square kilometer, shifting cultivation becomes wasteful and destructive. It is imperative in such areas that the system be either modified or else replaced by more conserving and, where possible, more intensive forms of land use. For most backward tribes this involves fundamental changes in their mode of life, religious ideas, and concepts of land rights.

In some instances it should suffice to adopt measures that eliminate the harmful or wasteful after effects of the system without requiring that this kind of land use be given up completely. Let us assume that we have a somewhat thinly settled region containing a forest that is not particularly valuable and the exploitation of which is impracticable or at present inexpedient. Deforestation and the spread of grasses are undesirable aspects

34 Nano: "Kaiñgin Laws," 1939, pp. 87–92.

of shifting cultivation in the area. In such a case reforestation would be assured by keeping fire out of the old clearings. This could be done by leaving relatively non-inflammable forest belts at least 60 feet wide between individual clearings. These forest belts would offer some protection against the spread of grass fires and at the same time would be a source of seed for further reforestation. Reforestation under such circumstances would be much easier and cheaper than if large contiguous areas were to be cleared and used.

Where natural reforestation is particularly slow, cultivators should be required to practice artificial reforestation. In scattered regions in Southeastern Asia tribesmen do this on their own initiative. Pendleton reports such a case in Burma.[35] Certain Dyak tribes in Borneo plant seeds of the candlenut tree (*Aleurites moluccana* Willd.) about 3 meters apart in a freshly cleared ladang before it has even been burned. In the eastern part of the Malay Archipelago shifting cultivators plant such trees as *Aleurites moluccana* Willd., *Erythrina lithosperma* Miq., *Albizzia falcata* Backer, *Hibiscus similis* Bl., *Casuarina equisetifolia* L., *Macaranga* species, *Euphorbia* species, *Sesbania grandiflora* Pers., and *Desmodium timoriense* DC. for the purpose of restoring the forest.

In Sumatra Deli tobacco planters initiated artificial reforestation after alang-alang had regularly invaded their fallow fields and caught fire so frequently that second-growth forest could not get a start. In Deli the tobacco fields are in use only 1 year and lie fallow 7 to 10 years. They should be protected by forest growth during that period. Therefore, in order to make this possible, the planters controlled fires by planting belts of teak along the highways and roads. These belts served as fireguards and at the same time provided lumber for tobacco-drying sheds. On the tobacco fields themselves they planted especially *Albizzia falcata* Backer, *Cassia siamea* Lam., *Pithecellobium saman* Benth., and *Leucaena glauca* Benth., which drive out the alang-alang[36] (Figs. 34–39).

Another way of speeding up reforestation is to spare a few of the larger trees at the time of the clearing. This is done by the Malays of Sumatra. They do not cut down *Koompassia parvifolia* Prain, *Arenga saccharifera* Labill., *Durio zibethinus* Murr. (durian), *Parkia* species, or *Garcinia grandifolia* Pierre.[37] The last three are fruit trees that are numerous in the forest, whereas the second is a palm whose flower stalk is tapped for its sugary sap and the first is preferred by bees as a hiving tree, which gives it its name, "bee-tree." This tree is preserved because the people want the honey. Worthington reports that in Africa "some tribes appear to realize the necessity of maintaining tree growth on their clearings and merely top the trees and lop the branches, so that the trees are enabled to

35 Similar cases are known in the residency of Benkoelen in Sumatra, in central Celebes, and in Flores. See Hagreis: "Ladangbouw," 1930, p. 631.

36 Jochems: "De begroeiing der tabakslanden," n.d., pp. 77–89.

37 *Ibid.*, pp. 30–32.

recuperate quickly when the area is again allowed to revert to bush."[38]

Several of the forest services of Southeastern Asia have made a rather successful use of the shifting cultivator in their endeavor to reforest the land. They have combined shifting culture and silviculture by hiring shifting cultivators to cut mature trees and to sow or plant new trees on cutover or burned-over land. For this the cultivators are paid a small wage. In addition they are allowed to grow their own crops for one or two years between the rows of young trees. This system, known as the "taungya forestry system," was developed in 1856 by the German forester, Dietrich Brandis, while he was Superintendent of Forests in Pegu, Burma.[39] It is ingenious in that it recognizes the habits of the shifting cultivator and fits them into a modern forest economy instead of persecuting the cultivator and forcing him to give up his traditional culture. This system has been used in the teak forests of Java since 1871. The forest service makes a contract[40] with the peasants of Java according to which the peasants are to prepare the land and plant and take care of teak trees (*Tectona grandis* L.) in return for a small sum of money and the right to plant such crops as maize and red pepper (*Capsicum annuum* L.) between the rows of teak. The cultivation of upland rice is discouraged, and the planting of cassava, peanuts, castor beans, and tobacco usually is not allowed (Figs. 40, 41).

The taungya forestry system is an excellent way of getting pure stands of commercial timber. The typical forest of the tropics is expensive to exploit because of the large variety of trees which it contains, many of which have little or no commercial value.[41]

As important as reforestation is soil erosion control. Where shifting cultivation is practiced in mountainous regions it often causes such acceleration in the rate of erosion that the rivers are choked with sediment and periodically do great flood damage. Flood control could be achieved in most cases by completely closing the headwaters and watershed areas to cultivators. On more gentle slopes the clearings should be rather narrow, should parallel the contours, and be separated by belts of forest. If at the same time the forest uphill were protected by fireguards, the clearings would have a chance to close quickly again so that the soil would be

---

38 Worthington: Science in Africa, 1938, pp. 378–379.

39 *Taungya* is the Burmese term for the temporary field of the shifting cultivator, a term that the foresters have adopted. E. P. Stebbing in his work, The Forests of India, Vol. 1, 1922, p. 376, writes:

"Brandis, moreover, recognized that the very system which, till then, had destroyed teak forests might be utilized to produce them, . . . and as early as 1856 an attempt was made in Prome district to sow teak in regular rows, together with the rice scattered by the toungya [*sic*] cutters. Twenty years afterwards this attempt had grown into a regular system, by which areas were being annually stocked with teak at a cost far below that of regular plantations." See also *Ibid.*, Vol. 3, 1926, pp. 415–428.

40 Becking: "De djaticultuur op Java," 1928, pp. 50–52.

41 Heske: "Der tropische Wald," 1939, p. 414. Heske's article gives an excellent short account of the forests of Southeastern Asia and points out that they are on the whole much easier to exploit than are the forests of tropical America because of the abundance of cheap labor.

protected against abnormal erosion. Terracing would be the most effective method of checking soil erosion, but this requires a good deal of labor, which is scarce in areas of shifting cultivation.

The Forestry Service of the Indies started an interesting soil-erosion experiment near Tjiwidej in the mountains south of Bandoeng at an elevation of 1800 meters. Two adjacent parcels of forest land, lying on rather steep slopes, are drained separately—out of each parcel comes a little brook—and thus form separate drainage units. Both parcels were used by shifting cultivators who had cut and burned the forest. In one case the slope was left undisturbed; in the other the slope was terraced. At the foot of each slope catch basins were constructed to retain the sediment that was washed down by rain water and transported by the brooks. The amount of water that drains out of each unit was measured and the sediment weighed. The purpose of this experiment was to get quantitative measurements of the amount of erosion that is caused by shifting cultivation and to study the effect of terracing in reducing soil loss. During the first year both parcels lost very little soil, because the humus layer protected it; but in the second year the unterraced parcel lost 5.065 kg. of silt per square meter whereas the terraced parcel lost only one-half as much, or 2.524 kg.[42] (Figs. 42–45).

Under the conditions that we have discussed so far, no radical change in the system of cultivation is required. If, however, the population increases to such an extent that there is no longer enough available land to allow for a sufficiently long rotation cycle, more drastic remedies are necessary.

Such a situation arose in parts of the Outer Islands and attracted the attention of government specialists. Legal experts familiar with Indonesian customary law, foresters, and agricultural advisers collaborated to seek a solution.[43] Realizing that shifting cultivators are usually reluctant to give up their traditional form of land use, the specialists approached the ladang problem by searching for means of reforesting ladangs quickly. Experiments were made with the planting of quick-growing leguminous and non-leguminous trees, bushes, and shrubs to shorten the fallow period, restore the humus, protect the soil against accelerated erosion, eliminate the grasses, and, in the case of leguminous species, improve the soil by adding nitrogen.[44] We have already seen, however, that artificial refor-

[42] I was able to visit the Tjiwidej experiment with Mr. L. Gonggrijp, under whose supervision the work was carried out. For a detailed report on the project see Gonggrijp: "Het erosie-onderzoek," 1941, pp. 200–220.

[43] H. J. Friedericy of the Department of the Interior and H. Vonk and E. J. Wind of the Department of Economic Affairs, for example, worked together on the problem of ladang culture in the residency of Palembang.

[44] Experimental plantings included, among others, *Erythrina lithosperma* Miq., a light-loving tree which is typical in Southeastern Asia of dry, rather open climax forests and of second-growth forests; *Hibiscus tiliaceus* L.; *Albizzia falcata* Backer, one of the fastest-growing trees of Southeastern Asia, used widely as a shade tree on coffee plantations; *Aleurites moluccana* Willd., a tree at home in the drier parts of the East Indies; *Leucaena glauca* Benth.; *Pterocarpus indicus* Willd., a large tree frequently used along roadsides as

estation is not a discovery of Western-trained scientists but has long been practiced by certain groups in Southeastern Asia. It was the plan of the government to spread and enforce this practice of artificial reforestation wherever advisable. This practice, whereby the cultivator determines the vegetative cover during the fallow period instead of abandoning the ladang to chance growth, brings him to a transition stage between shifting cultivation and permanent cultivation. From there the step to permanent cultivation with crop rotation is but a short one.

Permanent cultivation requires repeated tilling and therefore more labor and tools than the shifting cultivator is accustomed to. Work animals are also desirable. An important obstacle to intensification of agriculture in wide areas of Southeastern Asia, however, is the fact that hill and mountain regions [up to about 1500 meters in elevation] are frequently infested with malaria so that the energy of the people is greatly reduced.[45] Moreover, the acquisition of such implements as plows, harrows, and hoes, and especially of work animals, calls for financial resources or credit, both of which the average shifting cultivator lacks. Some shifting cultivators do own livestock, such as cattle or carabaos, but utilize neither their labor nor their manure. Groups of the Mois in the mountains of Annam have carabaos, but they do not use the animals in their agricultural activities because they "cannot talk" to them.[46] On the Lesser Sunda Islands livestock is kept because it gives social prestige to the owner. Overstocking of the pasture areas on these islands has led to overgrazing and has cut into the forest area until there has developed a scarcity of land suitable for ladang cultivation. The same has happened in large areas in Africa. There it has been necessary to teach the native peasant the use of cattle and make him familiar with the plow. This was done successfully in northern Nigeria by combining the animal husbandry of the nomadic Fulani with the indigenous shifting cultivation of the Hausa peasant, thus developing "a system of permanent farming whereby a man with cattle of his own, say two working oxen and a few cows, can plough and maintain in fertility an area of land sufficient to supply all the requirements of himself and his family, and to obtain a surplus balance as a return for his labour. . . . Thus mixed farming represents a gradual evolution from the shifting methods of cultivation to a static type of agriculture, by the introduction of animal husbandry, and with the use of animal cultivation."[47] Above all, the peasant learns to preserve the animal manure and

a shade tree; *Dolichos lablab* L., a climbing herb which is valuable as a vegetable; *Crotolaria anagyroides* H.B. and K., a bush of vigorous growth which attains a height of from 12 to 14 feet and forms an excellent green manure; *Tephrosia candida* DC. and *Tephrosia vogelii* Hook., low shrubs useful as green manures; and *Mimosa invisa* Mart., a leguminous creeper which suppresses grasses.

45 Gourou: L'utilisation du sol, 1940, pp. 181 and 350.

46 *Ibid.*, p. 354.

47 King: "Mixed Farming in Northern Nigeria," 1939, p. 271. See also Faulkner and Mackie: West African Agriculture, 1933; and by the same authors: "The Introduction of Mixed Farming in Northern Nigeria," 1936, pp. 89–96.

to apply it regularly to his fields in order to keep up the fertility which formerly was restored through long forest fallow. This animal fertilizer is more important than the help the peasant gets from his animals in the way of labor. Thus the area to be cultivated may be limited by the amount of manure available, rather than by the extent of land that can be plowed.

Van Beukering, in his summary of the results of the ladang *enquête*, is of the opinion that similar practices should be introduced in the Outer Islands of the Netherlands Indies; but the pasture problem must first be carefully studied before mixed farming can be widely established in the Indies. Where animal manure is not available, the composting of vegetable waste, or even of crops grown for this specific purpose, supplies an excellent fertilizer.

The intensification and modification of shifting cultivation are less difficult in flat country than on sloping land where there is the hazard of soil erosion. Perhaps the most obvious solution would be to move the population from mountainous land to the plains and valley floors, but this can be done only in areas where population density is low. In Southeastern Asia the density of population on the plains is usually so high that, should it be possible to bring malaria under control, a movement from the plains to the uplands and mountains would be desirable rather than a migration in the opposite direction. Another solution would be to have the shifting cultivator plant permanent tree crops to protect the soil against accelerated erosion. Steep slopes which now produce but a single moderate rice crop every few years could be made more productive if planted with perennials. The cultivator could then buy his staple food with the money obtained from the sale of the crops harvested from his trees.

The speed with which the change from shifting cultivation to permanent cultivation may be made will depend both on the geographical environment and the cultural level of the community. Shifting cultivators with permanent villages and a regularly established rotation of fields will respond much more readily to a reform brought from the outside than a group that is always ready to abandon its dwellings and move to a new locality.

–All the photographs in this book except those for which acknowledgement is made were taken by the author.)

(upper left)—*Caiñgin* freshly cut and burned in second-growth forest. Cagayan Province, Luzon. (Courtesy of the Bureau of Forestry.)

(upper right)—Cutting down the larger trees has been avoided by girdling. The caiñgin is ready for planting. Bataan Province, Luzon. (Courtesy of the Bureau of Forestry.)

(center left)—Burning a freshly cut caiñgin. Cotabato Province, Mindanao.

(center right)—Burning *ladangs*. Residency of Palembang, Sumatra.

(lower left)—Ilongot's caiñgin planted with upland rice. Nueva Vizcaya Province, Luzon. (Courtesy of the Bureau of Forestry.)

(lower right)—Ladang planted with upland rice. The second-growth forest in the background will be cut in a year or two. Residency of the East Coast of Sumatra.

Fig. 10 (upper left)—The temporary tree-house in this *caiñgin* provides an excellent lookout as well as protection wild animals. Cotabato Province, Mindanao.

Fig. 11 (upper right)—Shifting cultivator in a *ladang*, about to plant upland rice seeds. Residency of the Lampoeng Di Sumatra.

Fig. 12 (center left)—The man on the right is driving a planting stick into the ground, while the man on the left pla upland rice seeds.

Fig. 13 (center right)—The holes made with the planting stick are ready for seed.

Fig. 14 (lower left)—Abandoned caiñgin invaded by *cogon*. Bukidnon Province, Mindanao.

Fig. 15 (lower right)—Abandoned ladang. Residency of the Lampoeng Districts, Sumatra.

(upper left)—New *ladangs* planted with upland rice and maize on a steep slope. Notice the young second-growth forest on the second slope and the older forest in the background. Residency of Tapanoeli, Sumatra.

(upper right)—New *caiñgins* on slopes in second-growth forest. West of Baguio, Mountain Province, Luzon.

(center left)— Ladang of upland rice and maize on level land. Residency of the East Coast of Sumatra.

(center right)—New caiñgin ready for the first planting. The flimsy hut will be abandoned, together with the caiñgin, after one harvest. South of Sagnay, Camarines Sur Province, Luzon.

(lower left)—Guarding the ripening rice against birds from a temporary shelter in a ladang. The man has a permanent house in the village. Residency of the East Coast of Sumatra.

(lower right)—Well built village house of a shifting cultivator who raises pepper in his old ladangs. Residency of the Lampoeng Districts, Sumatra.

FIGS. 22 and 23 illustrate undesirable conditions that may result from shifting cultivation.

FIG. 22—*Caiñgins* cut and burned out in a valuable commercial forest before the holder of the timber concession had a chance to exploit it. Negros Occidental Province, Negros. (Courtesy of the Bureau of Forestry.)

FIG. 23—*Cogonales* in the foreground and patches of second-growth forest in the hills. The *cogon* is gaining the upper hand because of repeated fires, which prevent natural reforestation. Near Port Banga, Zamboanga Province, Mindanao. (Courtesy of the Bureau of Forestry.)

FIG. 24—Filipino foresters investigating the extensive cogonales of eastern Nueva Ecija Province, Luzon. (Courtesy of the Bureau of Forestry.)

5 and 28 (upper left and upper right)—Pepper vines growing in an old *ladang*. Residency of the Lampoeng Districts, Sumatra.

(center left)—Young rubber tree planted in a rice ladang. No more rice will be planted after the second rice harvest. Western Bali.

(lower left)—Young rubber trees in an old ladang that has produced two rice crops. Residency of Palembang, Sumatra.

(lower right)—Closer view of a young rubber tree in the ladang shown in Fig. 27.

FIG. 30 (upper left)—Batak benzoin collector beside a benzoin tree planted in an old *ladang*. Residency of Tapanoeli, Sur (Courtesy of the Department of Economic Affairs.)

FIG. 31 (upper right)—Peasant-owned rubber garden. Residency of Tapanoeli, Sumatra.

FIG. 32 (center left)—Houses abandoned by shifting cultivators in an exhausted coffee garden. Residency of Palem Sumatra. (Courtesy of the Department of Economic Affairs.)

FIG. 33 (center right)—Rubber mangles and sheets of rubber drying in a peasant-owned rubber garden. Residency of Dj Sumatra.

FIG. 34 (lower left)—Teak belts lining the road through a tobacco plantation. Between Medan and Arnhemia, Residen the East Coast of Sumatra.

FIG. 35 (lower right)—Teak belt partially cut down to prepare for a tobacco crop. Between Medan and Arnhemia.

6 (upper left)—Eight-year-old second-growth forest on a tobacco plantation, being cleared for the raising of a single tobacco crop. Residency of the East Coast of Sumatra.

7 (upper right)—After the tobacco crop has been harvested the local population is allowed to raise one upland rice or maize crop. Notice the bushes and trees, which the rice-growers must not destroy lest reforestation be retarded. In the background a tobacco-drying shed. Residency of the East Coast of Sumatra. (Courtesy of the Experiment Station of Deli Planters Vereeniging.)

8 (center left)—Upland rice following a tobacco crop. In the background a second-growth forest consisting of *Albizzia falcata* and in front of it a strip of *Leucaena glauca*. Residency of the East Coast of Sumatra. (Courtesy of the Experiment Station of Deli Planters Vereeniging.)

9 (center right)—Two-year-old second growth on a tobacco plantation. Residency of the East Coast of Sumatra.

0 (lower left)—A section in a teak forest which has been cut by shifting cultivators working under the *taungya* forestry system. Residency of Madioen, East Java.

1 (lower right)—Maize and vegetables between pine trees planted at an altitude of 1900 meters by shifting cultivators working for the Forestry Service. Slope of Goenoeng Tangkoebanprahoe, Residency of Batavia, West Java.

This group of photographs was taken at the Soil Erosion Experiment Station near Tjiwidej, Residency of Priangan, West Java.

FIG. 42 (upper left)—Unterraced slope under cultivation.

FIG. 43 (upper right)—Terraced slope under cultivation.

FIG. 44 (lower left)—View of the catch basins.

Chapter III

THE SEDENTARY CULTIVATOR

Sedentary, or permanent, cultivation of land is much more vital to the economy of the countries of Southeastern Asia than is shifting cultivation. The majority of peasant cultivators obtain their livelihood by tilling the same piece of land year after year. The continuously utilized arable land, however, is not all cultivated in the same way throughout Southeastern Asia. We may distinguish three main categories of arable land: garden land, surrounding the houses and villages; wet land; and dry land. These three classes of land are devoted to different crops and vary so much from one another in size and reliability of yields, and therefore in value that in Java, for example, the social status of a man depends upon the type of land that he cultivates. There are districts in Java where only those peasants who own wet land, houses, and garden land have the full rights of a villager.

The Garden Plot

Garden culture on land surrounding peasant houses in Java is devoted to a large variety of crops, ranging from fruit trees to plants that are not required in large quantities but only to supplement the diet and add flavor to the daily rice meal. Ochse and Terra, who have made exhaustive studies of garden culture, are of the opinion that it is a remnant of "primitive grainless hoe culture," in which all the useful plants found in the forest were concentrated in plots near the dwellings. Originally, the number of plants grown in the garden was greater than it is today, for some plants have been taken out of this economy to be grown on a larger scale in fields. The origin of garden culture explains, according to Ochse and Terra, the truly amazing variety of plants that are found on garden land.[1] Sollewijn Gelpke, a leading expert on native agriculture in Java during the past century, remarked on the richness of the gardens:

> Whoever strolls around a compound and takes things in with a botanical eye will see before him a variety of plants of which the uninitiated can have no conception. . . . Such wealth of vegetation is still more impressive to him who approaches it from an economic point of view. He sees palms, bamboo, bananas, and other trees that resemble each other closely, bearing fruit and giving support to various climbing plants.[2]

Often one is able to distinguish three levels, or stories, in a garden. The ground story consists of low-growing plants, especially tuberous plants that tolerate shade, such as elephant yam (*Amorphophallus campanulatus* Bl.), white arrowroot (*Maranta arundinacea* L.), "purple arrowroot" (*Can-*

---

[1] Ochse and Terra: "Het onderzoek," 1934. See also Terra: "De voeding der bevolking," 1932, pp. 552–593.

[2] Sollewijn Gelpke: Naar aanleiding, 1901, p. 203.

na edulis Ker), sweet potato (*Ipomoea batatas* Poir.), taro (*Colocasia esculenta* Schott), *Curcuma* species, Goa bean (*Psophocarpus tetragonolobus* DC.), bitter cucumber (*Momordica charantia* L.), snake gourd (*Trichosanthes anguina* L.), Lima bean (large) (*Phaseolus lunatus* L.), Goa potato (*Dioscorea aculeata* L.), wild yam (*Dioscorea hispida* Dennst.), and sweet yam (*Dioscorea alata* L.). The middle story contains taller-growing plants, such as cassava (*Manihot utilissima* Pohl), citrus varieties such as *Citrus nobilis* Lour. and *Citrus aurantifolia* Swingle, papaya (*Carica papaya* L.), banana (*Musa* species), rose apple (*Eugenia aquea* Burm.), and soursop (*Anona muricata* L.). The upper story is formed by tall fruit trees, such as coconut (*Cocos nucifera* L.), jackfruit (*Artocarpus integra* Merr.), sapodilla or chico (*Achras zapota* L.), pomelo (*Citrus maxima* Merr.), of which numerous forms, differing in size and color, are known in the Indies, *belinju* (*Gnetum gnemon* L.), mango (*Mangifera indica* L.), durian (*Durio zibethinus* Murr.), candlenut tree (*Aleurites moluccana* Willd.), tamarind (*Tamarindus indica* L.), *pete* (*Parkia speciosa* Hassk.), stinkbean tree (*Pithecellobium lobatum* Benth.), and *santol* (*Sandoricum koetjape* Merr.).

This list does not by any means include all the plants that appear in the gardens. The plants vary in number and type from region to region in accordance with differences in soil, climate (elevation), and the taste of the owners.

Estimates indicate that about 15 per cent of the cultivated land in Java and Madura is taken up by gardens. Unfortunately, there have been no detailed studies of garden culture outside of Java, although the products of the garden play an important role in the diet of all the peoples of Southeastern Asia. Terra, who compared garden culture in various districts of Java, found that the share of land allotted to gardens, together with the intensity of their cultivation, increases as the total amount of crop land per head decreases, down to the point where the average amount of crop land per head is only 0.15 hectare. Where the amount of land is even less than this the share of garden land decreases, although there is a tendency at first to increase the intensity of the garden culture. Close to the larger cities, the amount of crop land per person, together with the amount devoted to gardens and the intensity of the latter's cultivation, falls off to a minimum since the population here has supplementary sources of income.

With the development of a money economy appears a specialization that is directed toward the urban centers. There has always been some natural specialization inasmuch as some plants do not grow well everywhere. The coconut, for example, does best in the lowlands. But the new specialization depends not only upon such natural factors as soil and climate, but also upon economic factors such as marketing facilities and the development of communications. This specialization can be observed generally in Southeastern Asia.

Ochse and Terra found that in calories of food value the returns of the gardens proved to be slightly higher per hectare than those of the irrigated fields. One of the great advantages of garden culture over field culture is that there is always something ready to harvest from the former and therefore something to sell when money for daily household needs becomes scarce. Another advantage is that the garden supplies the family with a great variety of side dishes rich in vitamins; the vegetable proteins of the garden produce are also of high value.

As to the possibilities of improving garden culture, these two authors state:

> It is very evident that an improvement of the compound horticulture is a very difficult problem through the highly varied and dense planting of these compounds as a result of which they come to resemble a very complicated symbiosis. But for all this it must be possible by systematic investigation to improve not only the quality but also the quantity of the production. It may be said that the compounds with their story or layer formation superficially considered will find no favour in the eyes of western agriculture, but they are an essential link in the native production apparatus.[3]

The alarming decline in the welfare of the people of Java in the face of a mounting population prompted Netherlands Indies officials to examine every possibility of improving the peasants' condition. This accounts for the studies of Ochse and Terra. Lacking detailed studies outside Java, we are obliged to depend upon scattered references of uneven value.

Credner states that in Thailand practically every house has a few trees on the lot, and the villages are often hidden by tree clusters. The greatest intensification of garden culture is found on the right bank of the Mae Nam Chao Phraya (the Menam) in Bangkok Noi, just opposite Bangkok. Here irrigation is employed in garden culture, a technique not usually found in Southeastern Asia. The gardens are surrounded by walls 1 meter high and are traversed by trenches 2 meters apart. The beds between the trenches are elevated. The trenches work both as drains and as irrigation ditches. Small sluices control the water. In the gardens there is a great variety of fruit trees and palms which climbers, such as *Piper betle* L., use as supports. In the trenches the peasants grow lotus and water chestnut (*Trapa natans* L.).[4] Zimmerman, in his survey of rural Thailand, points out that in most regions the peasants rely too much on one crop, rice, and he recommends in several instances improvements in the cultivation of the home lots so that families may have a better supply of fruits and, above all, of vegetables. In the Khlong Rangsit section he observed that "on account of the lack of homestead and village sites among the people there is no gardening and home production of food."[5] This observation may be applied to Southeastern Asia in general. Wherever the dominant form of land tenure is tenancy the flimsy shack of the tenant has no mixed fruit and vegetable garden around it, for his status is much too uncertain. A tenant

[3] Ochse and Terra: *op. cit.*, pp. 371–372.
[4] Credner: Siam, 1935, p. 246.
[5] Zimmerman: Siam, 1931, p. 307.

who does not know where he will be the next year cannot be expected to plant fruit trees. Under these conditions the diet is reduced to the simplest foods and is usually deficient.

Zimmerman criticizes some of the characteristics of garden culture in the Thai part of the Malay Peninsula. The improvements that he recommends show that he has approached the gardens with a Western eye, looking for fruit trees "in rows at proper spaces, not scattered as is the case now." [6] He notices lack of tilling, an observation that could be duplicated in every country of Southeastern Asia. As a matter of fact, it is fortunate that the gardens—especially those that have many fruit trees and palms—are not clean-cultivated; otherwise erosion would have caused the same damage here that it has caused in the Westerners' rubber, tea, and coffee plantations—these were planted in rows and kept clean until the discovery was made that soil erosion is extremely severe under such conditions. Pendleton remarks:

> The average Occidental, accustomed as he is to clean cultivated orchards and vineyards in temperate zones, is inclined to keep a plantation in the tropics likewise cultivated. This is now quite generally recognized as a mistake. . . . Near Songkla [in southern Thailand] . . . are considerable plantings of fruit trees, planted as if they were forest trees. . . . Annually, before harvesting the fruit, the undergrowth is slashed and left on the ground, but there is no stirring of the soil or other cultivation. The results are excellent, and there is no soil deterioration, no catastrophic erosion, no expense for cultivation.[7]

There is little literature on the importance of garden culture among the Malays of British Malaya. According to Grist,

> fruit cultivation is of little more than local importance. The total area under fruit, other than pineapples, is estimated to be 63,000 acres. . . . This area does not take into account the numerous small compounds surrounding Malay homesteads, in which a few fruit trees are planted and which, in the aggregate, must amount to a very considerable area. Although certain fruits acquire a reputation in particular localities, the distribution of the fruits is scattered throughout the country. For instance, North Johore is noted for its durians, Malacca for dukus [*Lansium domesticum* Jack], Penang for rambutans, but in point of fact, equally good quality fruit of these kinds can be found in other areas.[8]

Malaya lacks the detailed information about gardens and their economic importance that is found in Java, for instance, where the garden area of each village is known, because in Malaya there is no population pressure and therefore no need for exact field studies and careful inventories of the source of food supply.

The Philippines are like Malaya in respect to garden culture. Some districts around Manila Bay have specialized in such fruits as mangoes,

6 *Ibid.*, p. 163.
7 Pendleton: "Some Interrelations," 1939, p. 50.
8 Grist: An Outline of Malayan Agriculture, 1936, p. 228.

lanzones, papayas, and chicos, which find a good market in Manila, but on the whole we rarely meet in the Philippines such intensive garden culture as in Java. The diet of the average Filipino is not well balanced and is particularly deficient in fruits and vegetables. In recognition of this fact the Bureau of Education and the Bureau of Plant Industry have both done much to impress the peasants with the need for fruit and vegetable gardens. Although the situation has improved during the past forty years and a large variety of fruits is now produced, the country still depends upon imports, whereas with a little effort most of the imported fruits could be produced locally and would furnish an additional source of income. It would be well to encourage the peasants to grow more fruits and to advise them on the most suitable varieties. In the other countries of Southeastern Asia, also, with the possible exception of Java, such advice would be valuable.

The published material available on the importance of gardens in Indochina is similarly limited in quantity. Even Gourou in his study of land utilization restricts himself chiefly to listing the fruits and vegetables found in the gardens and to stating that certain districts are renowned for the quality of their garden fruits. Gourou's list of fruits and vegetables is as varied and extensive, however, as similar lists for any of the other countries of Southeastern Asia.

Wet Fields

Throughout the countries bordering the South China Sea most of the cultivated land is used for the growing of lowland, or wet, rice. For this purpose the fields must be leveled and surrounded by earth dams to impound rain water during the rainy season or to confine the water that is carried to the fields through irrigation canals or lifted to the fields by mechanical means. If a field is properly leveled the surrounding dyke will keep the water everywhere at the same depth. The size of the wet-rice, or paddy, fields, and the amount of labor that is required to make them, depend upon the topography of the region. In the lowlands, where no terracing is required, there is relatively little labor involved in the making of paddies, which are frequently large. A hilly, or mountainous, terrain necessitates terracing, which demands greater effort and reduces the size of individual fields; the steeper the slope, the narrower the paddies. We have only to see the rice terraces of the mountains of northern Luzon, southern Bali, the Preanger districts of West Java, or Sumatra's West Coast in order to realize how much effort has been expended in building these terraces and how much energy must be devoted to their upkeep. The terraces, as well as the wet-rice fields of the plains, represent the accumulated efforts of generations of peasants, each of which has helped to extend the area devoted to the cultivation of wet rice (Figs. 50–58).

Not all people show the same skill in the making of paddies. According to Credner, the Thai are quite helpless on sloping land that requires terracing before wet rice may be planted. They would prefer to migrate to a region where level land is available rather than undertake the terracing of a hillside; on the other hand, they have been successful in building innumerable canals to tap rivers and spread the water over the plains.[9] However, they do not seem to be able to fight great floods. In contrast, the Annamese of Tongking have built huge dykes against destructive floods and the attack of the sea.

The problems presented by the cultivation of wet, or lowland, rice change from country to country, from region to region, and sometimes even from village to village. Lowland rice demands a soil thoroughly soaked with water and worked into "a fine soft mud by hoeing, plowing, trampling, or harrowing, and . . . free from weeds." [10] The field must remain under water until the ears of rice begin to fill and harvest time approaches. The amount of water required to grow lowland rice depends upon the variety of rice, the length of its growing period, and the nature of the soil. There are innumerable varieties of lowland rice. One way of grouping them is according to the number of days they take to mature, which ranges from 60 to 300 days. In the Philippines varieties that require less than 150 days are considered early maturing, those that need from 150 to 190 days medium-late maturing, and those needing more than 190 days late maturing.[11] The longer the growing period, the greater the water requirements and, generally, the higher the yields. Normally the water covering the field has a depth of from 10 to 15 centimeters. But "floating rice" can keep pace with a slowly rising flood and reach the amazing height of 4 meters or more. The stems settle on the mud when the water recedes, and new roots start at the upper nodes, while the lower parts of the stems rot away. Floating rice is especially well known in Cochinchina, and I have seen it in the Philippines. In 1939 the Department of Economic Affairs in Batavia sent a rice expert to Cochinchina to study floating rice and the possibility of introducing it into southern Borneo.

Lowland rice can be grown on many types of soil as long as a heavy clay subsoil is present to prevent the loss of water through percolation. Copeland states that lowland rice does best in heavy soils that hold water securely, and he adds that it is desirable "that there be, at a depth below the reach of the roots, but not too much further down, an impervious substratum-subsoil or hard-pan." [12] We distinguish between two main types of wet-rice fields: those that depend exclusively upon the rain that is impounded by their dykes, and those that are supplied by an irrigation system and are therefore no longer subject to the irregularities of rainfall.[13]

[9] Credner: Siam, 1935, pp. 203 and 217.
[10] Wickizer: "Rice and Wheat," 1941, p. 272.
[11] University of the Philippines: Handbook of Philippine Agriculture, 1939, pp. 7–11.
[12] Copeland: Rice, 1924, p. 45.
[13] An important variation of the second type is the cultivation of wet riceland that has been flooded by rivers rising above their natural levels. This mode of cultivation, which is found especially in the residency of Palembang in Sumatra, in southeastern Borneo, in Thailand, and in Indochina, is not so dependable as irrigation because of the irregularities of the river floods.

Irrigation systems also fall into two categories: peasant irrigation works and technical irrigation works. The former have existed throughout Southeastern Asia for more than a thousand years. In recent times scientifically trained engineers have designed large-scale systems with concrete weirs and locks. Fields solely dependent upon impounded rain water produce one rice crop a year, whereas irrigated fields often yield two crops. If even with irrigation only one crop can be produced, yet the irrigation system assures proper water control, whereas without it irregularities in the course of the rainy season may damage, or destroy, the crop.

Wet fields are protected against erosion by leveling, terracing, and dyking. But how does continued cropping affect the fertility of wet-rice fields? Diverse environments and cultural practices in Southeastern Asia create considerable differences in this respect. Much of the paddy land of the Philippines, Indochina, Thailand, Malaya, and Burma has lost a good deal of its original fertility because the peasants have not used manures, fertilizers, or crop rotation. In many instances, however, soil fertility has been preserved by the mineral matter dissolved in the irrigation water and by the silt deposited on those fields that are supplied with irrigation water coming from streams. The value of the water and silt depends, of course, upon the nature of the parent material from which it is derived. Rivers from a young volcanic region supply much richer water and silt than do those that drain a mountain region consisting of sedimentary rock or granite. For this reason, the Irrigation Service of the Netherlands Indies before making final plans for a new irrigation system takes samples of water and silt carried by rivers at different times of the year. The samples are analyzed in order to determine their influence upon the soil to be irrigated. A river found to be carrying water and silt that would lower the value of paddy land would be eliminated as a source of irrigation water, or its use would be limited as much as possible.

Irrigated fields of which the water supply does not contain minerals in quantities necessary for continuous satisfactory plant growth and paddy land that depends exclusively upon rain, both require other techniques for maintaining fertility. In many regions cultivators cut the rice stems just far enough below the ear to allow bundling. This practice makes possible the return of the rest of the plant matter to the soil. From this it is only one more step to work household refuse into the ground together with ashes of burned rubbish, rice husks, leaves, and manure.

As stable manure is not available in adequate quantities, intensive studies of the application of green manures to the cultivation of rice have been made in a number of countries of Southeastern Asia. A recent article by W. Bally reviews the results of these studies.[14] In a few rice-growing regions the peasants, on their own initiative, have learned to use legumes to improve rice soil. In wet fields the decomposition of green manure takes place under anaerobic conditions, and the final product is ammonia. Since

14 Bally: "The Employment of Green Manures," 1936, pp. T10–T31.

rice takes up nitrogen in the form of ammonia, the value of green manures on wet-rice fields is apparent.[15] In the Indies, *Crotolaria* species are the most frequently used green manures, the favored one being *Crotolaria juncea* L.; but *Phaseolus lunatus* L., *Dolichos lablab* L., and *Pachyrrhizus erosus* Urban are also used. The peasants of the Red River delta in Tongking are in the habit of working into the ground the wilted matter of a number of legumes that have been grown primarily for seed. The peasants of the province of Nam Dinh have developed a technique that may be unique. They push handfuls of indigo branches into the ground between the rows of rice a few days after the latter has been transplanted, for they believe that indigo is superior to pig manure or human excreta. In Indochina rice-growers frequently cultivate green manure, especially *Crotolaria* species, on dry fields and carry it to the rice fields at the time when these fields are being prepared for the seedlings. A green manure used exclusively in a few districts of the delta land of Tongking is the small aquatic plant, *azolla*.[16] Certain villages, thanks to favorable circumstances, have specialized in the growing of azolla to sell to the surrounding country, while other villages reproduce the plant until sufficient quantities of green manure have been obtained to cover all needs. The life cycle of azolla is completed in the months from November to April or May. Only the villagers of LaVan know the secret of maintaining and multiplying the plant during the hot season. They guard this secret jealously since it is the basis of their livelihood.

Further studies are needed to establish which varieties of green-manure plants are best suited to the various environmental conditions. It is important to find plants that grow rapidly and can be sown when the rice is still on the field, just after the water has been drained off to allow the grain to ripen. Experiments should also be made in the timing of the application of green manure. Should it be early, or just before the transplanting of the seedlings? Should it be on dry or on wet land?

Crop rotation also has its place as a means of keeping up the fertility of wet-rice fields. This practice seems to be better developed and more prevalent in Java than in any other country of Southeastern Asia. Extensive areas of *sawah*, as the Javanese call their wet riceland, are planted with dry-field crops such as maize, legumes, and other vegetables during the dry season, a practice that has a beneficial effect upon the soil. This form of double cropping is probably better than raising two rice crops in one year, because—although a different variety of rice is chosen for each season—both make essentially the same demands upon the soil; whereas a peanut crop, for example, grown during the dry season will enlarge the nitrogen content of the soil. In some areas of Java lowland rice is rotated

15 Imperial Bureau of Soil Science: Tropical Soils in Relation to Tropical Crops, 1936, p. 11.

16 For a map showing the extent of the use of azolla, see Gourou: Les paysans, 1936, p. 392. For further discussion see Nguyên-cong-Tiêu: "L'azolle," 1930, B, pp. 335–350 (also in *Proc. Fourth Pacific Science Congress, Java, 1929*, Vol. 4, 1930, pp. 355–370).

with sugar or tobacco grown under plantation management. But is the high fertility of the soil due to this crop rotation, or is it rather the result of the heavy application of commerical fertilizer and the thorough cultivation of the soil under Western management?

In no country of Southeastern Asia is there so great a variety or so intensive a use of commercial fertilizers as in Japan.[17] In general, the peasants of Southeastern Asia lack money for commercial fertilizers. In Indochina local phosphates are within reach of some of the Annamese, but the costs of imported fertilizers are prohibitive. In Java, although attempts to make possible the use of fertilizers through the agency of public credit institutions were successful in some special instances, not more than one per cent of the cultivated area received any commercial fertilizer in 1939. On the other hand, estimates indicate that more than 1,000,000 hectares of sawah have a phosphate deficiency and that a moderate application of phosphate would augment the yield by 25 to 50 per cent.[18]

The Agricultural Advisory Service was able to raise the productivity of wet-rice fields in West Java by about 10 per cent. It encouraged the peasants to use fertilizer on seedbeds, to sow rice more thinly on the beds (thus producing stronger seedlings), and to plant the seedlings in rows so that weeding could be done with a special instrument, the *landak* (Fig. 59). This weeder was developed in Japan. During the years immediately preceding 1939, these improvements in West Java resulted in an increase in output of unhusked rice amounting to 300,000 quintals, or 30,000 metric tons, with a value of nearly 1,000,000 guilders. This shows that some improvements in cultivation may increase production without a great capital outlay.[19]

Java may be compared only with southern China and with Japan in the cleanness of its rice fields. The Filipino does not pay much attention to weeding. The Malayan, Thai, Cambodian, Cochinchinese, and Filipino rice farmers have more flooded riceland per capita at their disposal than the people of Java and can thus afford to cultivate their land less intensively.

Irrigation Problems

Peasant irrigation works usually take care of the fields of only one village or of a small group of people who have joined together for the specific purpose of building and maintaining an irrigation system. These works consequently are small and usually inadequate, but nevertheless they de-

17 "The chief commercial fertilizers used in Japan are pressed herrings, pressed sardines, dried sardines, other fish cakes, bone meal, soybean cake, rapeseed cake, cottonseed cake, other vegetable-oil cakes, ammonium sulphate, calcium cyanamide, calcium phosphate, and potassium phosphate. Calcium phosphate and ammonium sulphate are probably the most representative chemical fertilizers."—Wickizer and Bennett: The Rice Economy, 1941, p. 252, footnote 33.

18 Departement van Economische Zaken: Voedselproblemen, 1940, p. 28.

19 *Ibid.*, pp. 27–28.

mand respect and admiration for the ingenuity displayed.[20] "The serious imperfections which characterize native irrigation works and make rice culture uncertain, particularly in the plains, can assuredly not be charged to the account of these who constructed them." [21] The weirs, by means of which the peasants raise the water of small rivers, are especially weak. The construction material consists of bamboo, trunks of coconut and other trees, and boulders. At times great bamboo baskets are filled with heavy stones and placed in the river, or large stones are laid loosely across the channel. Where the rivers are wide the people drive two rows of tree trunks into the river bed and fill the space between the rows with whatever debris they have at hand. These weirs frequently break because of river floods as they have no provision for allowing passage of flood waters; yet repair is often impossible for the duration of the rainy season, so that the crop is ruined. Another constructional disadvantage is the open intakes. During floods large quantities of water flow through these and cause an excessive soaking of the fields and damage to the canals. Wherever a canal has to cross a depression, a road, or another watercourse, bamboo or hollowed tree trunks are used to carry the water overhead. This often leads to large losses of water, especially during floods. Poor drainage is another weakness of peasant irrigation systems.

Scientifically trained engineers have come to the aid of peasant communities in Southeastern Asia and have designed irrigation systems based on technological principles and free from the weaknesses of peasant irrigation works. One of these technical irrigation systems may take care of tens of thousands of hectares of riceland belonging to hundreds of villages. In many instances the engineers have improved existing peasant irrigation works; more often they have made irrigation possible where peasant communities did not have the financial and technical resources to overcome natural obstacles. Technical irrigation works are usually government-financed.

In the Philippines most of the wet-rice fields are dependent exclusively upon the rain that falls directly on them. One centimeter of rain a day is said to be necessary for a healthy rice growth. But there is always the possibility that, particularly in the months of May and June, there may be so little rain that the peasants cannot prepare their fields. A shortage of rain in October is also dangerous as it may seriously damage, or even destroy, the crop, even though the precipitation during the rainy season may have been enough for the production of a satisfactory crop. To guard against an insufficiency of water for the wet-rice crop, therefore, the fields should be supplied by irrigation systems. The Philippines, however, do

20 The *subaks* of Bali are agricultural cooperative societies that have been formed to construct, maintain, and administer irrigation systems. The achievements of the Balinese subaks are very remarkable in a number of ways. Thus, the men of certain villages are especially skillful builders of irrigation works and are able to construct long canals and even tunnels of considerable length. These men are hired by the subaks and help them to solve their irrigation problems.

21 Van der Meulen: "Irrigation in the Netherlands Indies," 1940, p. 146.

not have adequate irrigation facilities, although in certain localities the peasants have built systems and the importance of irrigation is generally recognized. In Spanish times the irrigated area was estimated at 170,000 hectares.[22] Some of these irrigation systems were constructed by friars on their estates, but the works eventually fell into disrepair and only a few of them can still be used.

Not until 1922 did the government start a campaign of irrigation-system construction. In that year the Philippine Legislature authorized an issue of bonds amounting to 20,000,000 pesos. A survey at the time estimated that a total area of 560,000 hectares could be brought under technical irrigation. By 1938 the Bureau of Public Works had supplied about 75,000 hectares of land with irrigation water at a cost of 17,000,000 pesos. The expansion of irrigation was very slow, "owing to the difficulty of collecting charges, to complaints of unsatisfactory services, and to the low price of rice, which offered no inducement to incur the added expense."[23] The costs were to be charged against the land and were supposed to be paid by the owners over a period of from 20 to 40 years. They average 195 pesos per hectare. Miller wrote in 1932:

> The annual charge for the redemption fund and interest, as well as the maintenance of the irrigation works, is from ₱8.72 to ₱22.06 per hectare, depending upon the per hectare cost of the particular irrigation works and the number of years over which the cost is amortized. After the irrigation works have been paid for, a charge of about ₱3.00 per hectare for maintenance will accrue annually.[24]

After 1932 changes were made in the terms of the financial settlement. Payments were reduced, and it was suggested that the government remain owner of the systems. In 1934 the Philippine Economic Association stressed the need for an enlargement of the technically irrigated area, but it did not offer any suggestion as to how the costs of construction, maintenance, and operation might be recovered should the farmers insist upon very low charges.

In addition to the irrigation systems operated by the Bureau of Public Works, other government agencies controlled an irrigated area of 7,900 hectares. The *hacenderos*, private companies, and municipalities brought another 200,000 hectares under technical irrigation, thus raising the total area of land under this form of irrigation to about 300,000 hectares. If we subtract this figure from the total irrigated area, which the census of 1939 put at 670,000 hectares, we find that approximately 370,000 hectares were cared for by peasant irrigation systems.

It is generally estimated that about 80 per cent of all rice is grown on wet fields. Therefore, if we take 80 per cent of the total rice area of 1,936,200 hectares given in the census of 1939, and subtract from it the

---

22 Miller: Principles of Economics, 1932, p. 182.
23 Dawson: "Philippine Agriculture," 1940, p. 408.
24 Miller: *op. cit.*, p. 183, footnote 1—all figures as of 1932.

total irrigated area, we find that some 880,000 hectares depend exclusively upon impounded rain water.

In Java and Madura the possibilities of extending the sawah area are strictly limited. In 1936 the Spit Commission estimated that another 300,000 hectares of hitherto uncultivated land could be brought under cultivation, but most of that land would have to be converted into *tegalans* (permanent dry fields), and not sawahs. From 1928 to 1938 the sawah area increased from 3,253,878 to 3,367,879 hectares, or by 114,001 hectares. Most of this new sawah area had formerly been tegalan area. But at the same time the latter, despite the fact that it had lost some land to the sawah area, grew by 193,000 hectares as a result of new clearings. Thus the cultivated area of Java and Madura between the years 1928 and 1938 was enlarged by 307,000 hectares. If the cultivated area should continue to expand at the same rate, the remaining cultivable land on these two islands would be exhausted by 1948.

Netherlanders, experienced in the handling of difficult water problems in the mother country, were well prepared to help the Indonesians with their irrigation problems. They were able to harness the larger rivers of Java, a feat for which the engineering skill of the peasants was inadequate. The Indonesians had been unsuccessful in the irrigation of wide plains. The first European irrigation system was laid down in Buitenzorg during the regime of Governor-General van Imhoff (1743–1750). In the early days of the culture system, two irrigation projects were built in the delta of the Brantas River to supply water for sugar and indigo crops, both grown for export. The first project designed by Western engineers to benefit the peasants of Java was built in the 1880's in the residency of Semarang,where famines had been frequent. Since then the irrigation systems of Java have been steadily enlarged. There have been periods when it seemed as if the costs were greater than the benefits and progress was consequently slowed down. From 1893 to 1898 some 17,000,000 guilders were spent on the Solo River irrigation project, which finally had to be abandoned. A period of great expansion began in 1910 and was interrupted only by the economic depression.[25] By the end of 1939 the government had spent 250,000,000 guilders on irrigation construction and flood protection, of which only 42,000,000 guilders represented expenditure before 1900. These sums do not include outlays for the repair and maintenance of existing systems or for the salaries of the government personnel. Nor do they include the cost of irrigation systems constructed by sugar plantations, especially in the principalities of Soerakarta and Jogjakarta.

At the end of 1938 the total wet-rice area of Java and Madura amounted to 3,368,000 hectares. Of these, 1,204,000 hectares were technically irrigated, an additional 1,500,000 hectares were cared for by peasant irrigation works, and the remaining 664,000 hectares depended upon rain water

---

[25] See the graph presented by Van der Meulen: *op. cit.*, p. 149.

impounded on the fields.[26] The construction of technical irrigation systems had required an average investment of 190 guilders per hectare. This expenditure has greatly benefited the peasants of Java and the European planters who rented land from Indonesians. Yields were increased; two rice crops could be raised where formerly only one could have been raised. Unfortunately, the possibilities of extending technical irrigation by means of large projects were becoming fewer; for the difficulties and the costs were increasing.

In recent years a number of great reservoirs (*waduks*) have been built to store excess water during the rainy season for irrigation during the dry season. Most of these waduks were constructed in sugar-raising areas, since the value of the sugar crop justified a large expenditure. The distribution of water is controlled by the authorities in order to prevent disputes between cultivators. In the past, difficulties arose over the conflicting needs of the European sugar estates and the peasants whose rice fields adjoined these estates. Sugar cane is irrigated not by flooding but by scooping the water out of the ditches and pouring it on the plants by hand—something that can be done only by daylight. Whenever there was not sufficient water to take care of both enterprises at the same time the peasants were allowed to direct the water into their rice fields only at night. Under pressure of severe criticism from fair-minded persons this partiality was done away with; small field-reservoirs were filled with water during the night, and the water was discharged into the rice fields in the daytime.

The costs of constructing and maintaining these irrigation works are not charged against the fields, as the government in the Philippines tries to charge them, nor are there any water charges. According to Netherlands Indies law, however, the government, and not the Indonesian, is the owner of the land. The latter pays a tax for the use of the land, the amount of which is determined by the land's productivity. At regular intervals the fields are reassessed, and any improvements, especially the introduction of technical irrigation, put the field into a higher class and thus raise the tax. In this way, "expenditure on irrigation works is more or less off-set by an increase in receipts from land taxes due to improved results obtained by cultivators, and by more revenue from general taxes owing to enhanced general prosperity." [27]

In contrast to Java and Madura, the Outer Islands present possibilities for greatly increasing the area of wet riceland, either by the clearing and cultivating of virgin forest land or by the converting of dry fields into sawahs. In 1910 the government began the construction of irrigation works in the Outer Islands; but between 1910 and 1938 a total of only 17,000,000 guilders was spent there. In recent years expenditures have been greatly increased in conjunction with the migration of Javanese. Im-

26 These statistics exclude the principalities of Soerakarta and Jogjakarta.
27 Van der Meulen: *op. cit.*, p. 159.

portant projects were started in the residencies of Palembang and of the Lampoeng Districts of southern Sumatra and in the Celebes.[28]

British Malaya is the only country in Southeastern Asia that depends heavily upon rice imports for the feeding of its inhabitants. The greater part of the population is made up of immigrants who normally do not raise food crops but work in plantation or mining industries or live in cities. Rubber and tin enterprises producing for distant markets completely overshadow the cultivation of food crops for local consumption, just as the immigrants greatly outnumber the native Malays.

As early as 1896 Sir Frank Swettenham called attention to the desirability of expanding rice cultivation by constructing irrigation works "to free us from our dependence on foreign countries for the supply of our staple food," [29] and in 1897 he wrote:

> Above all things I should like to see a vastly increased area, under a well-devised system of irrigation, yielding a rice crop sufficient, not only for the people of the Malay States, but also for the large native population of the neighbouring Colony [Straits Settlements]. We have the land and the water, and what we want is a scientific distribution of the water over the vast uncultivated areas that would unquestionably produce a rice crop more than sufficient for all our needs, provided we had the funds necessary to construct the irrigation works.[30]

Shortly afterwards, the rubber boom hit Malaya. Within a few years hundreds of thousands of acres of potential riceland were planted with rubber. Even land used for rice cultivation was turned over to King Rubber when the Malay peasant found it more profitable and easy to grow rubber and buy his rice. Increasingly large quantities of rice had to be imported for the rapidly growing population. During the World War of 1914–1918 this extremely vulnerable food position of Malaya brought hardship to the people and a great financial burden to the government. Little was done, however, to improve Malaya's agriculture or to increase the production of rice, since the difficulties were forgotten as soon as there was an improvement in the rubber market.

The depression of 1929, however, forced the authorities to give serious study to the food-production problem. In 1929 a trained agriculturist became Director of Agriculture, a post previously held by a member of the administrative service. In 1930 the government appointed a committee to study the possibility of extending rice cultivation. The committee reported that the country had 600,000 to 1,000,000 acres suitable for rice growing, over and above the 700,000 acres already devoted to it. As a considerable part of the potential riceland was already planted with rubber, the committee recommended that the government prevent any further expansion of the rubber area on land suitable for rice. It also recommended

---

28 *Jaarverslag van het Departement van Verkeer en Waterstaat over 1939*, 1940, pp. 243–248.
29 Quoted in Haynes: "Memorandum," 1933.
30 *Ibid.*

the establishment of a Drainage and Irrigation Department in place of the old Hydraulic Branch of the Public Works Department. This new department began to function in 1932. Its main tasks were to help the Malay peasant with his problem of maintaining proper drainage and irrigation systems, to open new areas, and to distribute them among Malay pioneers. According to Grist,

> Malay *padi* growers have an inherent gift for water control, so that dams and irrigation channels devised and constructed by the local community are a common feature of the smaller *padi*-growing areas.[31]

Drainage and Irrigation Department engineers revealed, however, that in spite of this skill the Malay peasants downstream from the tin mines or rubber plantations were having increasing difficulties in keeping their rice fields in production, and year after year many hectares of riceland were going out of cultivation. The tin mines added their discharge of solid matter to the silt already carried by the rivers, and the development of rubber plantations led to accelerated erosion, especially in the days of clean weeding, so that the rivers of Malaya became choked with enormous quantities of silt.

> The pollution of a river by mining is very obvious in the discolouration of the water by the suspended solids of the mine effluent; the rise of a river bed by the uncontrolled discharge of tailings from hill mining and its disastrous effect on the valley is spectacular, but the silting of a river from soil erosion is not so obvious.
>
> The advance of the swamp as it spreads up the valley is so slow that it often takes on the appearance of a natural phenomenon of which man is the innocent victim.
>
> In the Malacca River valley, some 7,000 acres of *padi* land have gone out of cultivation from this cause. It is perhaps within living memory that most of this area was planted; to-day it has the appearance of a primitive swamp, but remove the reed cover by draining the swamp and there are the *batas* [little field dykes] of the old cultivators.
>
> To them the deterioration of their *padi* lands appeared as an act of God if they thought about it at all. Their loss came upon them very slowly, they found year by year how much more difficult it was to build their bunds to keep out the river, their crops were more frequently destroyed, the results of their labour got poorer and poorer. . . .[32]

Some thirty years ago Balun Bidai was a thriving village near the junction of the Kinta and Perak rivers.

> It is said that more than 2,000 people had their homes there and cultivated extensive *padi* lands. . . . The progressive silting of the Kinta River in its lower reaches gradually blocked up the mouth of the Sungei Tumboh and its waters found their way through the jungle and by swampy channels and lagoons reaching the Kinta at various points. The result was that the Balun Bidai *bendang* [rice

[31] Grist: An Outline of Malayan Agriculture, 1936, p. 121.

[32] *Annual Report of the Drainage and Irrigation Department of the Federated Malay States and the Straits Settlements 1937*, 1938, p. 95.

fields] became subject to severe flooding and some were permanently under water. The natural access to the area by water became so difficult that only small *sampans* could negotiate it whereas formerly there was a service of river steamers . . . and the population drifted away till it was reduced to less than 400.[33]

Extensive agricultural development for rubber planting in the beginning of this century caused the Batang Padang River to silt and by the year 1920 it had risen above the surrounding land in Lower Perak; its waters poured over the country far and wide to the valleys of the Kroh and Chikus rivers. The coarse silt blocked the mouths of its tributaries . . . it could no longer maintain its channel which became choked up with elephant grass growing prodigiously in the silt and this fine river with a normal channel 100 feet wide became so overgrown that it was impossible to pass down it even in a small boat.[34]

The Drainage and Irrigation Department undertook a great number of projects designed to prevent further excessive silting of rivers, to clear up silt-choked river channels, or to prevent destructive floods. In addition, it improved existing irrigation systems by cleaning and correcting drainage canals and by replacing the short-lived brushwood dams of the Malays with modern permanent dams which include provisions for the passing of floods. The department also opened up new lands and divided them into small rice farms. In the Sungei Manik Irrigation Scheme, for example, each settler received four acres of irrigated land and two acres of *kampong* (village) land. A very large number, if not the majority, of the new settlers were immigrants from Sumatra and Java. These immigrants are racially and culturally so close to the Malays of Malaya that they do not encounter the hostile attitude that Chinese and Indians experience when they attempt to acquire riceland in the Malay States.

Despite the efforts of the Department of Agriculture and of the Drainage and Irrigation Department to expand rice production during the 1930's, at the outbreak of the war in 1939 Malaya was still depending for two-thirds of its rice consumption upon the neighboring rice granaries: Burma, Thailand, and Indochina. During the season of 1932–1933 Malaya's rice area had amounted to 767,000 acres, compared with 753,000 acres during the season of 1938–1939. This slight decrease in rice area between 1932 and 1939, which occurred in spite of the establishment of new irrigation projects, was probably due not only to a shift of Malay peasants from marginal riceland to newly irrigated areas, but also to some degree to an improvement in rubber prices which once more lured the Malay away from rice cultivation. It is easier to tap rubber than to cultivate rice. It is estimated that 93 per cent of the Malay rice area produces wet rice and 7 per cent produces dry rice. The bulk of the wet-rice land is served by unimproved peasant irrigation systems or else depends upon rain falling directly on the fields; in 1939 only about 160,000 acres belonged to government-constructed irrigation projects.

33 *Ibid.*, p. 87.

34 *Ibid. 1936*, 1937, p. 17.

When the outbreak of the war in 1939 brought the rice problem of Malaya again to the fore, the government proposed that Chinese and Indians be permitted to grow rice. Sir Shenton Thomas, the Governor and High Commissioner, took the stand: "Rice production in Malaya must now be considered in terms not of Malay nationalism, but of national emergency." [35] Over the protest of Malay nationalists, a small amount of land in the states of Johore, Selangor, Perak, and Pahang was actually settled by Chinese.[36] The government furthermore attempted to make rice cultivation on new land more attractive by paying for the clearing and planting and by guaranteeing a minimum rice price.

Notwithstanding the abundance of the country's potential riceland, all the attempts of the government to make Malaya self-supporting in food were bound to fail because of the ratio of Malays to non-Malays in the population. As long as there is no substantial reduction in the Chinese and Indian populations of Malaya, and as long as the Malays successfully object to the establishment of Chinese and Indian communities engaging in rice cultivation, Malaya will continue to depend upon imported rice.

Whereas the riceland of Malaya is broken up into patches scattered along the coasts of the peninsula and the alluvial banks of its short rivers, the rice area of Thailand is concentrated in two great plains. The mighty Mae Nam Chao Phraya, with the Mae Nam Mae Klong, the Mae Nam Tha Chin, and the Mae Nam Bang Pakong, forms one enormous alluvial plain suited to the cultivation of lowland rice. This area has somewhat the shape of a triangle whose base at the head of the Gulf of Thailand is about 160 kilometers wide. Some 160 kilometers inland it is 80 kilometers wide, but northward it fingers out into narrow bands between the mountain chains of northern Thailand. The second important rice area is found upon the Khorat plateau along the rivers of the Mae Nam Mun system draining into the Mekong (Mae Nam Khong).

Thailand's economy rests almost exclusively upon the cultivation of lowland rice. Only a small part of the country receives a sufficient amount of rain (1800 millimeters or more during the rainy season) to grow lowland rice on fields that are dependent upon rain water falling directly on them. At most observation points of the central plain the average annual rainfall is from 1,000 to 1,500 millimeters. Therefore irrigation is required to produce one satisfactory crop a year. The additional water is supplied by rivers coming out of the mountains of northern Thailand and by those that drain the Khorat plateau. These rivers rise during the rainy season and almost every year flood wide sections along their banks. However, the floods depend upon rainfall in the headwater areas. Insufficient rain in the mountains means inadequate flooding or none at all and, consequently, a poor harvest. Exceptionally heavy or prolonged floods, on the other hand, can be just as disastrous for the rice crop. The Thai have long

35 Quoted in Barber: "British Malaya," 1940, p. 3.
36 Thompson: Postmortem on Malaya, 1943, p. 37.

been very skillful in harnessing these floods in a network of canals that distribute the water throughout the plains. From these canals the water either enters the fields by gravity or is lifted into them by human power or, occasionally, by mechanical power. But without technical irrigation the success of the harvest is always at the mercy of the height of the flood. A record of the flood levels of the Mae Nam Chao Phraya at Ayutthaya (Ayutia) for a period of 99 years shows that in only 32 of these years were the levels entirely satisfactory; in 22 years the levels were barely satisfactory; and in 30 years they were much too low. In 15 of the years the flood level was so high that the crops suffered from too much water. These figures indicate clearly the necessity for technical irrigation in Thailand,[37] which has been generally recognized since the turn of the century.

Because of the Thai peasants' limited financial resources and their lack of technical facilities it has been necessary for governmental or private agencies repeatedly to come to their aid. In 1892 the Siam Canals, Land and Irrigation Company obtained a concession to develop the Khlong Rangsit area, an extensive swamp region northeast of Bangkok between the Mae Nam Nakhon Nayok and the Mae Nam Pa Sak. The company dug the Khlong Rangsit, which is wide enough for boat traffic, in order to drain the swamp, and this canal also served for irrigation purposes when necessary. The flow of water and boat traffic is regulated by locks. The company failed, and the government called in a Dutch irrigation engineer, Homan van der Heide, who remained in Thailand from 1899 until 1909. He drew up plans for an extensive irrigation system, but, because of the expense it would entail, this was not constructed. However, Van der Heide supervised the building of a number of long canals that run in east-west lines from the Mae Nam Bang Pakong to the Mae Nam Mae Klong. Like the Khlong Rangsit, these canals are shut off from the rivers by locks. They serve as traffic arteries, as well as irrigation canals.

Van der Heide was succeeded by Sir Thomas Ward, who suggested a comprehensive irrigation system, consisting of several projects to be undertaken by the newly formed Irrigation Department "only as Siam acquired a population sufficiently numerous to occupy all the land thus made available." [38] The cost of the projects was estimated at 150,000,000 ticals. Ignoring Ward's proposal suggesting the Mae Nam Suphan Project (Map, Fig. 46) as the starting point, the government began with the South Pa Sak (sometimes called Prasak) Project, adjacent to the old Khlong Rangsit Project and forming a unit with the latter. This disregard of Ward's proposal was due to political pressure exerted by landlords who had interests in the Khlong Rangsit area. The South Pa Sak Project was finished in 1924 and entailed an expenditure of 15,780,768 ticals. The

---

37 Credner: Siam, 1935, pp. 213–218 and map 4 opposite p. 72.

38 Thompson: Thailand, 1941, p. 517. See also *Ibid.*, pp. 514–520, and Credner: *op. cit.*, pp. 218–222, for the following discussion.

irrigable area amounts to 83,000 hectares, of which some 60,000 to 70,000 hectares are actually under cultivation. The Chiang Rak-Bang Hia Project also has been completed. When the Nakhon Nayok Project is finished, the whole plain between the Mae Nam Chao Phraya in the west, the

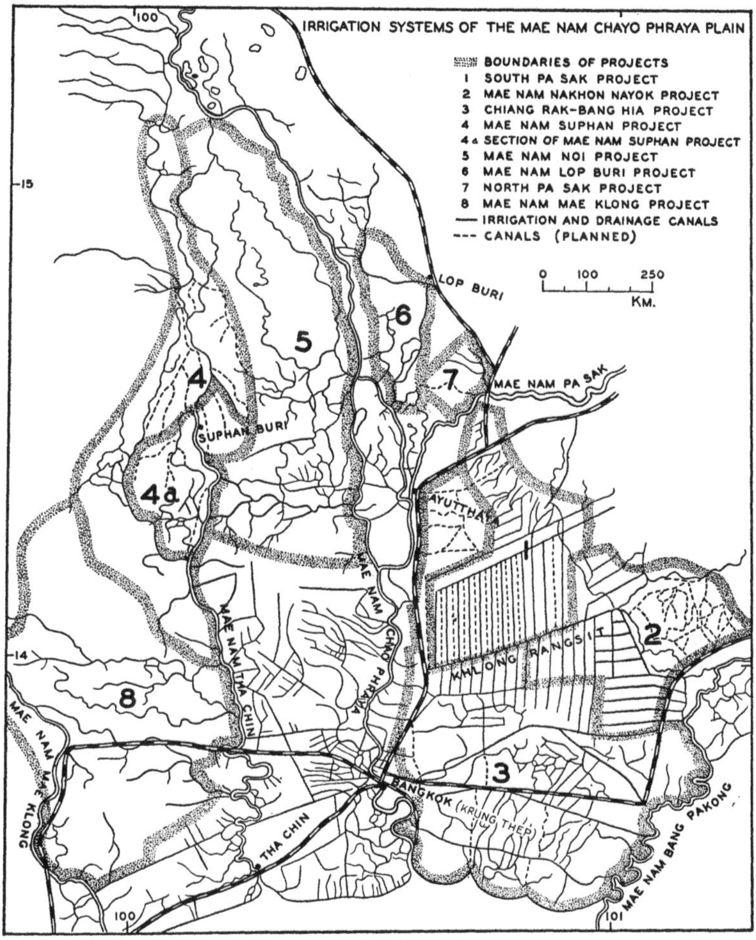

FIG. 46—Irrigation systems of the Mae Nam Chao Phraya Plain, Thailand. (Based on Credner: Siam, 1935, Fig. 26, p. 216.)

Mae Nam Pa Sak in the north, and the Mae Nam Bang Pakong in the east, embracing a total area of 540,000 hectares, will be under technical irrigation. This region should then be safe from destructive floods and also will be supplied with shipping facilities vital to rice export. Practically no development work has been done west of the Mae Nam Chao Phraya; only a part of the Mae Nam Suphan Project has been finished. The main task of this project is to canalize the Mae Nam Tha Chin, or Mae Nam Suphan, an old branch of the Mae Nam Chao Phraya that has silted up. The plain of the Mae Nam Suphan was once an important rice-producing region, which, in the fourteenth century, supported the capital of the state of U Thong. Apparently opportunities for cultivation have diminished in this district.

The silting of canals and rivers is a serious problem in Thailand. On the peninsula most of this silting has been due to tin mining. On the plains, however, much of it is the result of deforestation, while shifting cultivation has been the principal factor in the mountain regions.

Ninety-four per cent of the cultivated area in Thailand is devoted to rice. Twenty per cent of this is now under technical irrigation.

In Indochina the cultivation of lowland rice presents each geographical region with its own particular set of problems. The wide delta lands of the Red River (Song Ka) and the Mekong form the two chief rice-producing areas of Indochina, in comparison with which the small deltas and coastal plains of central and southern Annam—squeezed between the Annamese mountains and the China Sea—and the old alluvial plains of Cambodia southeast of the Tonle Sap (Grand Lac) are of only secondary importance.[39]

The life of the Tongkingese peasant is profoundly influenced by the torrential character of the Red River. Every year the level of the river rises rapidly during the summer months and endangers man and his work with widespread and destructive floods. The only means by which the inhabitants of the delta land are able to defend themselves and their fields is by constructing dykes to keep the Red River and the other rivers of Tongking within bounds. These dykes must be so long and firm that their construction and maintenance is possible only under the supervision and with the support of a well-organized state. The earliest records show that during the Trân dynasty, in the thirteenth century, the state was in control of dyke construction, but it is quite certain that dykes were built even before that time. From the Trân dynasty onward Annamese rulers and their officials paid a great deal of attention to the defense of the lowlands against river floods.

The delta land must be protected not only from the rivers but also from the sea and its tides. The Red River and the other rivers that cross the plains of Tongking carry enormous quantities of sediment and are extending the delta farther into the Gulf of Tongking. As soon as the slowly growing mud flats attain a height suitable for cultivation, they are protected from the sea by dykes. Though the system is similar to that practiced in

---

39 For detailed information on the distribution of wet-rice fields see: "Carte d'ensemble des rizières d'Indochine, 1:2,000,000"; this map accompanies Henry and De Visme: Documents de démographie, 1928. The map is not quite up to date because of the extensive improvements that have been made since 1927, but it can be supplemented by using Robequain: L'évolution économique, 1939, and Gourou: L'utilisation du sol, 1940.

the Netherlands, the danger of submersion is not so great in Tongking as it is in the polder areas of the Netherlands, for these lie below sea level, and furthermore the tides are less strong in the Gulf of Tongking than along the coasts of the Netherlands. Nevertheless, the tides are noticeable in the rivers far inland. The land along the sea shore and along the lower courses of the rivers, unless protected from salt water, becomes unsuitable for the cultivation of rice, except for salt-loving varieties.

Although the Annamese state had concentrated its efforts for centuries on the construction of dykes, the French found grave weaknesses in the system when they occupied Tongking in 1885. The Annamese flood control and irrigation systems represented an intermediate stage between local peasant and modern technical irrigation. The dykes were usually no higher than 9 meters, although floods of 11 meters and more are not exceptional. The dykes were irregularly designed and formed salients and coves, which prevented an even flow of floodwater. Frequently they were badly constructed. As the Annamese lacked the technical knowledge and facilities that would have enabled them completely to master the rivers, they could not prevent breaks in the dykes, and these developed annually in some parts of the delta.

The French continued the work of the Annamese emperors. At first the colonial administrators thought that a strengthening of the large dykes of the Red River and its effluents would suffice to control the flood waters. But despite the improvements, breaks continued until 1926. In the province of Hung Yen 26 breaks occurred between 1806 and 1900, destroying the crop in each instance. Since 1900 this province has experienced 13 dyke ruptures in a period of 26 years. In 1926 the Inspector General of Public Works, Albert Pouyanne, started a large-scale program of dyke construction, which follows a master plan covering the whole delta. In 1885 the dykes had a volume of about 20,000,000 cubic meters; by 1915 it had grown to 32,000,000, by 1926 to 48,000,000, and by 1938 to 87,000,000 cubic meters. The dykes along the Red River can now hold the highest floods. The highest recorded level reached by the river at Hanoi is 11.93 meters, and the dyke there is 13 meters high. The alignment of the dykes has been improved, and they have been strengthened to such a degree that there have been no breaks since 1926, although the floods of 1936 and 1937 reached record heights.

The dykes prevent floods, but they also make drainage of the excess water difficult. The Red River, the Canal des Rapides, and the Canal des Bambous, the three principal channels, contribute but slightly to the drainage of the delta. Because of the excess of rain water during the wet season, drainage canals had to be constructed. The canals lead into the Thai Binh and smaller rivers. Near the coast, the drainage is further regulated by a series of floodgates that allow an outflow at low tide. Before the construction of the drainage systems, much of the land was so flooded during the rainy season that it could be cultivated during the dry season only. This was true in the section from Ha Dong to Phu Ly, south of Hanoi, which today, with proper drainage, produces two crops.[40]

The third water-control problem that had to be solved in the Tongking delta region was complex. There are three types of riceland—high, middle, and low fields—and there are two possible crops, those of the fifth and of the tenth months of the Annamese calendar. The crop of the fifth month is the dry-season crop, which is harvested in June, and the other is the wet-season crop, harvested in November. The latter does not require irrigation. During the rainy season there is an abundance of water in little brooks, ponds, and depressions, from which the peasants can lift it into their fields should there be a lack of rain water. But this water is usually stagnant and not sufficiently rich in oxygen; furthermore, it does not contain silt, which would enrich the field. Yet, although a technical irrigation system would improve even the wet-season rice crop, the peasants might be reluctant to pay for irrigation water, as they consider water to be a gift from heaven, a gift of which there is an abundance around them. During the dry season, irrigation is necessary in many parts of Tongking if a second rice crop is to be grown. The advantage of a technical irrigation system is obvious when it makes possible cultivation of land that would otherwise lie fallow during the dry season. The desirability, however, of establishing irrigation on land that is now planted with dry-field crops during the dry season is debatable since the peasants require these dry-field crops for a more balanced diet. Furthermore, the soil profits by crop rotation; as soon as rotation is abandoned in favor of two rice crops, the yields from the November rice harvest decline.

The French administration has invested large sums in irrigation systems, and doubt has been expressed as to whether in some cases the increased yields justify the expenditure, above all since the money was advanced by the French and is supposed to be paid back with interest.[41]

Gourou estimates that the Tongking delta region contains 1,100,000 hectares of irrigated land, of which 250,000 hectares produce only a dry-season rice crop and 350,000 hectares only a wet-season rice crop, whereas 500,000 hectares of the delta produce two crops a year. Thus the harvested area amounts to 1,600,000 hectares.

In Annam the plains are small, with the exception of those in the province of Thanh Hoa which is crossed by the Song Chu and Song Ma. The rivers of Annam, like those of Tongking, are subject to annual flooding, but, because of their small drainage basins, their floods cannot compare with those of the Red River. It is relatively easy to construct storage dams at points on the rivers just within the mountains and thereby to facilitate irrigation during the dry season. Such dams and irrigation canals have been built in the deltas of Thanh Hoa and Nghe Tinh in northern Annam and of Phu Yen and Phan Rang in southern Annam. A total of

40 See Robequain: *op. cit.*, pp. 249–253 and especially the map on p. 251.
41 Gourou: Les paysans, 1936, p. 107.

90,000 hectares is irrigated by these systems. Further projects are under consideration.[42] Gourou estimates that Annam has 560,000 hectares of irrigated land yielding one crop and 246,400 hectares yielding two crops. The area devoted to wet rice amounts to nearly 807,000 hectares and the harvested wet-rice area to about 1,050,000 hectares.[43]

Water control in Cochinchina, the delta land of the mighty Mekong, presents different problems. Here the peasant does not require the shelter of strong dykes, for the annual floods of the Mekong come very slowly and are never as high as those of the Red River. Furthermore, there is only one rise, after which the water recedes, whereas the Red River may have three or four high floods during the course of the rainy season. Each year the Mekong covers vast stretches of the delta with a sheet of water and turns into swamps those parts that have poor drainage.

When the French took over Cochinchina the delta was very thinly settled, although Annamese had already started to come in as colonists. Recognizing the possibilities of large-scale rice cultivation in Cochinchina and the need for better drainage, the French began to construct a system of canals which became the decisive factor in the development of the country. These rectilinear canals serve at the same time for drainage, navigation, and irrigation.

This irrigation is helpful in Cochinchina since the rainfall, although generally sufficient for rice growth, is extremely variable. The tide makes itself felt throughout the delta region, in the river as well as in the canals. The automatic oscillation of the water level allows drainage at low tide and irrigation of the fields at high tide. Unfortunately, but a small portion of the riceland profits from this daily rhythm, for Cochinchina has only primary canals and lacks the subsidiary canals that would bring the advantages of the system into all corners of the plains. Now, only the riceland along the main canals is irrigated. The government expected that the landowners would graft upon these a system of secondary and tertiary canals, but this has been neglected by the great landowners "in that haste to produce and that fever of speculation which has characterized the history of western Cochinchina for 50 years."[44]

Thanks to large-scale canal digging and dredging, the area of wet riceland in Cochinchina increased from 522,000 hectares in 1880 to 1,175,000 hectares in 1900, and 2,200,000 hectares in 1937.[45]

Three types of rice cultivation may be distinguished in Cochinchina: rice that is transplanted once, rice that requires two transplantings, and floating rice.[46] These three types are adaptations to environmental conditions and differ in labor requirements, length of growing season, and

42 Robequain: *op. cit.*, p. 250.

43 Gourou: L'utilisation du sol, 1940, p. 255.

44 Robequain: *op. cit.*, p. 125.

45 *Ibid.*, p. 243.

46 For a map showing the geographical distribution of these three types, see Gourou: L'utilisation du sol, 1940, p. 290.

yields. On the whole, the farmers of Cochinchina cultivate their fields much less intensively than do those of Tongking and Annam. Hardly any double cropping of rice is practiced in Cochinchina, nor is much of the riceland cultivated during the dry season.

In 1941 changes in the political boundary between Thailand and French Indochina reduced the importance of Cambodia as a rice-surplus region of Indochina. Cambodia lost the Battambang area, west of the Tonle Sap, which in recent years had developed into a prosperous rice-growing district, possessing one of the most modern rice mills of Indochina.[47] The province of Battambang has a rice area of 134,000 hectares, whereas Cambodia, before the boundary change, had about 800,000 hectares of riceland.[48] Most of this is cultivated during the rainy season and depends entirely upon the rain caught in the fields. Around the Tonle Sap and along the Mekong a good deal of floating rice is grown, and some riceland is planted after the summer flood has receded. The crop from this land is harvested in the dry season and may be considered a dry-season crop, although it begins to grow during the latter part of the rainy season and profits from the floodwater that has spread over the land.

The deltas of the Red River and of the Mekong, the two "rice baskets" of Indochina, differ greatly when it comes to the final disposal of the crop. The contents of the Tongkingese basket barely meet the needs of the local population, whereas a large part of the Cochinchinese rice may be exported, since the production far exceeds the local consumption. The delta of the Red River is densely settled; but Cochinchina, especially the western part, is still a pioneer country with a relatively low population density. The importance of Indochina as a rice-exporting region is based entirely on the production of rice in the south, particularly in Cochinchina and, to a lesser extent, in Cambodia.

Dry Fields

Dry fields[49] represent the third type of land under permanent cultivation in Southeastern Asia. The steady increase in the foreign demand for tropical products raised on dry fields, combined with the growth in local food requirements, has led to a great expansion of the area under permanent cultivation. The continuous use of unirrigated land, however, offers one of the most troublesome problems of tropical agriculture, because of the difficulty of preventing soil depletion. Erosion is especially active in a

---

47 Thompson: "War and Further India's Rice," 1941, p. 185.

48 Gourou: L'utilisation du sol, 1940, p. 309.

49 In the following discussion we shall exclude non-irrigated or dry land that comes under the regime of shifting cultivation. Sometimes a non-irrigated field lying fallow may have the same appearance as a recently abandoned ladang. But the essential difference is that under permanent culture land tenure is more highly developed than under shifting culture; so that a permanent field lying fallow belongs to an individual who, after a short period, will again cultivate it, whereas an abandoned ladang gives the former cultivator no claims for the future.

climate characterized by heavy rainfall often concentrated in one season, and it will, unless checked, lead to the exhaustion of the fertility of the soil at a comparatively rapid rate in the tropics. In Southeastern Asia permanent dry fields are usually found on sloping ground exposed, much more than in the case of garden or wet-rice land, to the danger of erosion. Garden land is protected by dense stands of perennials, while terracing and dyking guard wet fields against destructive erosion. The terracing of dry fields is one of the most effective ways to control the loss of soil, but, since terracing requires so much labor, we find terraced dry fields only in densely settled regions, and even there the peasants often will not undertake this additional labor unless they are urged to do so by the authorities. As long as there is enough land available, the peasants prefer to let dry land recuperate by lying fallow for short periods rather than to protect it by terraces; but with the increasing density of population and the consequent shrinkage in the size of individual farms this practice is no longer possible (Figs. 61–64).

As early as the second half of the nineteenth century, K. F. Holle, the agricultural adviser in the Netherlands Indies, constantly stressed the need of controlling erosion on dry fields. Following his advice, the government, in Article 5 of the 1874 Ordinance on Alienation of Domain Forest Land, stipulated that permission to clear domain land on sloping terrain be given to the peasants only on condition that they terrace the slopes and thus control erosion.[50]

The need for the terracing of slopes became especially apparent in Java in the coffee lands that the peasants had to plant under the forced-culture system. Many of these lands were completely exhausted after a few years, and Holle showed that this rapid exhaustion occurred because the topsoil had been washed away by heavy rains. In 1873 he wrote detailed instructions for the terracing and cultivation of coffee lands.[51] However, Holle's instructions and the stipulations of the Ordinance of 1874 were subjected to criticism and often ignored. In 1889, therefore, the Government of the Indies asked Forester Berkhout to investigate the practical results of the application of terracing to tegalans. Berkhout's report summarizes the arguments for and against terracing; these still make interesting reading.[52] The arguments against terraces on tegalans were put forth as follows:

1. The people are not used to terracing dry land.
2. Terracing demands more labor.
3. It reduces productivity in the first year if the land is rather steep and the humus layer thin.

---

50 Article 5 reads: "Dat bij hellend terrein zoo er gevaar bestaat voor afspoeling van de bouwkruin, de aanleg geschiede terrasgewijze, en voorts onder zoodanige voorwaarden als plaatselijke omstandigheden wenschelijk maken." Ontginnings-Ordonnantie, *Staatsblad 1874*, No. 79.

51 Nederlandsch-Indische Maatschappij: Handleiding, 1873.

52 Berkhout: "Terrassenaanleg," 1892, pp. 139–141.

4. It ties the native more to the soil; so that he can less easily escape the Treasury and the control of native officials. (This is a disadvantage for the individual but an advantage for the government.)

Berkhout lists as arguments in favor of terracing:

1. Terracing makes permanent cultivation of the land possible—because the soil is not washed away—and tillage becomes easier.

2. If the terraces are made narrow at the start and are widened each year, until on steep slopes they have reached a width of from 3 to 4 feet and on less steep slopes of from 5 to 6 feet, the increased labor will be compensated for by the advantage of easier tillage of the land in following years than would be the tillage of virgin soil.

3. The cultivation of crops is much easier on land that is under permanent culture than on land that has been lying fallow for several years and has been invaded by malignant weeds such as *alang-alang* (*Imperata* species).

4. The people do not have to move constantly. Whereas non-terraced land has to be abandoned nearly every year, well terraced land can be utilized for from 10 to 20 consecutive years and even for much longer if fertilization and crop rotation are applied.

5. The deep tillage and increased airing of terraced soil improve its physical condition and raise productivity.

6. The rental and sales value of the land rises.

7. The land will be fenced and the peasant will build a simple hut on it, something that would not be profitable if the land were used for only one year. This reduces the danger that the crop will be damaged by wild animals.

8. The land can be manured without danger that the manure will be washed away before it has had time to be effective.

9. The people learn to appreciate the land and to till it better.

10. More perennials—such as areca-nuts and sugar-palms—are planted on terraced land than on land that requires fallowing every so often and on which the fruit trees would have to struggle with alang-alang.

11. The Treasury finds it easier to collect taxes from a sedentary, stable population than from a people that is constantly moving.

. . . .

14. It considerably reduces the land requirements of the native population in the mountains. It is therefore easier to expand private plantation agriculture and governmental coffee cultivation.

15. It reduces the difficulties of forest conservation. Forests whose conservation is not necessary for climatological reasons can be released for cultivation in accordance with the increase of population and of land requirements.

16. Terracing breaks the force of rain water. The tilled soil absorbs more water. Thus terraced land regulates the flow of the rivers and reduces the violence of floods.

17. It reduces the silting of rivers and their estuaries.

In a study published in 1894, Holle reports actual cases from the regency of Soekaboemi that clearly prove the advantages of terracing tegalans.[53] Nevertheless, the government did not require the peasants of Java to terrace tegalans already in use before the Ordinance of 1874, although officials in some districts, by bringing pressure to bear upon the peasants, compelled them to terrace. For several decades following the early 1890's, the authorities paid less attention to soil erosion. In many districts the terraces were not kept in good condition, and so much soil was washed away that in more recent years the authorities have become alarmed over the increasing signs of exhaustion of the tegalans and their decline in yields. A recent study states: "Another problem which on sloping dry land demands attention is the erosion of the soil. Terracing of these lands should be forcefully recommended, as otherwise the soil will become completely sterile." [54]

TABLE III—JAVA: GROWTH OF THE TERRACED TEGALAN AREA IN THE RESIDENCY OF PRIANGAN, 1937-1941

DATE	TERRACED TEGALAN AREA IN HECTARES
1937 (December)	1,370
1938 "	4,290
1939 "	12,880
1940 "	14,950
1941 (May)	16,930

Source: Joosten: Ontwikkeling . . . van den Landbouwvoorlichtingsdienst, 1941, p. 1067.

Holle was especially familiar with the Preangers, and many of his illustrations are taken from this part of West Java. I had the good fortune to spend some time in the field with H. A. A. M. Wirtz, the adviser for Indonesian agriculture in Priangan (formerly Middle and East Preanger). Mr. Wirtz was trying to improve the fertility of tegalans by means of terracing. In Priangan tegalans represent more than half the area cultivated by Indonesians. At the end of 1937 there were 1,370 hectares of terraced tegalans in Priangan, and by May 1941 this figure had increased to 16,930 hectares (Table III). Wherever possible Wirtz recommended building up the faces of the terraces with stones. Where stone is not available the faces of the terraces can be sufficiently protected by a grass such as buffalo grass (*Paspalum conjugatum* Berg.), which at the same time supplies a good cattle fodder. It is extremely important to protect the faces of the terraces with a cover of vegetation; otherwise erosion is accelerated through the increased cultivation of the soil, and the terraces will be washed away within one or two rainy seasons. Where *Paspalum* had not yet been established as a cover crop on the fronts of the terraces, the cover very often consisted of a spontaneous growth of common weeds and grasses. Unfortunately, the peasants like to dig out these weeds in order to pre-

[53] Holle: "Eenige gegevens omtrent den inlandschen landbouw," 1894.
[54] Departement van Economische Zaken: Voedselproblemen, 1940, p. 29.

vent their spread over the fields. Wirtz opposed this removal of weeds and grasses. Because of its deep roots, *Leucaena glauca* Benth. is an excellent protector of terrace banks when planted as a hedge along the edge of the terraces. It can be kept low so that it will catch the soil that otherwise might be washed down (Figs. 65–70).

In two instances during my stay in Priangan I saw remarkable proof of the extent of recent erosion on sloping unterraced land. The peasants of this region frequently grew tea on their tegalans. They kept the ground so free of weeds that the soil between the bushes was unprotected, and erosion had been so extensive that the roots were exposed for as much as a foot and a half, leaving the bushes standing upon "stilts." This was in all likelihood a repetition of what had occurred during the past century in the coffee lands. Had these tea lands been terraced, as Holle recommended, they would not have lost so great a quantity of soil in the 10 or 15 years that they were in production.

The work of the Agricultural Advisory Service in encouraging the terracing of tegalans in West Java has had pronounced effects. It has led to a rise in land prices—even of tegalans not yet terraced—for the farmers began to realize the potentialities of dry land. As soon as the danger of erosion has been conquered, the peasants have begun to apply more stable manure and green manure to the land and to plant cash crops such as tobacco, onions, and tomatoes. The building of terraces has also created employment for men especially skilled in terracing. Whereas formerly a good deal of the tegalan area had been leased, now the owners have started to cultivate a greater portion of the land themselves.[55]

The island of Madura is probably the most remarkable area in Southeastern Asia in respect to the cultivation of tegalans.[56] The island is very densely settled, but the natural resources are rather limited. Because of the topography and soil, over 80 per cent of the arable land consists of tegalans, mostly on slopes. The tegalan area has been extended to such a degree that the island is practically deforested. A large part of the tegalans is terraced, including much land that from the viewpoint of agricultural economics has not been worth the labor that went into the construction of the terraces. A portion of the dry land, however, was left unterraced after the forest was cut down, and its cultivation resulted in such severe erosion that the land has become worthless for farming. Even terracing and other methods of soil improvement could not rehabilitate this land. The agricultural officer of Madura told me in 1940 that he believed some 120,000 to 150,000 hectares of tegalan area should be bought from the Madurese and reforested by the forest service. However, there is no doubt that the situation would be much more serious today had the Madurese not done his best by terracing and other soil-saving methods of cultivation to pre-

55 Joosten: Ontwikkeling . . . van den landbouwvoorlichtingsdienst, 1941, p. 1071.

56 I spent two days on the island with J. F. Dekker as my untiring guide and interpreter.

vent erosion and rapid exhaustion of the land. Otherwise practically all the tegalans would now have been washed into the sea, and many of the people would have been forced to abandon the island (Figs. 71, 72).

The Philippines supply us with a striking contrast to Madura in the island of Cebu, which, like Madura, consists largely of limestone. Cebu, too, is densely settled. The island has the highest density of population of any province in the Philippines, that is, 219.4 persons per square kilometer in 1939 (as compared with 357 per square kilometer in Madura in 1930). Cebu is also very mountainous. As the population increased, more and more mountain forest was cleared, and the land was planted with annual food crops. However, the Cebuanos, in contrast to the Madurese, did not terrace their dry fields. Erosion ran its course, with the result that today in many parts of the island there is no longer any soil covering the limestone, while over wide areas the layer of soil has become so thin that terracing would no longer be possible. The peasants try to retard erosion by laying dry cornstalks across the slope—futile attempts. The continued cultivation of annual crops on the steep hill slopes of Cebu will finally destroy what remains of the dry fields and force the Cebuanos in increasing numbers to migrate to the neighboring islands, especially to Mindanao and Negros (Figs. 73–76). Years ago C. F. Baker, Dean of the College of Agriculture, University of the Philippines, recommended combining forestry with agriculture by planting narrow belts of trees along the contours between parallel wider strips of crop land, but so far no steps have been taken to test the value of this suggestion.[57]

A beginning in reforestation was made in Cebu when, in 1916, the Bureau of Forestry was given jurisdiction over a tract of about 4,000 hectares of mountainous land to the west and northwest of Cebu City. During the following years reforestation was carried out with *Tectona grandis* L., *Vitex parviflora* Juss., *Aleurites moluccana* Willd., *Aleurites trisperma* Blanco, and *Leucaena glauca* Benth. Wherever the plantings could be protected against firewood-seekers and against fire they did well on these steep hills, with their thin soils, and sent their roots into the crevices of the limestone. By 1928 the older plantings were supplying wood and oil-seeds. An economic survey of the land revealed that 95 per cent of the reforested area should be considered as permanent forest land, unsuited to agriculture. In spite of this the Bureau of Lands, under political pressure from the local population, demanded and obtained control over the land, with the exception of the forest experiment station and its seed plots at Camp 7 on the road from Cebu City to Toledo (Figs. 77, 78). Soon thereafter the reforested land was sold to Cebuanos on liberal terms, which required only a small first payment. As soon as the down payments had been made the new owners cut the trees and sold the timber and firewood, the value of which was estimated in many cases to be over 100 pesos per

57 Pendleton: "Glimpses of Hinterlands," 1935, pp. 831–832.

hectare. In many instances the land was abandoned as soon as the forest crop had been harvested; in others the land was cultivated for a short period. The soil that had formed during the time the land was being reforested eroded away. Granted that the population pressure in Cebu is great, it was nevertheless shortsighted to give way to local politics and allow the destruction of the new forest. The best course for Cebu would probably be for the peasants to give up growing three maize crops a year on steep slopes and to plant instead perennial tree crops such as kapok, mangoes, rimas, and other fruit trees.

In Cebu the Bureau of Forestry lost control to the Bureau of Lands. Even in Madura the forest service found that reforestation plans were not especially favored by officials of the Department of the Interior, who would have preferred an increase in the crop area to a decrease.[58]

So far we have discussed only one aspect of dry-field cultivation, the need for erosion control through terracing. In addition to erosion we also face the problem of a decline in fertility because of the fact that each crop takes away a certain amount of mineral matter vital to the growth of plants. Thorough tillage, crop rotation, the application of stable manure and commercial fertilizer, the growing of green manure, and mixed cropping—that is, the growing of more than one crop on the same soil at the same time—are some of the means by which the peasants of Java try to prevent depletion of soil fertility on their dry fields. The method, or methods, used depends upon the local environment, local tradition, the industriousness of the people, and the degree to which they must rely upon crops coming from the dry fields. In many regions of Java the peasants are accustomed to plowing their land from six to nine times and then harrowing it several times before they plant. Crop rotation is generally practiced. The peasant, when asked why he practices crop rotation, says that a living crop does not like to smell anything of its kind that has died, and for that reason he changes the crop from season to season. The use of animal manure is limited by the small number of livestock. In many parts of Java the density of population is so high that peasants have to use most of the land for the growing of food crops rather than for the production of fodder with which to feed animals. But aside from this, peasants in some districts are not used to preserving animal waste. Although the application of commercial fertilizer has given good results in many districts, the peasants are slow to change, either because of their conservatism or because they do not have the capital for this intensification of agriculture. However, in districts where such commercial crops as tobacco and sugar cane are raised, too much fertilizer may be used. I found this excessive use of fertilizer in a tobacco-growing district of Madura, where the peasants apply fertilizer as late as two weeks before the crop is harvested. In such districts agricultural advisers were endeavoring to bring about a reduction in the quantity of fertilizer applied and a better timing of its use.

58 Mijers: "Herbebosching," 1941, pp. 909–942.

The Agricultural Advisory Service of the Netherlands Indies for many years carried on experiments with green manures and recommended a variety suitable for different conditions, but the main obstacle to the application of their research was the difficulty of introducing green manure into a crop rotation. Some of the green manures used in Java are *Crotolaria anagyroides* H. B. and K., *Crotolaria usaramoensis* Baker, *Tephrosia candida* DC. and *Tephrosia vogelii* Hook., *Calopogonium mucunoides* Desv., *Centrosema pubescens* Benth., *Centrosema plumierii* Benth., and *Mimosa invisa* Mart. Green manure requires a relatively large amount of water in order to produce the desired foliage. If there is sufficient water, the peasants can also grow other crops such as legumes, which supply food as well as fertilizer. Most of the peasants of Java have so little land that they cannot afford to sacrifice a single food crop from a field in spite of the probability of an increased yield after the green manure has been plowed under. Although it is preferable to grow green manure unmixed with other plants, the planting of a green manure crop among the food crops is recommended where land is scarce. This is done in Java, especially in maize and cassava fields. The specific green manure that is chosen should meet the following conditions:[59]

1. It should have a rapid growth,
2. An abundant production of leaves,
3. No excessive development that might harm the main crop,
4. An absence of climbing or trailing stems;
5. The plant should have deep roots with many nodules;
6. It should not become woody too easily—as herbaceous plants decompose much more quickly; and
7. It should be immune from the diseases and insects that attack the main crop.

Another agricultural practice common to many parts of the world, especially the tropics, is mixed cropping, which seems to bring in its train some of the advantages of rotation.[60] Nicol stresses the great importance of having leguminous plants grown among non-leguminous plants. He points out that the legume component in the mixture provides nitrogen to the non-legume.[61] In various parts of India—for example on the Bombay Deccan—bulrush millet (*Pennisetum typhoideum* Rich.) is grown with various legumes. This can be done year after year without greatly affecting the fertility of the soil. No crop rotation is practiced here. The proportion of bulrush millet to legumes varies according to the quality of the land; the poorer the land, the smaller the proportion of millet and the larger the proportion of legumes. Wallace wrote in 1887:

> The growth of mixed crops is a widespread practice which is well worth consideration and study. . . . The advantages . . . are distinctly great. . . . There is

59 International Institute of Agriculture: Use of Leguminous Plants, 1936, p. 60.
60 Willis: Agriculture in the Tropics, 1914, p. 29.
61 Nicol: "Mixed Cropping," 1935, p. 189.

but one explanation of the existence of these practices, viz., that they have been found advantageous after long experience and much careful consideration on the part of a body of workers who, for power of observation and an intelligent interest in and knowledge of every day occurrences, would put to shame those classes which hold a corresponding position in educated Europe.[62]

Well known is the American Indian's practice of interplanting maize, beans, and squash. Chinese and Japanese peasants practice mixed cropping widely, and we may observe it almost everywhere in Southeastern Asia. Intensive mixed cropping in the garden culture of Java has already been described (see above, pp. 43–45), but there is relatively little interplanting of legumes with non-legumes. On dry fields the peasants of some parts of Java are accustomed to interplant cassava and maize—both, however, non-legumes. Whereas maize can be harvested after 100 days, cassava needs up to a year or more of growth. Cowpea (*Vigna sinensis* Savi), Goa bean (*Psophocarpus tetragonolobus* DC.), Lima bean (large) (*Phaseolus lunatus* L.), and hyacinth bean (*Dolichos lablab* L.) are also interplanted with cassava. On tegalans in Bali, in 1941, I observed the following crops grown with upland rice: *Eleusine indica* Gaertn., sesame (*Sesamum indicum* L.), and pigeon pea (*Cajanus cajan*). This last enriches the soil as a result of its large production of leaves as well as by the nitrogen in its nodules.

Kaserer has noted that there is little or no interpenetration of plant roots of the same species when they are grown together, whereas a significant degree of root interpenetration takes place when two different species are grown in close association, the degree of interpenetration increasing with the degree of dissimilarity. In one experiment, for example, a legume and a non-legume showed the maximum amount of root interpenetration, the explanation apparently being that one plant takes up the nutrients produced, or made available, by the other.[63] If these observations are correct we have here a scientific argument for the value of mixed cropping. Further studies are needed to determine the desirability of improving and extending it in specific areas.

Another important method of improving the fertility of dry fields is the use of compost. Vegetable composts are employed extensively in India and Africa. The masters at making compost from household refuse, vegetable waste, and human excreta are the Chinese, who cling to this method of fertilization wherever they practice truck farming in the neighborhood of urban centers in Southeastern Asia. Fungi and bacteria are used to break down suitable mixtures of vegetable and animal wastes. A method that was developed in India is the "Indore process" in which, "by arranging these mixtures in the proper way, and by watering and turning them, to supply moisture and air, the waste materials are transformed in about ninety days into a finely divided humus rich in the foods required by grow-

62 Wallace: India in 1887, 1888, pp. 196 and 198.

63 Kaserer: "Beobachtungen," 1911, pp. 1022–1030.

ing crops. The process can be adapted to climate by manufacture either in shallow pits or low heaps."[64]

Dry-Field Crops Cultivated by Peasants

The list of dry-field crops cultivated in Southeastern Asia is rather large, but there are regional variations. Cassava, for example, is commonly grown on tegalans in Java (2,300,000 acres), whereas it is unimportant in Sumatra, Borneo, the Philippines (40,000 acres), Malaya (24,000 acres), and Indochina (75,000 acres). The most important dry-field crops are upland rice, maize, cassava, sweet potatoes, peanuts, soy beans and other pulses, tobacco, and sugar cane, all of which are annuals. Of the first four, only upland rice is a native of Southeastern Asia, whereas the other three are natives of the New World and were introduced into Southeastern Asia by the Spaniards in the sixteenth century.

TABLE IV—CHEMICAL COMPOSITION OF CASSAVA AND IMPORTANT CEREAL FLOURS

	CASSAVA (PER CENT)	MAIZE (PER CENT)	WHEAT (PER CENT)	RICE (PER CENT)
Moisture	10.75	13.41	13.65	13.12
Protein	0.87	7.56	12.35	9.85
Crude fat	1.19	2.79	1.75	0.88
Ash	1.24	1.78	1.81	1.51
Crude fiber	6.12	3.19	2.53	0.63
Carbohydrates	79.83	71.27	67.91	74.01

Source: The Cassava Industry in the Philippines, 1939, p. 7.

The inhabitants of Southeastern Asia have long relied upon root crops, fruits, leaves of various plants, and wild roots and tubers gathered in the forests, as well as upon rice. Scheltema refers to a number of early statements which indicate that "rice was not yet the principal food for a large portion of the native population [of Java] as late as the first part of the 19th century and that the rice diet has become much more important in the course of the past hundred years" as a result of the great extension of the cultivation of irrigated rice fields.[65] If it is correct that the rice consumption per capita increased during the past century, the opposite trend has been evident during the present century. For a while the increase in the area of irrigated land was great enough to allow for a larger rice production, but with the steadily increasing pressure of population the peasants now have to depend more and more upon unirrigated land. Thus the production of dry-field crops is gaining in importance.

Authorities in Southeastern Asia have been especially worried over the increase of cassava production and consumption. Because of the exhaustion of the soil fertility on many dry fields, the growing of upland rice and

64 Worthington: Science in Africa, 1938, p. 387.
65 Scheltema: The Food Consumption, 1936, p. 10.

maize has become unprofitable, and many peasants have changed over to cassava, which makes lesser demands upon soil fertility. The per acre yields of cassava are high, and the plant is a source of carbohydrates, fat, and protein. From the standpoint of dietary needs, however, the ratio between carbohydrates on the one hand and protein and fat on the other is unfavorable. People who live on cassava merely fill their stomachs. The insufficiency of protein intake is detrimental to their health. Table IV compares the chemical composition of cassava, maize, wheat, and rice.

Maize, as the table indicates, compares quite favorably in protein content with wheat and rice and is especially rich in fat. It has gained an important position in the diet of the people of central and eastern Java, Madura, the island of Cebu, and the Cagayan Valley in the island of Luzon. Both maize and cassava are easy to store and therefore have a great advantage over sweet potatoes, which must be consumed soon after they ripen.

In Thailand maize is an important crop on the Khorat plateau and in other regions where dry land is available. Among the root crops cassava has gained a greater importance than taro and yams. The last two are grown mainly by Chinese gardeners in the neighborhood of large towns. Cassava, too, is grown commercially by Chinese, especially near the railroad on the peninsula.

Most of the so-called dry-field crops are grown in Indochina during the dry season on land that is flooded during the rainy season and at that time produces rice. Gourou has estimated that in Tongking 100,000 hectares of land are used exclusively for the production of dry-field crops and that 350,000 hectares of irrigated land are occupied by such crops during the dry season.[66]

This practice of growing dry-field crops on irrigated land during the dry season is especially widespread in Java, where population pressure is so great that the peasants have to use every possible means to enlarge their food supply. If there is not enough water available to grow a second wet-rice crop, they will always try to get a second harvest of quick-growing dry-field crops during the dry season.

In contrast to the peoples of Java, the Filipinos do not, as a rule, go in for the raising of dry-field crops on wet riceland during the dry season, although this should be possible if the land were cultivated. There is enough water in the soil a short distance below the surface for the growing of crops practically everywhere during the dry season. Repeated tilling would keep the soil soft and fine so that water could move upward by capillary action to replace that which is lost through evaporation. Cultivation would enable plants such as maize, peanuts, and soy beans to grow on land that otherwise would be baked dry and hard. Although the ratio between cultivated land and population in the Philippines does not make a dry-season crop a necessity in most regions, there are a number of districts where the peasants traditionally raise a dry-season crop on irrigated

66 Gourou: Les paysans, 1936, pp. 407–408.

land. An example is *mongo* (*Phaseolus mungo* L.), grown in Pangasinan and Batangas. Alarmed by the poor rice crops in 1940 and the difficulty of importing rice from Indochina and Thailand, the Philippine Bureau of Plant Industry, early in 1941, started to distribute seed of quick-growing dry-field crops, especially maize and legumes, in order to supplement the nation's rice bowl. But the average Filipino farmer is reluctant to accept this new practice, because "it is not the custom"—a powerful argument not to be easily overcome.

Dry-Field Crops Cultivated by Plantations

The discussion thus far has been limited to the native cultivation of dry land and to annual crops. However, extensive areas of non-irrigated land are occupied by perennial crops that are grown on plantations owned by Westerners. The crops produced include tea, coffee, cacao, quinine, rubber, palm oil, coconuts, kapok, sisal, and other fibre plants. Some of these crops were cultivated for a long time without the proper measures being taken to protect the soil against accelerated erosion. Although the detrimental effects of plantation agriculture in Ceylon and southern India are notorious in the plantation history of southern Asia, results were just as disastrous in Malaya and in the Netherlands Indies. Westerners, accustomed to clean cultivation or clean weeding in the temperate zone, carried this practice over into the tropics, with the result that soil erosion was very severe on steep slopes and led to great losses in soil fertility. Today, however, these mistakes have been recognized; there is hardly a plantation that has not adopted measures to protect sloping land against erosion and level land against soil exhaustion. The planters of Java were the pioneers in this field, introducing or developing most of the protective measures. These measures include contour drains, silt pits, contour ridges, terraces, hedges, shade trees, and especially cover crops. The use of these measures varies according to the crop, or crops, produced and according to climate, soil, and slope; but the purpose of all of them is to encourage an effective ground cover and to check the rate of water run-off. The financial depression of the 1930's—resulting in the need for very strict economy and a decrease in production costs—has helped considerably to reduce the amount of clean weeding and to convince conservative planters of the value of preserving a ground cover, litter, and mulch. Such measures help not only to retard erosion but also to enrich the soil. The first Western planter who broke away from the custom of keeping the ground of his plantation free from weeds was L. M. Tyl, who in the years 1890–96 developed the "weed-system" and thus introduced the idea of a ground cover into plantation circles of Southeastern Asia. But it took many years before his idea was generally accepted. It is interesting to note here that the Asiatic-owned plantations and the small holdings producing perennial crops, such as rubber, usually did not practice clean weeding. This, in the early days, was regarded by European observers as a sign of laziness and poor management, but, as has been so often the case, the Westerners were finally forced to admit the soundness of Oriental cultivation practices.

As we have seen, there are two principal means of protecting the soil between tree crops: by allowing the growth of indigenous plants or weeds (in other words the development of a natural mixed cover), or by establishing a pure cover of leguminous plants, many of which came originally from the New World. Rubber planters in Malaya seem to have had more success with the first method, which has been called "the forestry method of cultivation" or "rubber forestry." [67] Of course, the growth of indigenous plants or weeds must be controlled, and this is done by slashing the undergrowth from time to time. Frequently it is necessary to practice selective weeding in order to suppress undesirable plants and encourage more valuable ones.

Much research has been directed toward the study of leguminous cover crops which supply nitrogen, as well as foliage for green manure. This research was started in 1910 in the Botanical Garden at Buitenzorg. Some of the cover crops that have proved especially valuable and are widely used on modern plantations are *Calopogonium mucunoides* Desv., *Centrosema pubescens* Benth., *Centrosema plumierii* Benth., *Desmodium* species, *Tephrosia candida* DC. and *Tephrosia vogelii* Hook., *Mimosa invisa* Mart., *Cassia* species, and *Vigna oligosperma* Backer.[68] Several of these have already been mentioned in the foregoing discussion on the use of green manure crops in native agriculture.

Because a number of perennials such as tea, coffee, and cacao are commonly grown under shade trees, many writers seem to believe that these plants need shade. This may be true in part, but more important is the fact that the shade trees protect the soil against excessive losses of moisture and undesirable changes in soil composition, texture, and structure and thus prevent a reduction of the quantity of humus in the soil. Most of the commonly used shade trees are legumes which, in addition to shade, supply litter and nitrogen and help to increase the moisture-holding capacity of the soil, thus reducing the danger of erosion. *Erythrina micropteryx* Poepp., *Erythrina lithosperma* Miq., *Leucaena glauca* Benth., *Albizzia moluccana* Miq., *Albizzia chinensis* Merr., *Albizzia falcata* Backer, *Pithecellobium dulce* Benth., or Manila tamarind, and *Peltophorum pterocarpum* Backer are especially useful in this respect.

---

67 Haines: "The Uses and Control of Natural Undergrowth," 1934.

68 For a very extensive list of cover crops, green manures, and shade trees, see Heyne: De nuttige planten van Nederlandsch-Indië, Vol. 3, 1927. See also International Institute of Agriculture: *op. cit.*

Part II

AGRICULTURAL COLONIZATION IN THE PHILIPPINES AND THE NETHERLANDS INDIES

Chapter IV

LANDLESS FILIPINOS

The Population Problem in Southeastern Asia

The population in all the countries of Southeastern Asia is distributed very unevenly. It is concentrated mainly in the lowlands along the coasts, on the plains of great rivers, on intermontane plains, and on certain of the islands in the Philippines and the East Indies. Other large areas, less attractive for various reasons, be they permanent or temporary, have low densities of population. This great concentration is due chiefly to the fact that rice is the staple food. Growers of wet rice prefer extensive stretches of level land. The growth of population in any region leads to an expansion of the area under cultivation as long as there is suitable land available. Thereafter the size of the individual farm decreases. This decrease in the amount of cultivated land per peasant must be compensated for by a greater intensity of land use in order to bring about larger yields. There are limits to this, however, and when the stage has been reached where the agrarian population encounters great difficulties in obtaining subsistence from the land, in spite of the fact that its agriculture is very intensive, then we may speak of overpopulation.

There are various ways of relieving the pressure of population, such as emigration, industrialization, and a lowering of the birth rate. The remedy of the population problem that will be examined here is migration to regions with agricultural potentialities.

In Southeastern Asia the Indies, the Philippines, and Indochina are suffering most severely from a maldistribution of population. This and the following chapters deal specifically with the Philippines and the Netherlands Indies, where I was able to study the problem at first hand.

Density and Distribution of Population in the Philippines

The last census of the Philippines,[1] taken in 1939, showed a total population of 16,000,000, more than twice the number (7,635,000) [2] recorded in 1903. The density of population amounted to 53.8 persons per square kilometer in 1939, but there was, as has been already indicated, a great

[1] The statistical data in this chapter for which the sources are unspecified were obtained chiefly from the *Census of the Philippines 1939*, the annual reports of the Philippine Commission and later of the Governors General, and unpublished material supplied directly by government departments. The text of the acts and proclamations referred to may be found in the following compendiums: U. S. Statutes at Large, Vol. 32, 1903; Public Laws Enacted by the Philippine Legislature (31 vols.), 1903–1937; Public Laws of the Commonwealth (5 vols.), 1935–1939; and Messages of the President (5 vols.), 1935–1941.

[2] It should be pointed out that the census of 1903 was not so complete as that of 1939; so that the increase between 1903 and 1939 was probably not quite so large as the census figures indicate.

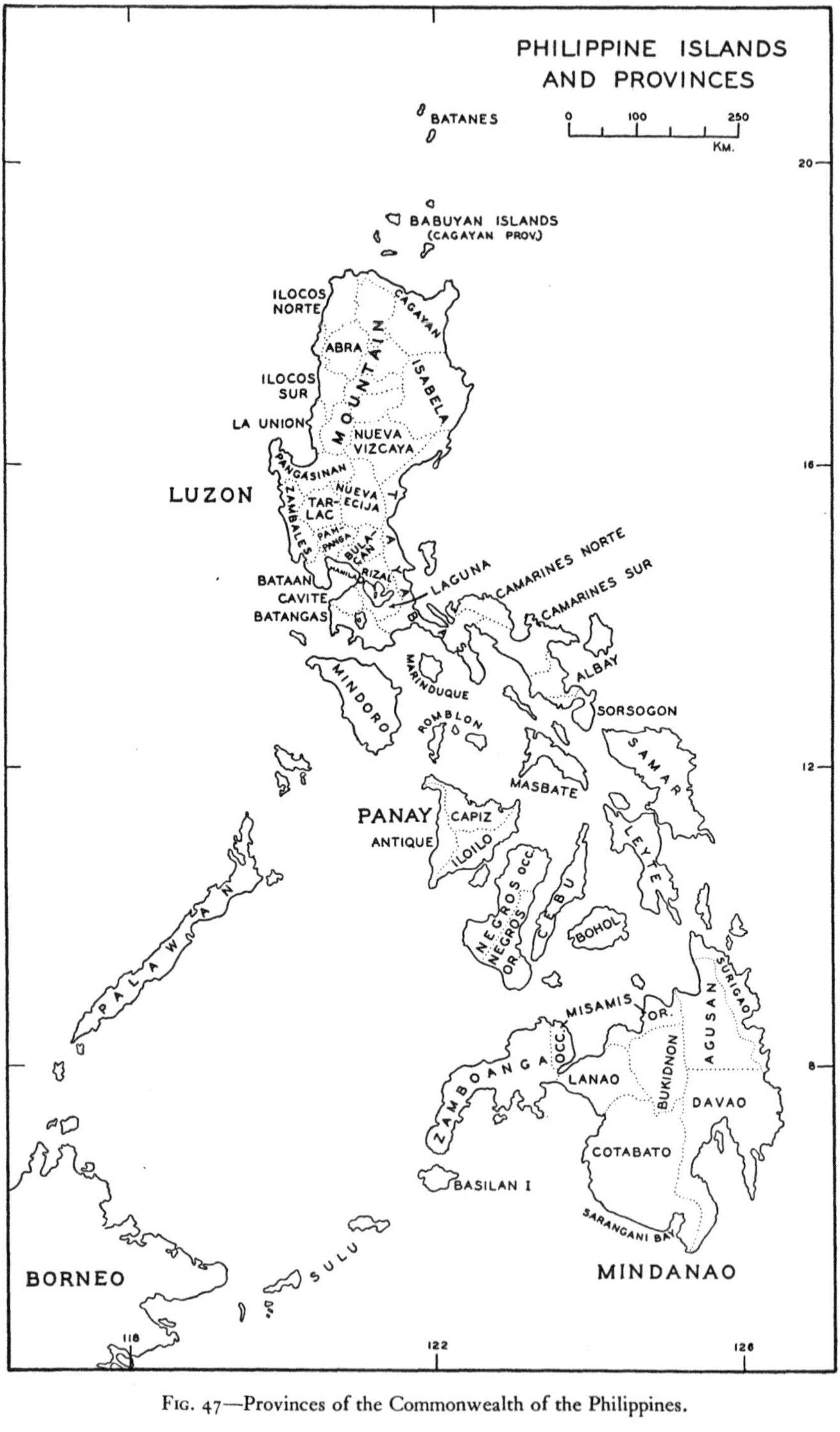

FIG. 47—Provinces of the Commonwealth of the Philippines.

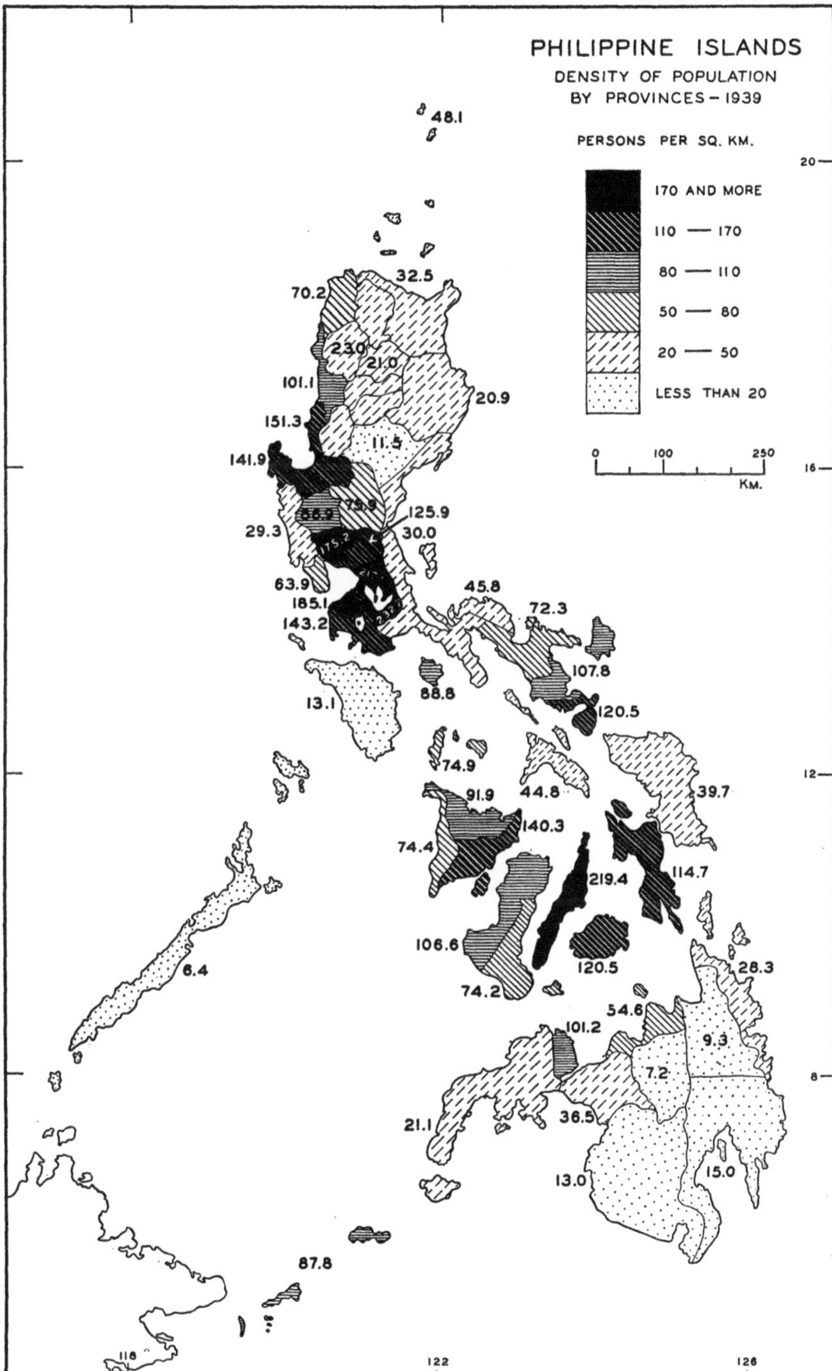

FIG. 48—Density of population in the Commonwealth of the Philippines, by provinces, as recorded by the census of 1939. (The map does not show the province of Manila.)

unevenness in distribution (Map, Fig. 48). Of the ten largest islands of the Philippines, Cebu had the highest density with 214.8 persons per square kilometer, followed by the rest of the Visayan group of Bohol, Leyte, Panay, and Negros; while Luzon, the largest of the islands, had a density of 70.4 persons per square kilometer. In contrast to these islands,

TABLE V—AREA, POPULATION, AND DENSITY OF POPULATION OF THE PHILIPPINE COMMONWEALTH AND OF THE TEN LARGEST ISLANDS, 1939

ISLAND	AREA (SQ. KM.)	POPULATION (THOUSANDS)	DENSITY PER SQ. KM.
Philippine Is.	297,410	16,000	53.8
Luzon	104,688	7,375	70.4
Mindanao	94,630	1,828	19.3
Samar	13,080	471	36.0
Negros	12,705	1,219	95.9
Palawan	11,785	44	3.7
Panay	11,515	1,292	112.2
Mindoro	9,735	117	12.0
Leyte	7,214	836	115.8
Cebu	4,410	947	214.8
Bohol	3,865	450	116.3

Source: *Census of the Philippines 1939*, Vol. 2, p. 42.

where the density considerably exceeded the average density of the Philippines, Mindanao, the second largest of the islands, averaged only 19.3 persons per square kilometer, Samar 36.0, Palawan 3.7, and Mindoro 12.0 (Table V).

A study of densities by provinces, excluding Manila, shows that Laguna (in Luzon) had the highest density in 1939, with 232.2 persons per square kilometer, followed by Cebu (the province of Cebu includes some smaller islands besides Cebu) with 219.4 and Rizal (in Luzon) with 217.1. In Luzon the provinces around Manila Bay, those in the central plain, the Ilocano provinces (La Union, Ilocos Sur, and Ilocos Norte), and those of the Bicol region (Sorsogon, Albay, and Camarines Sur) are the most densely settled. In the Visayas—in addition to Cebu—Iloilo (in Panay) and Bohol are very congested, and Leyte and Negros Occidental are less so. In Mindanao only one province, Misamis Occidental, has a density higher than 100 persons per square kilometer.

Figures for the density of population based on total land area are not particularly satisfactory. More significant is the ratio between population and cultivated area. This is certainly true in the Philippines, where most of the peasants cultivate small farms rather intensively and are, on the whole, crowded on plains and in valleys, while extensive areas of hinterland—lowlands, uplands, or mountain country—remain thinly settled or empty. The average density per square kilometer of cultivated land in the Philippines was 404.7 persons in 1939. However, in such provinces as

Ilocos Sur, Ilocos Norte, and Cebu, provinces that have little urban concentration, the densities were considerably higher: 804.3, 666.5, and 767.5 persons, respectively. As only a part of the arable land is used agriculturally, the ratio of population to cultivated area for the Philippines as a whole is not alarming, being considerably lower than that of Java and Madura or of Japan. Although more than half of the land in the Philippines is considered cultivable, only 13.3 per cent of the total area was actually under crops in 1938, with 22.4 per cent of the total belonging to farms. The difference between the area under crops and the farm area was made up by idle, pasture, forest, and "other" land.[3] In the mountainous Ilocano provinces, in Cebu, and in Bohol, however, practically all cultivable land is in use.[4] Moreover, the peasants of these provinces use a good deal of land on steep slopes that, since no safeguarding measures have been taken, are exposed to severe erosion (Figs. 73, 74). It is clear that the provinces with the highest densities per unit of cultivated land, far from having possibilities of expanding their crop area, are really in need of reducing the area under cultivation and carrying out a reforestation program to prevent further deterioration of their soil resources. There is already a constant stream of migrants from the Ilocano provinces and the crowded Visayan provinces to adjacent, or even to distant, parts of the country. This migration started in Spanish times but received its strongest impetus during the American regime, when extensive road building opened up areas that had been inaccessible.

TABLE VI—PHILIPPINE COMMONWEALTH: PERCENTAGE DISTRIBUTION OF THE CULTIVATED LAND ACCORDING TO THE SIZE OF FARM AND PERCENTAGE DISTRIBUTION OF FARMS ACCORDING TO SIZE, 1939

SIZE OF FARMS	PERCENTAGE OF CULTIVATED LAND	NUMBER OF FARMS	PERCENTAGE OF ALL FARMS
Less than 1 hectare	4.8	368,903	22.5
1–1.99 hectares	15.5	489,053	29.9
2–2.99 "	16.4	306,496	18.7
3–3.99 "	12.2	164,458	10.1
4–4.99 "	8.7	94,106	5.8
5–9.99 "	18.3	142,587	8.7
10–19.99 "	9.7	48,367	3.0
20 or more "	14.4	20,756	1.3
Total	100.0	1,634,726	100.0

Source: *Census of the Philippines 1939*, Vol. 2, pp. 927 and 928.

ECONOMIC CONDITIONS

The population problem in certain parts of the Philippines is consider-

[3] "Other" land signifies waste land or land occupied by houses, buildings, or improvements.

[4] In Cebu 28.6 per cent of the total area was under crops in 1938, in Ilocos Sur 12.6 per cent, in Ilocos Norte 10.5 per cent, in La Union 24.2 per cent, and in Bohol 23.6 per cent.

ably aggravated by such economic factors as land-tenure systems, widespread tenancy, usury, poverty combined with indebtedness, and the uneconomical size of many of the farms.

On nearly one-fourth of all farms in the Philippines less than one hectare of land was cultivated; these accounted for less than one-twentieth of the total cultivated area. On more than half of all farms less than two hectares

TABLE VII—LUZON: PERCENTAGES OF TOTAL FARM AREA AND OF ALL CULTIVATED LAND OPERATED BY SHARE, CASH, AND SHARE-CASH TENANTS IN CERTAIN PROVINCES, 1939

PROVINCE	PERCENTAGE OF TOTAL FARM AREA OPERATED BY TENANTS	PERCENTAGE OF CULTIVATED LAND OPERATED BY TENANTS
Bulacan	60.5	66.5
Nueva Ecija	57.3	67.8
Pampanga	52.9	67.0
Cavite	54.7	58.5
Tarlac	45.3	52.4
Laguna	41.3	44.3

Source: *Census of the Philippines 1939*, Vol. 2, pp. 983–984.

were cultivated, and these occupied only one-fifth of the country's cultivated land. On the other hand, on slightly less than one-twentieth of all farms 10 or more hectares were cultivated, but these farms included about one-fourth of the area under cultivation (Table VI).

One of the characteristics of Philippine agriculture, and of Oriental agriculture in general, is that land devoted to rice culture is worked in small units, each usually requiring the regular labor of not more than one family. Whenever the labor demand exceeds the capacity of the family unit, neighbors cooperate.

Although the Philippines have many small farmers who own the land they till, the proportion of landless farmers is astonishingly high for a tropical country. According to the census of 1939, only 49.2 per cent of all farmers were owners of the land they worked; an additional 15.6 per cent owned a part of the land; whereas more than one-third (35.1 per cent) of all farmers were tenants, the great majority of them being share tenants.[5]

The various provinces differ widely in the degree to which the land is worked by tenants. The map in Fig. 49 shows the percentages of cultivated land farmed by tenants. Comparing the number of tenant farmers, i.e. share tenants, cash tenants, and share-cash tenants, with the total number of farm operators, we find that in 1939 the percentage of tenancy ranged all the way from 1.8 to 70.4 per cent. Seven out of the 49 provinces had had a tenancy rate exceeding 50 per cent, 16 had a rate of more than 40 per cent, and 25 one of more than 30 per cent.[6] The prevalence of

[5] *Census of the Philippines 1939*, Vol. 2, p. 968.
[6] *Ibid.*, p. 970.

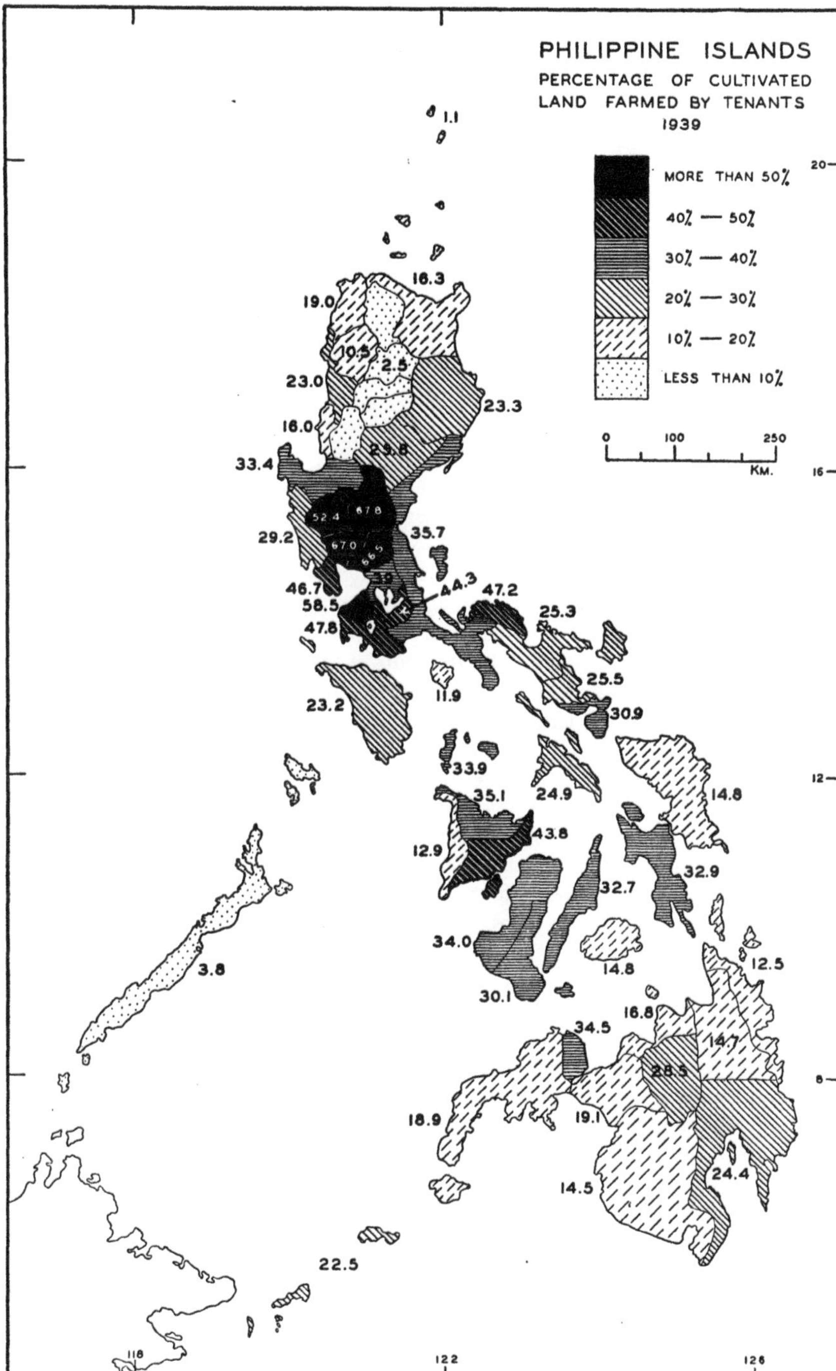

FIG. 49—Percentage of cultivated land farmed by tenants in the Commonwealth of the Philippines, by provinces, as recorded by the census of 1939.

tenancy in certain provinces in Luzon can be seen in Table VII, which gives the percentages of total farm area and of cultivated area operated by all types of tenants.

The percentage of tenants was low in La Union (17.7 per cent), Ilocos Norte (19.5 per cent), and Ilocos Sur (24.2 per cent), provinces that are among those having the highest ratios of population to cultivated area.[7] The explanation for this may be found in the character of the Ilocanos, who are notably industrious, thrifty, and attached to their land.

Historical Development of Land Tenure

Caciquism

In order to understand the land-tenure conditions of today in the Philippines one has to go back to early Spanish and pre-Spanish times. In those days, as now, the inhabitants of the lowlands and coastal regions of Luzon and the Visayas lived in villages. Each village, or *barangay*, as the Tagalogs called it, had its own chief, or *dato*. Under the aristocracy, which was represented by the dato and his family, there were three classes of people: free men, serfs, and slaves. The free-born members of a barangay usually paid no taxes or tribute to the dato but were obliged to follow him in war and to help with the cultivation of his land, the harvesting of his rice, and the construction of his house. In a tropical climate this last task is repeated every few years. The serfs were obviously what today we should call tenants or *aparceros*. They had their own houses and turned over to their masters the yield of half the land that they cultivated. The slaves lived in quarters provided by the master, could not marry without his permission, and could be sold at any time.

The cleared land of a barangay was divided among the free-born, each one recognizing the rights of the others. There were no titles. The forested land was not divided but was held by the barangay. No one could cultivate the land of any barangay other than his own unless he had purchased or inherited the right to do so. The position of dato was usually held by one family through hereditary right. This gave the family definite economic and social advantages over the other families of the barangay.

When the Spaniards established themselves in the Philippines they claimed ownership of all land for the Crown, but they recognized the rights of the native peasants to the land they were actually working. This was only a small fraction of what it is today. In order to attract Spanish settlers to the colonies, the Crown divided the conquered land into grants, or *encomiendas*, and forced the native population to pay tribute to the *encomenderos*. The encomenderos abused their power, exploited the people, and interfered with the religious work of the early missionaries, thereby arousing the enmity of the Church.

The encomienda system went out of existence at the beginning of the

[7] *Ibid.*

seventeenth century, and the country was then governed by colonial government officials who, in the political sphere, took the place of the datos of pre-Spanish days. The Spaniards were careful, however, not to disturb the social structure of the villages or the social and economic position of the datos.[8] The datos became known as *cabezas de barangay*. Several barangays made up a pueblo or municipality, headed by the *gobernadorcillo*, who originally was elected by all the household heads of a municipality, but later, in the nineteenth century, by the cabezas de barangay. The families from whom the gobernadorcillos and cabezas de barangay were drawn constituted the aristocracy and enjoyed special political and social privileges. The Spaniards introduced the term *cacique*, or chief, into the Philippines from Haiti and applied it to the datos, or cabezas de barangay, and the other leading families of a community. By recognizing them, the Spaniards helped the caciques to preserve their power over the people and gave them the opportunity of getting more and more land into their hands, of making more and more people financially dependent upon them (inasmuch as they were the tax-collectors); and of reducing freeholders to the status of tenants. Thus many of the leading families of pre-Spanish days kept their favorable social and economic position, and often strengthened it by intermarriage with the Spaniards.

This same group retained leadership after the coming of the Americans because it was the educated element in the country. It was able to monopolize most of the advantages derived from the American regime and to obtain control over national as well as local political life. Hence these caciques, the landed aristocracy of the Philippines, having all the privileges that the possession of money and land may give in an agrarian community, are today in a position to exploit a large part of the rural population. In some provinces this part amounts to more than half of all the farming families.

Although both Spanish and American authorities in the Philippines were aware of the pernicious hold that the cacique class had on the less fortunate part of the population, neither took effective measures to break it. Caciquism became deeply ingrained in the life and thinking of the Filipinos; and as long as the mass of the people remained illiterate and ignorant, it was relatively easy for the caciques to regulate affairs in the rural communities according to their own wishes. The Spanish Government made several half-hearted attempts to systematize land tenure by issuing land titles. As late as the second half of the nineteenth century two royal decrees were issued, one in 1880 and a second in 1894, urging land-

8 "The king . . . has ordered . . . that the honors of the chiefs be preserved to them as such; and that the other natives recognize them and assist them with certain of the labors that they used to give when pagans. This is done with the lords . . . of barangays, and those belonging to such and such a barangay are under the chief's control. When he harvests his rice, they go one day to help him; and the same if he builds a house . . ." Blair and Robertson: The Philippine Islands, Vol. 16, 1904, p. 155. Briefer and more recent studies covering the historical material in this chapter are Benitez: History of the Philippines, 1940, and Miller: Principles of Economics Applied to the Philippines, 1932.

holders to secure titles. The decree of 1894 granted a term of one year during which claims for free titles could be filed. After the expiration of that term all land not covered by title was to revert to the Crown, and no longer would it be possible to obtain a title free of charge. Only a few made use of this offer. To the great majority of peasants, accustomed to unwritten rules of land tenure, the land law was too involved, the idea of a land title too strange. In the opinion of the Spaniard and the American, most Filipinos were without individual legal rights to the land, but in their own opinion they were the actual owners of this land that their forefathers had cleared and that had been in constant use by their families. The comparatively few people who acquired legal titles were mostly persons belonging to the cacique group, and these often laid claim to more land than actually they had a right to. Thus in many cases peasants who had felt secure in the possession of their land and had not known or cared about titles were suddenly confronted with the fact that a wealthy person, with the law behind him, was claiming their land. These peasants were then driven from it or forced to become tenants.

The American administration inherited the results of the defective Spanish land system, which had been so unintelligible to the majority of the peasants and had encouraged land grabbing by a relatively small group of unscrupulous individuals. At the time of the American occupation the number of Filipino peasants without titles was estimated at 400,000. There existed no complete records of the issued titles, nor were these based on clear and accurate surveys. It is therefore no wonder that land disputes and agrarian troubles have persisted.

Friar and Church Lands

One of the most difficult problems facing the Americans concerned the large estates in the hands of religious orders, the so-called friar lands, which for more than 200 years had been a source of conflict and repeated revolts. During the centuries that the Roman Catholic Church and such orders as the Dominicans, Augustinians, and Recollects had spent in the Philippines, they had come into possession of large areas of the best agricultural lands, mainly in the provinces of Cavite, Laguna, Rizal, and Bulacan, that is, in provinces close to Manila. Most of this land was covered by titles issued in 1880 and 1894, although a part was covered by titles dating from the seventeenth and eighteenth centuries. In many instances the claims of the friars conflicted with the claims of those who were living on the land and who insisted that they were the true owners.[9] In such cases the tenants paid rent, or *canon*, to the administrators of the estates, but only under pressure.

During the Philippine Revolution the Malolos Government confiscated

9 *Fourth Annual Report* [of the Philippine Commission], 1904, Part 1, pp. 189 and 191.

all the friar estates. But the United States Government was bound by the Treaty of Paris to protect the property interests of the friars. Knowing that a return to the old relation between the friars, or their representatives, and the tenants would cause continual agrarian conflicts, the Philippine Government decided to buy the estates belonging to the Augustinians, Dominicans, and Recollects. After long-drawn-out negotiations, the government, on December 22, 1903, agreed to pay $7,239,784.66 for 23 estates aggregating about 165,000 hectares. The contract stipulated that the sum was to be reduced if surveys should reveal that the estates were actually smaller than they had been assumed to be. When the negotiations between the government and the friars were finally concluded on October 20, 1905, the purchase price was set at $6,934,433.36, a sum that was raised through the sale of Philippine government bonds authorized by the United States Congress.[10] This transfer brought the greater part of the friar lands with their tenants, estimated at more than sixty thousand persons, under direct control of the government. The friars, however, were not willing, in 1903, to sell all of their land; moreover, the Church itself also possessed a number of estates; so that, in spite of great efforts on the part of the American administrators, the question was not completely solved.

According to the census report, Roman Catholic religious organizations owned 41,782 hectares of land in the Philippines on December 31, 1938. It is doubtful, however, whether this figure is complete, since the four Church- or friar-owned estates of Buenavista, San Pedro·Tunasan, Lian, and Dinalupihan alone account for a total of 41,618 hectares[11] and the Church owns other estates as well. There is a possibility that the census excludes the 27,408 hectares of the Buenavista Estate since it was leased in 1939 by the government for a period of 25 years with an option of purchase. The American administration, inheriting this difficult friar- and Church-land problem from the Spaniards, has done much, especially through the labors of the Taft administration, to alleviate it.

Land Tenure Conditions at the Outbreak of War in the Pacific

At the beginning of the twentieth century the Church and the friars, together with the caciques, were the great landowners; today this group is enlarged by private corporations, which own more land than the Church and the friars ever did. In the sugar-producing areas especially, land has been concentrated in the hands of corporations. On December 31, 1938, business corporations owned 346,017 hectares of land: of these, 121,096 hectares belonged to Filipino, 106,473 to American, and 70,981 to Spanish corporations. A good deal of this land is worked by cash tenants, the remainder by laborers under farm managers.

---

10 U. S. Public Act, July 1, 1902, Secs. 63, 64, and 65.

11 Allen: "Agrarian Tendencies in the Philippines," 1938, pp. 54–55.

The census does not reveal to what extent land is owned by caciques, but it is known that a great deal of the riceland of the central provinces, the sugar land in the various sugar-producing regions, and the coconut and abaca land is owned by caciques, who divide their estates, or *haciendas*, into small units and have them worked either by *inquilinos* (cash tenants) or *kasamas*[12] (share tenants). The great majority of the cacique holdings comprise less than 100 hectares each; only a relatively few exceed 1,000 hectares. Since the Spanish days, these holdings have grown steadily and the percentage of tenant families has correspondingly increased. This is evident in the province of Nueva Ecija, which in large part was opened up and developed after 1918. In 1935 Nueva Ecija had 15 holdings of more than 1,000 hectares in extent; another 50 were from 500 to 1,000 hectares and about 2,000 were from 24 to 100 hectares.

Under the *inquilinato* system, which exists in all the provinces of Luzon and also in some of the other islands, an agreement is made between landlord and inquilino whereby the former turns his land over to the latter for a stipulated amount of annual rent, in cash or kind, called canon. The inquilino has his own animals and agricultural implements and must himself finance his agricultural operations. In densely populated provinces there is such a demand for land by people with the necessary funds to work as inquilinos that the landowners auction off the right to cultivate their land to the highest bidder, who must often pay the canon in advance. Local conditions, the pressure of population, the quality of the land, the type of crops to be raised, and other factors determine the amounts bid by the inquilinos. The cash rent therefore varies greatly from province to province and even within provinces.

Often an inquilino leases land without intending to work it himself. Instead he enters into an agreement with a share tenant who does the actual cultivation. Landlords, especially big absentee landlords, prefer to lease their land to a few inquilinos instead of to thousands of kasamas, for in the inquilinato system no labor or supervision is required of the landlord. It is of no practical concern to the landlord how good or how poor a crop the inquilino raises, since the former receives his fixed amount of canon, preferably before the beginning of the agricultural year. The Church and the friars used, and still use, the inquilinato system in preference to the kasama system. Modern corporations have followed their example.

In recent years there have been frequent conflicts between landlords and inquilinos because, as the pressure of population increases—creating a scarcity of good riceland—the landlords repeatedly raise the canon and refuse to renew the leases if the inquilinos will not pay the higher amount. For a time inquilinos may submit to successive increases in the canon, but finally they decline to agree to a new raise. They may have put a great deal of improvement into the land. Often they, or their forefathers, actu-

12 Also known as *aparceros*.

ally cleared the land and then leveled and dyked it so that it could be used for the cultivation of wet rice. It is natural that an inquilino should hate to give up the farm that he or his family has worked for a long time, and that he should refuse to leave the land if the landlord is unwilling to recompense him for the improvements he has put into it.

In a region where the competition for land is sharp the landlord has every opportunity to raise the canon and force the inquilinos to do extra work for him. According to a survey made by the Department of Labor, the *hacenderos* "oblige the inquilino to construct camarines [store houses], houses, dikes . . . without compensation. When a hacendero believes that it is necessary to construct a road within the hacienda, build a bridge and other improvements, he asks contributions from his inquilinos. . . . Those who do not give the contribution are expelled from the hacienda." [13] In 1939 there was a dispute between the owners and inquilinos of Hacienda Esperanza in Pangasinan Province. One of the complaints of the dismissed inquilinos was that

> some years ago they contributed towards the construction of a concrete dam which has cost ₱12,000.00. The contributions ranged from ₱7.00 to ₱10.00 a hectare. . . . The condition was that the contribution in question would be returned to them at the rate of ₱1.00 a year after the dam had begun operating. It was to be deducted from the canon they were to pay. Likewise, they were made to understand that if they contributed they would be allowed to remain on the land as long as they wished, if they paid the corresponding canon then prevailing. The dam began operating in 1935, but no refund to the inquilinos has ever been made. Not only this, but now [in 1939] they are being driven away.[14]

Such complaints have been frequent.

In 1939 the actual number of farms run by cash tenants (inquilinos) was small (17,111) as compared with the number worked by share tenants (547,933) and constituted only one per cent of all the farms in the Philippines. The percentage of farms run by inquilinos in the Luzon rice provinces of Cavite, Rizal, Pampanga, Pangasinan, and Tarlac exceeded this average, however.

Thus the prevailing form of tenancy, especially on riceland, is the kasama, or share-tenancy, system, which resembles the share-cropping system of the cotton-producing regions of the United States. The kasama system has a number of variations. Under a common form the landowner is obliged to furnish land, seed, and the cash needed for transplanting and harvesting the rice, while the kasama does the labor with his own work animals. The crop is divided equally, after the kasama has paid in kind half the expenses for transplanting and harvesting. The kasama will receive two-thirds, and the landlord one-third, of the crop if the former owns

[13] Commonwealth of the Philippines Labor Department: Fact-Finding Survey Report, (Ms. dated 1936), p. 37.

[14] Memorandum by Modesto Villalobos, executive officer of the Anti-Usury Board, to the Secretary of Justice, Manila, April 26, 1939.

the work animals and pays for seed, transplanting, and harvesting. Under still another arrangement, the owner furnishes the tenant with land, work animals, and seed. The harvest is divided equally, after the costs of the seed and the planting and harvesting have been deducted.

A characteristic feature of the kasama system is that the tenant has no funds and usually no rice for his family at the beginning of the contract. He therefore receives a *bugnos*, or advance, from his landlord. This sum, varying from 25 to 125 pesos, usually bears no interest but is returnable in money or in kind at the termination of the contract. In addition, the tenant is given a weekly or monthly *palay* (unhusked rice) ration to keep him and his family going. The palay received up to the time of planting is usually paid back without interest; whereas the rations delivered after the crop is once planted are subject to high interest, and here the door is opened to usurious practices, which will be discussed later. The kasama is completely in the hands of the landowner, who has to support him until the time of the harvest and the division of the crop. Often the tenants run so deeply into debt that they remain in debt after the crop has been divided and they have turned over their share to the landlord in payment of their loans. Thus, instead of being able to draw upon their own rice during the off-season between February and July, they must ask the landlord for new loans in money or palay and fall even more deeply into debt.

It is to the advantage of the landlord to have the tenant indebted to him, not only because of the high interest rates but also because then the tenant may be forced to do all kinds of extra work and may not leave his holding. The debt binds the tenant to the land and makes him almost a slave of the landlord, who thereupon determines every step to be undertaken—the crop to be grown and the time of planting and harvesting. This control at times goes so far that literate tenants, who have the right to vote, are not at liberty to choose their own candidate but must cast their vote for the landlord's candidate. An investigator for the Department of Labor reported a contract which compelled the tenant to attend the Roman Catholic Church and to work with his family for the landlord at fixed times without pay. Visitors were not allowed to enter the hacienda. Violations of rules listed in this particular contract gave the landlord the right to impose fines of from two to four *cavans* of palay.[15] (One cavan equals 75 liters.)

Usury, which presents the darkest aspect of the relation between landlord and tenant, is an old institution in the Philippines. Various interest rates are known by specific terms, such as *takipan*, *talindua*, and *terciahan*. Under the takipan rate the tenant pays back two cavans of palay for every cavan that he has borrowed. Should he be unable to settle this debt at the time of the first harvest, it increases to four cavans at the second harvest. The talindua rate demands three cavans for every two borrowed, and the terciahan, four cavans for every three borrowed. In all instances the

---

15 Commonwealth of the Philippines Labor Department: *op. cit.*, p. 226.

rates are exacted on loans over periods of six months and less. If a loan has been made in cash, it is settled in kind, but the landlord subtracts from the market price of palay on the day of delivery as much as 50 to 75 centavos. This arrangement is called *takalanan*.

Those who defend these interest rates argue that a tenant who borrows, for example, 10 cavans of palay at a time when a cavan costs 4 pesos owes not 10 cavans of palay but 40 pesos. If at the time of the payment of the debt the price for one cavan is only 2 pesos, then the tenant must deliver 20 cavans in order to meet his obligation of 40 pesos. The defenders of this usurious system intentionally overlook the fact that the landlord will not sell his palay when it is cheap but will wait until the price has improved, and thus realize excessive profits. It is evident that in many cases the same rice that was turned over by the tenant will be used to sink him more deeply into debt and to exploit him further. This practice of changing from kind into money and back into kind, in order to make full use of price differences and to force the debtor to pay in kind at what is for him the most disadvantageous moment, has been made illegal in the Philippines through legislation which will be discussed later.

The late Governor General Theodore Roosevelt investigated a case of usury in which a small farmer had borrowed 300 pesos ($150.). He never obtained any more cash. Each year, when the time came to pay his interest, the usurer offered him remission if he would sign a new document substituted for the original one and written in a language he did not understand. At the end of the fourth year he owed 3,000 pesos.[16]

Tenant indebtedness is desired by the landlords, who use various methods to involve their tenants in debt. One way is to force more advance money on the tenant than he actually needs. This procedure is called *pasunod*. Or the landlord may compel his tenant to buy clothing from him at exorbitantly high prices. Under the *cantina* system, common in Negros, tenants and laborers are obliged to make their purchases at the landlord's store, which charges more than other local stores. It is no wonder that kasamas rarely, if ever, have a chance to rid themselves of debt, since their landlords are so anxious to keep them in financial dependence in order to prevent their leaving the land.

If these kasamas, living as they are from hand to mouth and depending upon the landlord for their subsistence, are refused credit and rice rations, they become helpless and desperate. They are, on the whole, poorly clothed, poorly fed, and poorly housed. Very few of them have a chance to secure an education for their children. Hayden gives an excellent description of this section of the Philippine population,

> who as tenant farmers, renters or virtually landless workers eke out a miserable living from the rich Philippine soil. These people constitute a depressed minority which has been largely left behind in the march of Philippine progress. Astonish-

16 Roosevelt: "Land Problems," 1934, p. 200.

ingly ignorant, for the most part unable to use effectively any language save their local dialects, and economically helpless, it is they who are the most complete victims of the local *cacique*, the remorseless usurer and the exploiting political or religious charlatan.[17]

An important factor in the poverty of the kasama is the smallness of the average kasama farm, which in 1939 covered only 2.88 hectares, including uncultivated as well as cultivated land. Of this average farm the kasama was actually able to cultivate only 2.22 hectares. More than half of the kasamas (53.9 per cent) had farms of less than two hectares. These are averages for the Philippines as a whole. In the more crowded provinces conditions are still more serious. Because of the high percentage (54.8) of kasamas who are on rice farms, the following discussion will be limited to rice tenancy.

The cultivation of one hectare of riceland requires about 300 hours of man labor in addition to 150 hours of animal labor. The labor requirements are especially high during the planting and harvesting seasons; whereas in the intervals between planting and harvesting and between harvesting the old and planting the new crop the farmer and his family have little work to do unless they are engaged in a secondary industry. A farmer who tills less than two hectares of riceland does not make full use of his working capacity even during the busy season. With some additional help from other members of the family an efficient farmer can handle three hectares of riceland, especially if his land does not depend upon rain exclusively but is supplied by an irrigation system. Here, then, lies one reason for the poverty of the kasama: his labor is only partly employed—the equivalent of 75 eight-hour days a year.

Another cause of poverty is to be found in the yields. Most of the Philippine riceland produces only one rice crop and is usually not used for the raising of secondary crops during the dry season. The average yield of rice (both wet and dry) for the Philippines as a whole was 23 cavans per hectare in 1935. Yields vary greatly from province to province and depend upon the fertility of the soil, the weather, the variety of rice grown, and the methods of cultivation. The highest yields for certain lowland rice varieties come to 60 cavans per hectare in the best years. The average per-hectare yield in 1935 of lowland and upland rice for Nueva Ecija was 38 cavans, for Nueva Vizcaya 35 cavans, and for Pangasinan and Bulacan 30 cavans.

Most of the kasamas cultivate lowland rice and may expect an average yield of about 40 cavans per hectare, of which at least half goes to the landlord. Out of his share the kasama has to pay half the cost of transplanting and harvesting the rice, which on a farm of 2 hectares amounts to the equivalent of 5 cavans; thus, in this case, if the price of rice at the time of harvest is 2 pesos per cavan, the kasama's total income will be

17 Hayden: The Philippines, 1942, p. 378.

only 70 pesos. Velmonte reports that tenancy surveys made by the College of Agriculture show that the income of a rice tenant is not more than 130 pesos on a farm three hectares in area. Tenant families try to supplement their meager incomes with work outside the farm during the dry season, but sometimes there is not enough work for all the tenants.

TABLE VIII—PERCENTAGES OF FARMS WORKED BY THEIR OWNERS, ACCORDING TO SIZE OF FARM, IN THE PHILIPPINE COMMONWEALTH AND IN CERTAIN PROVINCES, 1939

PROVINCE	LESS THAN 1 HECTARE	LESS THAN 2 HECTARES	LESS THAN 3 HECTARES	MORE THAN 3 HECTARES
Philippines	24.6	52.1	68.4	31.6
Cebu	47.8	77.0	87.7	12.3
Ilocos Norte	53.2	85.6	94.3	5.7
Ilocos Sur	51.1	83.9	93.7	6.3
La Union	42.0	75.5	89.0	11.0
Pangasinan	37.2	69.0	84.4	15.6

Source: *Census of the Philippines 1939*, Vol. 2, pp. 975, and 989–995.

Yet if a kasama is fortunate enough to obtain work as a laborer when there is nothing for him to do on the farm, he may, according to Velmonte, raise his annual income to 200 pesos.[18] But even such an income is hardly sufficient to support the average tenant family of five or six members.

One way of helping tenants to increase their annual income would be to do away with the old custom of dividing the land into such small farms and instead to give each family at least three or four hectares. The tenant farms of the less congested areas, such as Nueva Ecija, are somewhat larger than the average in the overpopulated provinces of Bulacan, La Union, Ilocos Sur, and Ilocos Norte. In the latter provinces the only way to increase the size of tenant farms would be to encourage large-scale emigration, thus reducing competition for the land of the caciques; this might result in an enlargement of the remaining individual tenant farms. At present, tenants in the crowded provinces practically beg for land to farm, and whenever a cacique dismisses a tenant there are others eager to take his place.

Another way of adding to the income of the kasamas would be to raise the average crop yields through seed selection and to encourage the growing of secondary crops, such as beans, maize, peanuts, and sweet potatoes, during the months between the rice harvest and the next rice planting. The Philippines have extensive areas that can be brought under technical irrigation (see pp. 50–52). In many instances this might enable more farmers to raise a second rice crop and to make fuller use of their working time.

Many farms that are operated by their owners are also uneconomically small, although these farmers are better off than the kasamas since they

18 Velmonte: "Farm Tenancy Problems," 1938, p. 130.

own their whole crop. Slightly more than half of all the farms in the Philippines worked by their owners have less than two hectares of land, and slightly less than one-fourth have under one hectare (Table VIII). Farms operated by peasant owners are especially small in Cebu and in the Ilocano provinces, as Table VIII shows. In Cebu, Ilocos Sur, and Ilocos Norte about half of all the peasant owners have farms of less than one hectare, whereas more than three-fourths cultivate less than two hectares.

The several factors responsible for the low economic level of the major part of the rural population of the Philippines clearly demand special attention if help is to be given to the poor and depressed peasant. Two main types of remedy suggest themselves. The first would be to check such abuses as usury. This would not, however, go to the root of the Philippine problem. The second, and more thoroughgoing, type of remedy would be to effect a redistribution of the population by encouraging migration to hitherto unoccupied lands.

Rice Share Tenancy Act

In February 1933, under the administration of Governor General Theodore Roosevelt, the Philippine Legislature passed "An Act to promote the well-being of tenants (*aparceros*) in agricultural lands devoted to the production of rice and to regulate the relations between them and the landlords of said lands, and for other purposes." [19] This law was to protect the kasamas against abuses by the landlords. The caciques were, however, strong enough to make the law ineffective by inserting the provision that the law should go into effect "only in provinces where the majority of the municipal councils shall, by resolution, petition for its application to the Governor General who shall make the law effective by proclamation." This provision prevented the law from being enforced, since the caciques control the municipalities.

When increasing unrest in the rice regions of the Philippines forced the National Assembly to reconsider the case of the share tenant, it passed Commonwealth Act No. 178 which amends the former act. This act was approved on November 13, 1936. Again the cacique group prevented its general application by inserting the provision, "This Act . . . shall take effect . . . by proclamation to be issued by the President of the Philippines upon recommendation of the Secretary of Labor, when public interests so require, in the municipalities and on the date designated in such proclamation; and once enforced, its effects may similarly be suspended. . . ." The lawmakers were apparently of the opinion that the kasamas deserved protection only after conditions had become so bad that the kasamas began to disturb law and order and came to open clashes with the landlords.

Hayden observed in 1931 that Philippine government officials were most anxious to prevent an investigation of the Tayug uprising on the

[19] Act No. 4054, Feb. 27, 1933.

part of American members of the governor general's staff of advisers since "such an inquiry would inevitably be pushed into the whole realm of the oppression of the poor peasant by the local boss [cacique], the usurer, the Constabulary, and the local official." [20]

By issuing a proclamation on January 20, 1937, President Quezon made the Rice Share Tenancy Act effective in the five provinces of Bulacan, Nueva Ecija, Pampanga, Pangasinan, and Tarlac,[21] and by February 1941 it had been extended to Albay, Bataan, Capiz, Cavite, Iloilo, Laguna, and six municipalities of the province of Isabela. However, the application of the act to these provinces did not prevent serious clashes and disputes between kasamas and landlords.

The act is rather long and contains so many ambiguous sections that it cannot be understood by a simple, uneducated kasama. Its main provisions are:

1. Contracts between landlord and tenant must be written in the local dialect (Section 4).

2. Contracts are good for one year unless there is an agreement to the contrary (Section 6).

3. The cost of transplanting, harvesting, threshing, irrigation, and fertilizer, if any, are to be borne equally by the two parties. The crop shall be divided equally (Section 8).

4. All advances obtained by the tenant in connection with cultivation shall bear not more than 10 per cent interest per agricultural year. On all loans other than money—such as rice—interest up to 10 per cent may be added to the invoice price of the loan (Section 10).

5. Fifteen per cent of the crop must be left to the tenant regardless of any indebtedness (Section 14).

6. The landlord must not dismiss his tenant except for good reasons. The tenant must not leave the landlord before the expiration of the contract except for good causes (Sections 19 and 25).

In 1938 and 1939 reports reached authorities in Manila that some of the landlords of the provinces in which the Rice Share Tenancy Act was effective threatened to dismiss at the end of the agricultural year any ten-

---

20 Hayden: *op. cit.*, p. 380.

21 "Whereas, extensive agricultural lands in the Provinces of Bulacan, Nueva Ecija, Pampanga, Pangasinan and Tarlac, are dedicated, under the 'kasama system,' to the rice industry; and

Whereas, time and again serious controversies due to conflicting interpretations of the terms of verbal contracts and other matters affecting the relationship between landlords and tenants in the aforesaid provinces have arisen, thereby menacing public peace and order:

Now, therefore, I, Manuel L. Quezon, President of the Philippines, by virtue of the powers vested in me by law, and upon recommendation of the Secretary of Labor, do hereby declare the provisions of Act Numbered Four thousand fifty-four, otherwise known as the Philippine Rice Share Tenancy Act, as amended by Commonwealth Act Numbered One hundred seventy-eight, to be in full force and effect . . . in all the municipalities of the Provinces of Bulacan, Nueva Ecija, Pampanga, Pangasinan and Tarlac . . ." (Proclamation No. 127).

ant who dared to demand observance of the act. In order to prevent the wholesale dismissal of share tenants the National Assembly passed Commonwealth Act No. 461, approved June 9, 1939, which states that a tenant may be dismissed only for reasons specifically mentioned in the Rice Share Tenancy Act and then only with the approval of a representative of the Tenancy Division in the Department of Justice. This act was replaced on August 22, 1940, by Commonwealth Act No. 608, which also provides compulsory arbitration of all disputes by a representative of the Department of Justice and gives landlord as well as tenant the opportunity of appealing to the Court of Industrial Relations in case either objects to the decision of the Department of Justice.

The landlords hold that Commonwealth Act No. 461, amended by No. 608, is unconstitutional because it deprives them of their right to dispose of their property as they please.[22] One of them wanted to dismiss three of his tenants. These refused to vacate the land, claiming that they could not be dispossessed or ejected without the approval of a representative of the Department of Justice. Both parties brought their case before the Court of First Instance of Pampanga. The plaintiff held that the above-mentioned acts were unconstitutional. But in the court's decision of January 24, 1941, Acts Nos. 608 and 461 were declared valid and constitutional, and the parties were directed to submit their controversy to arbitratioı before the Department of Justice.

In the summer of 1940 there were a number of serious clashes between tenants and hacienda guards. On June 4, 1940, the *Manila Daily Bulletin*, an American newspaper, stated:

> What had been feared during the past several months has occurred in central Luzon. Bloodshed due to tenancy troubles was reported from Nueva Ecija and Pampanga last night while Bulacan was also seething with clashes threatening in the rice farms.
>
> Fear of famine, actual hunger and the new tenancy law are reportedly largely responsible for all the troubles in the rice region of Luzon. In some towns and barrios people are actually suffering from lack of food with the refusal of many of the landlords to advance money to tenants. In other farms, hundreds of tenants are afraid they would starve during the coming year as the planting season is now upon them and they are still waiting for the definite solution of the ejectment cases filed against them. If they cannot continue tilling the lands they have been working on for many years, their families will have nothing to eat.

This report claimed that Act 608, which forbade landowners to dismiss tenants of long standing without the approval of the Department of Justice, was largely responsible for the unrest among the tenants, who did not know that the landlord had the right to refuse a new contract if the tenants had not fulfilled their obligations under the old one. The leaders of tenant organizations, of which there were several in central Luzon, in-

22 Ramos: "Discrimination against the Landowners," 1941, p. 7.

structed their followers to stay on the farms under any circumstances, whether or not their ejection was justified under the very tenancy act which they had invoked for their protection.

Not only private landowners but also the government had difficulties with tenants whom they wanted to dismiss. In 1940 the manager of Buenavista Estate, which was leased by the government, evicted a number of tenants because of their failure to pay the rent. When a lawyer and tenant leader claimed that Commonwealth Act No. 538 prohibited the eviction of a tenant despite non-payment of rent, the Department of Justice ruled that this act affected only estates owned by religious orders and could not be applied to Buenavista since this estate had been leased by the government from the Hospital de San Juan de Dios before the passage of the act. So on June 11, 1940, constabulary soldiers arrested 125 tenants of Buenavista Estate for tilling lots from which they had been ejected.

These cases demonstrate that the Rice Share Tenancy Act, with its numerous amendments, failed to regulate relations between landlord and tenant. The root of pre-war agrarian unrest in central Luzon was not so much the tenancy contract and its stipulations about division of costs and yields of production (although usury and oppression played their role) as the fact that the income of a tenant who worked less than two hectares of land was not enough to give him and his family a bare living and keep him out of debt. There was nothing new in this economic situation. By 1941, however, tenants were organized and led by lawyers, labor organizers, and agitators, who constantly admonished and encouraged them to insist upon their rights granted by law, and who instructed them concerning these rights. In those regions where the tenants were organized in unions the old relationship between landlord and tenant was rapidly disappearing.

Public statements by President Quezon and other high officials also encouraged the kasamas in their struggle with the caciques. At the first convention of the National Commission of Peasants on July 16, 1939, at Cabanatuan, Nueva Ecija, Quezon declared:

> The rights of the *kasama* are no less sacred than the rights of the landowners. . . . The landlords must realize by this time that the *kasama* is not a slave. The word *kasama* means partner or associate and it is as such that the law contemplates him to be. Under the law, the *kasama* is almost the owner of the land and cannot be ejected except for causes as enumerated in the statute. . . .[23]

After listening to such statements the kasama was apt to remember only that he "is almost the owner of the land and cannot be ejected" and to forget the qualification, "except for causes as enumerated."

Government Purchase of Landed Estates

In recent years the Commonwealth Government has purchased a num-

23 Commonwealth of the Philippines Labor Department: *Labor Bulletin*, Vol. 2, 1939, pp. 377 and 378.

ber of the estates on which disputes were most frequent. Such a policy had been promised in the platform of the coalition between the *Nacionalista-Democrata* party and the *Nacionalista Democrata Pro-Independencia* party in 1935, which stated in Article XI:

> When the resources of the country so permit, we shall begin the expropriation of great estates, so that they may be divided into lots and sold to private citizens, preferably their actual occupants. We shall encourage the formation of small land ownership, which is the bulwark of democracy, the guaranty of public order, and a stabilizing force. It is our desire that every Filipino should own his land, the house in which he lives, and the farm which he tills.

TABLE IX—HOMESITES UNDER THE ADMINISTRATION OF THE RURAL PROGRESS ADMINISTRATION, JUNE 30, 1940

NAME	AREA (HECTARES)	FORMER OWNER	DATE OF ACQUISITION	PRICE (PESOS)
Marikina, Rizal Province	108.6731	Angel M. Tuason *et al.*	Dec. 29, 1938	295,791.48
Tunasan, Laguna Province	216.7999	Colegio de San José	Expropriation started Sept. 28, 1936; sale on Aug. 31, 1939	414,855.00
Dinalupihan, Bataan Province	99.6261	Monte de Piedad	Expropriation started Aug. 25, 1937; sale, June 30, 1939	93,250.82
Lian, Batangas Province	36.5212	Colegio de San José	Expropriated Mar. 4, 1937	—
Capellania de Tambobong, Rizal Province	90.0532	Archbishop of Manila and Philippine Trust Co.	Expropriated June 9, 1939	—
Laryuhan, Nueva Ecija Province	1.0565	Cajetano de Borja	Expropriated June 19, 1939	—

Source: Commonwealth of the Philippines Rural Progress Administration: Report of the General Manager on the Progress of its Activities. Unpublished.

This promise in the party platform of 1935 found its way into the Constitution of the Philippines in Article XII, Section 4, which reads: "The National Assembly may authorize, upon payment of just compensation, the expropriation of lands to be subdivided into small lots and conveyed at cost to individuals."

In one of its first acts[24] the National Assembly authorized the President to start expropriation proceedings or negotiations for acquiring those portions of large estates that were occupied as homesites by the tenants. This

24 Commonwealth Act No. 20, July 11, 1936, authorizing an appropriation of 1,000,000 pesos. To this sum was added another 2,000,000 pesos under Commonwealth Act No. 260, April 18, 1938.

act was hailed by the press as a definite step that would lead to the breaking up of vast tracts of land formerly owned by a few individuals and would give tenants a chance to become owners, at least of their home lots.

On October 7, 1937, Administrative Order No. 49 of President Quezon created the Landed Estates Survey Committee, which was to work out a long-range program with regard to the future treatment of large estates. This committee apparently was never particularly active. It did, however, recommend the acquisition of Marikina homesite in the municipality of Marikina in Rizal Province.

On August 23, 1938, the President approved Commonwealth Act No. 378, which authorized the leasing of large estates by the government for 25 years. A few months later the Rural Progress Administration (R. P. A.) was organized. This was a government corporation with a capital of 1,500,000 pesos, divided into 15,000 shares of 100 pesos each. All shares were held by the Commonwealth Government, with the exception of six, which were issued in the names of the members of the Board of Directors. The R. P. A. was entrusted with the execution of the policies set forth in Commonwealth Acts Nos. 20 and 378.[25] Homesites which had been acquired after 1936 were transferred from the Bureau of Lands to the R. P. A.

According to the first report of the general manager, covering the period from March 5, 1939, to June 30, 1940, the R. P. A. had control, on the latter date, over six homesite sections of the estates of Marikina, Tunasan, Dinalupihan, Lian, Capellania de Tambobong, and Laryuhan (Table IX). In addition to acquiring the residential parts of these estates, the administration had also leased the large Buenavista Estate, of 27,408 hectares, belonging to the Hospital de San Juan de Dios. The tenants of this estate had been organized into the Buenavista Farmers' Coöperative Association under the management of José G. Sanvictores, one of the six directors holding shares in the R. P. A.

Between June 1940 and the summer of 1941 the R. P. A. acquired the Guadelupe Estate in Rizal and Hacienda Bahay Pare in Candaba, Pampanga. The latter had been the scene of long-standing agrarian unrest. The owner, Don Ramon Santos, had bought Bahay Pare's 2,189 hectares from the Order of the Augustinians in 1926 for 900,000 pesos. Whereas the Augustinians had operated the hacienda under the inquilinato system, Santos changed to the kasama system, with its greater possibilities for exploitation. This created animosity on the part of the tenants.

The surveying and resale of the homesites were not so easy as might have been expected. The government had bought the sites in order to help tenants acquire home lots from which they could not be evicted even though dismissed as tenants from the farmland owned by the estates. Nevertheless, difficulties arose. On the Marikina homesite some *Ganap*[26]

---

25 Executive Order No. 191, March 2, 1939.

26 A peasant organization.

followers were reluctant to pay for the lots, as they had been led to believe that the government would eventually give them the lots free of charge. The people of the municipality of San Pedro, Laguna, in which the Tunasan Estate is located, were divided on the issue of the survey of the homesite, in spite of their oft-repeated petitions to the government to buy the residential part of the estate. The people obstructed the survey of their lots, refusing either to give their names or to indicate the boundary lines. Finally, the R. P. A. had to call in the constabulary in order to make the survey possible. The R. P. A. and the occupants of the residential lots on the Dinalupihan homesite differed over the value of these lots. The occupants protested in meetings held in the Dinalupihan public plaza.

The cases cited above as well as the difficulties on the government-managed Buenavista Estate show that the Filipino tenant has become more and more conscious of the power that unions give him. This creates new problems. In the words of one Philippine authority: "The forgotten man refuses to be forgotten any longer. He now demands recognition of his rights; in fact, he is demanding more than what is his right." [27]

It is difficult to see how a government policy of buying one hacienda after another can solve the problem of the little kasama, who, because of the great density of population in the rice-growing regions, cannot buy more from the new owner, the government, than he was already farming when his landlords were the friars or the caciques. The government, however, could improve the lot of the kasama if it would construct irrigation systems wherever possible and would develop the hitherto unused parts of the haciendas that it had acquired. But it has not done this. As a matter of fact, as late as December 31, 1940, about one-fourth (22.9 per cent) of the friar land purchased by the government under Taft in 1905 was still vacant.

Public Land Policy

American Land Policy in the Philippines

In 1902 the United States Congress granted the Philippine Government power to administer the extensive public lands for the benefit of the Filipinos. These lands had become the property of the United States by the Treaty of Paris.[28] Congress limited to 16 hectares the amount that could be acquired by any one person. This limitation discouraged individual Americans from acquiring public lands in the Philippines. The upper limit for corporations and associations was set at 1,024 hectares. These restrictions were demanded by American Democrats who were opposed to the Philippine policy of the Republican administration.

The defenders of a less restricted land policy argued that the public

27 Sanvictores: "The Agrarian Problem in Buenavista," 1939, p. 191.
28 U. S. Public Act No. 235, July 1, 1902, Sec. 12.

domain in the Philippines was so extensive that it would take the Filipinos hundreds of years to develop their country with their own financial resources. It would therefore be to the advantage both of the Philippines and the United States if outside capitalists could be induced to share in the development of the islands by establishing plantations of tropical crops.

The opposition, however, feared that a policy of giving large grants to individuals or corporations would turn the Philippines into a plantation colony and would be detrimental to the political and economic future of the Filipinos themselves; it would reduce them to a nation of plantation laborers. Large-scale investments of American capital, by impeding the political progress of the Filipinos, would be an obstacle to eventual independence. This view was supported by General Arthur MacArthur. When asked by the Senate Committee on the Philippines about his attitude toward the proposal to sell public land in blocks of 5,000 acres to American corporations, he answered: "I deprecate the idea . . . that we are going to sell public lands there in blocks of 5,000 acres with a view to rapid exploitation. That I should regard as disastrous both to American and Filipino interests absolutely." [29]

On an earlier occasion General MacArthur was asked what effect exploitation of the Philippines would have upon the Filipinos. His answer was: "One of the great apprehensions of the people there is that . . . exploitation will . . . deprive them of their resources and relegate them eventually to a position of social inferiority, about which they are very sensitive." [30]

In its fight against large land grants and extensive franchises the Congressional opposition was strongly supported by a powerful lobby of American beet- and cane-sugar interests, which looked with alarm upon any competition from a modern Philippine sugar industry working with cheap labor. Additional support came from the Anti-Imperialist League, with headquarters in Boston, which fought against a strong entrenchment of American plantation capital in the Philippines since it would operate against the interests of the Filipinos.

As it turned out, the sugar industry developed in spite of the limitations placed on alienation of public land. The government-owned Philippine National Bank financed several sugar mills, or *centrals*. The rubber industry, however, was virtually kept out of the Philippines, although Mindanao, which lies south of the typhoon belt, is perfectly suited for the growing of rubber. American rubber capital, confronted by the 1,024-hectare limit, turned instead to British Malaya, Sumatra, Brazil, and Liberia, where extensive land tracts were available.

Equipped with the power to dispose of the public domain, the Philippine Government passed the Public Land Act on October 7, 1903. This

29 Hearings before the Committee on the Philippines, 1902, p. 1379.
30 *Ibid.*, p. 876.

act introduced the homestead system into the Philippines, prescribed regulations for the sale or lease of public lands, and provided for the confirmation of the titles that were still unestablished and the granting of free patents to occupants and cultivators of public land who had cultivated this land for a certain length of time. It permitted the sale or lease of public land not only to citizens of the Philippines and of the United States, including other insular possessions, but also to corporations organized in the Philippines or the United States, regardless of the nationality of the stockholders. It was under this law that Japanese corporations were able to acquire a part of their holdings in the province of Davao.

The Philippine Commission, which had passed the Public Land Act in 1903, was never satisfied with the land law and up to 1913 continued recommending its liberalization to Congress—supposedly in the interest of a more rapid economic development. In his annual reports Dean C. Worcester, a member of the commission and Secretary of the Department of the Interior until 1913, emphasized the harm that was inflicted on the Philippines by the restrictions upon which Congress had insisted in 1902. According to Worcester, 10,000 acres, or nearly 4,050 hectares, was the minimum amount of land upon which a corporation could safely afford to establish an up-to-date plantation equipped with modern machinery and implements. Others clamored for 25,000-acre tracts. But neither the President nor the Congress of the United States ever responded favorably to the insistent recommendations of the American administrators in the Philippines. One advocate of a more generous attitude toward capitalist interests called the reluctance of Congress "a pernicious interference in a matter which is of purely local interest, which should be left to the discretion of those on the spot who are familiar with the local conditions." [31]

In some other colonies, however, too much was left to those "on the spot who are familiar with local conditions," and the result in those places was that Western capitalistic enterprises overshadowed everything and reduced the local people, if they were willing to work for wages, to the status of agricultural laborers. If they were unwilling, labor was obtained from the outside, as in Malaya, which came to depend almost entirely upon Chinese and British-Indian laborers. Large-scale importation of Chinese labor into the Philippines was proposed in 1902 but was forbidden by Congress.

The first really important change in the Philippine land law came in 1919 when the Philippine Legislature passed Act No. 2874. Filipinos consider this law as the first important step toward the nationalization of their land resources, since it limited the exploitation and utilization of public lands to Filipino and American citizens, except in the case of citizens of other countries whose laws give the same rights to Filipinos as to

31 Ireland: The Far Eastern Tropics, 1905, p. 214.

their own citizens. Such people were also permitted to acquire public lands in the Philippines; but no foreigner was able to take advantage of the reciprocal provision of the law. These restrictions were directed especially against Chinese and Japanese. The act further provided that at least 61 per cent of the capital stock of corporations must belong to citizens of the Philippines or the United States.

The new law thus made it virtually impossible for aliens to acquire public lands legally; at the same time it tended to relax the restrictions on individual holdings. By the new law individuals could obtain 24 instead of only 16 hectares through homesteading procedures, and 100 hectares instead of 16 through purchase. Later amendments of Act 2874 increased to 144 hectares the amount of public land purchasable by an individual. At no time, however, was the Philippine Legislature willing to raise the limit for corporations above that originally set by Congress in 1902. As late as 1928, 1929, and 1930 various governors general appealed to the Philippine Legislature to encourage the investment of American capital in the islands. On July 16, 1928, Governor General Henry L. Stimson in his message to the legislature said:

> I believe that the chief obstacle to our development has lain in certain unwise limitations of law, most of them introduced here a generation ago from the United States under the influence of the theories then prevailing but which have long since been exploded in the land of their origin. . . .
>
> I believe that American capital has learned the unwisdom and the danger of attempting unfairly to exploit the land in which it enters. . . .
>
> I, therefore, recommend . . . a thorough revision of your laws governing the creation of corporations and the regulation of their activities, including the control over your public utilities and including a wise and conservative revision of your land laws . . . [32]

But in his final report covering his term in the Philippines, Stimson had to admit that the resentment against large landholdings "is one of the strongest political sentiments" of the average Filipino.

> The existence of this native sentiment has not been generally recognized in the United States, but the events of my own year's experience brought it to my attention as one of the deepest and most controlling currents of public opinion in the islands and one which it would be folly to disregard or attempt to defy. . . . The existence of this strong feeling against land monopolization added to the ever-present racial sensitiveness toward foreign exploitation, whether American or Mongolian, tends to make the problem of stimulating the introduction of the much needed capital for economic development from America a very delicate one. . . .[33]

Governor General Dwight F. Davis complained about the tendency of Filipinos to "raise the bogey of 'exploitation' " as soon as the words "eco-

32 *Annual Report of the Governor General of the Philippine Islands 1928*, 1930, pp. 41–42.
33 *Ibid.*, p. 4.

nomic development" were used.[34] Yet in spite of all arguments put forward by Americans interested in attracting capital to the Philippines, the Filipino remained firmly convinced that it was better for his country to be developed slowly and gradually by small farmers than to be rapidly and efficiently exploited by a few large companies working the land with tenants or hired laborers.

Governor General Theodore Roosevelt, Davis's successor, praised the Philippine Legislature for having "very wisely prohibited individuals from acquiring" public land in large blocks.[35] Frank Murphy, the last American governor general in the Philippines, summarized the American land policy: "The public domain has not been exploited. Large estate grants —a sore spot in other days—have not been tolerated. The natural inheritance of the Filipino as the owner of his fatherland has not been violated."[36]

Filipinos have frequently expressed their gratitude for the conservative land policy initiated by Congress at a time when they themselves were not yet in a position to prevent the exploitation of the natural resources of their islands.

What would have been the result for the Philippines if Mindanao, for example, had been turned into one large plantation worked by laborers imported possibly from China, Java, or southern India, if the supply of Filipino labor had proved inadequate? Would American capital have been willing to consent to the independence of the Philippines as a whole, or would it have insisted on a separation of Mindanao from the rest of the country? It was certainly fortunate for the development of Philippine unity that the American people were very much aware of the danger of the monopolization of the public domain by powerful corporations. Had these been allowed to open up the islands, the pace would have been much too rapid for the Filipinos. The country had suffered severe losses of population in the fighting against Spain and the United States, and these had resulted in the abandonment of agricultural land. At the same time rinderpest had wiped out a large number of the carabaos needed to cultivate the soil. In the early years of this century the Filipinos were not yet ready to move as pioneers into hitherto unused land.

The Problem of Land Titles

One of the most difficult problems that faced the early American administration was to determine the extent of public domain and of private holdings. Only a relatively small number of persons held Spanish land titles. American officials were convinced that clear titles, guaranteed by the government, were absolutely necessary to stabilize conditions and prepare the way for economic improvements.

34 *Ibid. 1929*, 1931, p. 22.
35 *Ibid. 1932*, 1934, p. 12.
36 *Ibid. 1935*, 1937, p. 14.

In 1902 the Philippine Commission passed a law providing for the registration of land ownership under the Torrens system, which grants government-guaranteed titles after the land has been surveyed and ownership proved in court. As this procedure was as foreign to their way of thinking as the Spanish land system had been, Filipinos were extremely slow to make use of it. During the first seven years after the passage of this law, only about 4,000 Torrens titles were issued, almost all of them covering large and valuable private holdings. The ordinary peasant did not know about the law and, if he had known, could not have afforded to apply for a Torrens title because of the incidental costs of surveys, attorneys, and court fees. While only a few individuals obtained titles, the number of occupied lots, which was estimated at 2,300,000 on January 1, 1903, increased annually by 20,000 during these first seven years. Thus the government was not making any headway.[37]

In order to speed up the registration of titles, a new system was developed in 1910 to provide for cadastral surveys of whole municipalities. For three years the two houses of the legislature could not agree on the Cadastral Act, but it was finally passed in February 1913. In accordance with this act all the parcels of a municipality were surveyed and then presented to the court in groups of 1,000 as one case. Here the government brought *pro forma* suit against the occupants because they lacked titles, and ordered them to vacate the land. If the occupants could prove their claim of ownership to the satisfaction of the court and if their claim was not contested, they received the necessary titles.

The advantages to the government of these cadastral surveys were immediately apparent. The receipts from land taxes increased by an average of 25 per cent, much of this increase being undoubtedly due to the fact that the people could no longer underestimate the amount of land they were occupying as had previously been the case. The possession of these Torrens titles, on the other hand, made it possible for the owner to borrow money from banks at reasonable interest rates, whereas without titles they had had to go to usurious moneylenders.

However, the cadastral survey, the hearing of the cases, and the issuing of titles did not always progress as rapidly as had been hoped. Cases piled up in the courts because of lack of personnel. Moreover, the owners of land which had been surveyed and for which titles had been prepared, often refused to call for the papers because of the payments that were required for the certificates. On December 31, 1932, for example, there were 159,230 duplicates of original certificates of title awaiting delivery. As a total of 500,210 decrees had been issued between 1903 and the end of 1932, it is clear that nearly 32 per cent of all certificates had failed as yet to reach the hands of the landowners.

Between 1909 and June 30, 1937, the Bureau of Lands made cadastral

---

37 *Annual Report of the Philippine Commission . . . 1910*, 1911, p. 10.

surveys for 1,040,793 lots, amounting to 3,105,763 hectares; and an additional 93,261 lots, amounting to 248,358 hectares, were surveyed by private surveying companies. It was estimated at the same time that there were also 4,096,974 parcels, amounting to 25,708,136 hectares, awaiting cadastral surveys.[38] They could be handled by 30 cadastral survey parties in the course of about 60 years. These figures indicate that the cadastral survey of the Philippines was still far from complete before war broke out.

Filipino Response to American Land Policy

What was the attitude of the Filipino to the homestead system—a system that leads to a form of dispersed settlement, the isolated farm, which has become the prevalent type in the United States and Canada? There was no common reaction, for the homestead system was only one of several settlement patterns in the Philippines. Where wet rice is grown in old, densely settled areas, the peasants live in hamlets and villages. Some of the non-Christian folk in the mountain and forest areas live in scattered dwellings; and the Christian Filipinos who grow such crops as maize, coconuts, or abaca also frequently prefer to live in isolated houses on their land rather than in villages. But, on the whole, villages predominate, certainly in those regions that need relief from population pressure. The Spaniards had a great deal to do with the concentration of the population in large villages, but they did not introduce the village system into the Philippines. The Malay inhabitants of the islands were already living in hamlets, or barangays, at the time of the arrival of the Spaniards. The Spaniards did, however, encourage this tendency to settle in groups rather than on isolated farmsteads, as they found that the work of Christianization and supervision was much easier when people were brought together in large villages "under the bells."

This colonizing technique happened to harmonize with the Malay custom of congregating in villages on higher ground and turning all the irrigable low-lying land into rice fields. The village system facilitates group work and cooperation, which are essential in the cultivation of wet-rice land, in the construction of irrigation dams and canals, and in the performance of other tasks that require the labor of more than one family. Thus custom, together with the close ties of relationship among families, the gregarious nature of the people, their agricultural technique, and the Spanish influence, built up a preference for the village system that is difficult to overcome.

The average Filipino of the central provinces, accustomed to the comforts of village life and to a dependence upon the help of his family, friends, and neighbors, was reluctant to move out into a pioneer region where he would have to live alone. Hence, to the surprise of American officials in the Philippines, who had expected a ready response, Filipinos applied for

38 Mills: Planned Surveys, 1937, pp. 172, 196, and 297.

homesteads only in small numbers during the early years. In 1912 the *Cablenews-American* commented editorially on this failure of the Filipino to take up public land:

> The framers of the homestead law took it for granted that the agricultural people of the Philippines were accustomed to settle on isolated farms and live in dwellings far removed from their neighbors. They ignored the very patent fact that the Philippine farmer does not live on the land he tills and cannot be persuaded to do so. The folly of this assumption has been abundantly proved during the years that the homestead law has been on the books by the very few entries that have been made under it.[39]

In his report of 1911 Worcester argued that formerly the Filipinos had had to settle in villages because of the lack of security for life and property, but that this could no longer be the reason for their reluctance to move out to homesteads, since order had been established. He blamed the caciques for trying to prevent the landless from applying for homesteads through fear of losing them as tenants.

After 1912 the number of homestead applicants grew slowly but steadily. In the period from 1912 to 1925 there was an annual increase ranging from 5,000 to 10,000. From 1926 to 1930 an average of about 12,000 persons a year filed applications. After that the number declined slowly to about 7,000 in 1934, the last year before the inauguration of the Commonwealth Government.

From July 26, 1904, to November 14, 1935 (the end of the American regime), a total of 212,094 homestead applications had been received by the Bureau of Lands, of which 81,086 were rejected and 21,577 cancelled. Thus 48.4 per cent of all applicants had failed in their attempts to get homesteads. Another 19,160 cases were still pending in November 1935; in the whole period 55,450 had been approved and only 34,821 applications, or 16.4 per cent of the total number, had been patented.[40]

This large number of rejections and cancellations of homestead applications is surprising and indeed alarming in a country like the Philippines where there is so great a need for a better distribution of the population. Why did nearly half the persons anxious to acquire homesteads fail in their attempt? A lack of surveys of the public domain was chiefly to blame for the rejections. For years homesteaders were not guided in their search and had to decide for themselves on the location of the land for which they wished to apply. Only upon receipt of the application did the Bureau of Lands determine whether the land belonged to the public domain, whether the Bureau of Forestry considered it to be agricultural rather than forest land, and whether the land was free from other claims. Even an approved claim did not protect an applicant from losing the results of

---

39 *Cablenews-American*, May 9, 1912, quoted in Forbes: The Philippine Islands, Vol. 1, 1928, p. 325, footnote.

40 *Annual Report of the Director of Lands 1935*, 1936, pp. 66–67.

his labor if some other claimant came forward after the former had already worked for a period on the land; for the government did not clear the title on public land in a court procedure until it was applied for by homesteaders.

The failure to meet certain requirements regarding residence and cultivation likewise led to the cancellation of applications. If a homesteader was unable to cultivate at least one-fifth of his land within 5 years, he lost his claim and it could be given to another applicant. In such cases it would have been more humane to have reduced the homestead to correspond to the amount actually cleared rather than to deprive the peasant completely of the fruits of his labor.

Not only did the attachment of the Filipino to the village type of settlement, the machinations of caciques, and the lack of organization that prevented thousands of would-be homesteaders from reaching their objectives constitute obstacles to a more rapid settling of the public domain by homesteaders, but the lack of roads was also a serious obstacle. Without roads it was practically impossible to reach those districts where most of the public land was located. Whenever a road was to be constructed through an undeveloped section, homesteaders flocked in rapidly, to find only too often that influential persons who had been privately informed of the construction even before it was begun had taken up the choice land on both sides of the road. These people then held the land for speculative purposes, using hired labor to meet the bare minimum requirement for improvements. Recognizing this abuse, Governor General Theodore Roosevelt issued an executive order in 1932, providing "that for two kilometers on either side of any new road that might be opened through the public domain no tract of land of more than 350 acres could be taken up [through purchase or lease], that tracts of that size must be only sparingly granted, and that they could not have more than a limited frontage on the road. Under no circumstances were they to be contiguous." [41]

That the evil nevertheless persisted was admitted by the former Secretary of Agriculture and Commerce, Benigno S. Aquino, who told the Convention of Bureau of Lands Officials in November 1940: "When I travelled through the provincial roads from Davao to Cotabato and thence to Bukidnon I was shown those extensive parcels of land which have been registered in favor of rich people from the Visayas and Luzon." [42]

In the absence of feeder roads that would make it possible for them to market their surplus products, small homesteaders find it difficult to develop land 10 or 15 miles or more from the highways. It is therefore important that they should have a chance to settle near such highways. Wealthy individuals who are farming extensive areas can more easily overcome the transportation problem. They have the means to cut roads

---

41 Roosevelt: "Land Problems," 1934, p. 202.

42 *Proceedings of the Convention of Bureau of Lands Officials*, 1940, p. 85.

through to the highways, and it is profitable for them to use trucks to market their crops.

Malaria, together with the lack of medical care, represents another grave obstacle to homesteaders, many of whom come from parts of the islands where malaria is practically non-existent, as on the low coastal plains. In such regions as the interior of Mindoro, Mindanao, or the foothills on both sides of the Cagayan Valley, the homesteaders are exposed on arrival to this disease at a time when they must work hard and live under pioneer conditions and their resistance is lowered. The practice of clearing the banks of little streams benefits certain species of anopheles; this results in a rapid spread of malaria, which is found throughout the Philippines in elevations that are above the coastal plain and below 2,000 feet.[43] During a visit to the Mallig plain in the province of Isabela in 1941, I met settlers who belonged to a group of some 60 Ilocanos, including women and children. Since their arrival in this section two years before, a third had died of malaria. Malaria is without doubt the greatest scourge of the homesteader who moves from the coastal plain into the hilly interior anywhere in the Philippines.

In certain parts of the islands the local non-Christian tribes resent the infiltration of lowland Filipinos and attack the settlers. Like all primitive communities they claim exclusive rights to the territory they inhabit. Any invasion of it is considered by them a violation of their sovereign rights and is answered by the killing of the trespassers.[44] Between 1938 and 1940 the Ilongots of the Sierra Madre along the boundaries of Tayabas, Nueva Vizcaya, and Nueva Ecija killed about 30 persons in an attempt to prevent Christian settlers from invading their hunting and gathering grounds. These killings resulted in a withdrawal of the settlers who had pushed farthest into the territory of the Ilongots.[45] Where such troubled situations exist, the government might set aside reservations for groups like the Ilongots and then prevent settlers from encroaching upon these reservations.

In spite of all the defects of the homestead system, the voluntary extension of agricultural settlement in Luzon progressed, even though haltingly. The homesteaders moved from the Ilocos coast and from the densely settled parts of the central plain into the provinces of Camarines Norte and Sur, Tayabas, Nueva Ecija, Nueva Vizçaya, Isabela, and Cagayan. The development of a net of trunk highways in Luzon, especially from the central plain across the mountains of Nueva Vizcaya into the valley of the Cagayan River, and from Ilocos Norte along the north coast to Aparri, brought thousands of Ilocanos into Nueva Ecija, Nueva Vizcaya, and the two provinces of the Cagayan Valley. Now Nueva Ecija, with the exception of the mountainous part, is fairly well settled. At the time of the out-

---

43 Holt and Russell: "Malaria," 1932, pp. 305–369.

44 Thurnwald: Economics in Primitive Communities, 1932, p. 189.

45 *Manila Daily Bulletin*, March 19, 1941.

break of the Pacific war the movement into Nueva Vizcaya was still going on, although this province lacks the extensive ricelands that are characteristic of Nueva Ecija. Migration into the Cagayan Valley was also in full swing.

On the other hand, spontaneous, large-scale migration of homesteaders to Mindanao proved impracticable in the early days. Their lack of funds for the long sea trip, their desire to remain in touch with the home province and village, and their fear of Mindanao's Moro population were strong deterrents to the migration of land-seeking peasants from Luzon. From the most crowded parts of the Visayas, however, people had been migrating of their own initiative to the northern coastal regions of Mindanao ever since Spanish times. But this movement was insufficient to relieve the population pressure in the Visayas, and by 1913 the Philippine Government had recognized the necessity for providing financial assistance in the colonization and development of Mindanao.

–Ifugao rice terraces near Banaue, Mountain Province, Luzon.

(lower left)—Section of the Ifugao rice terraces. Note the height of the terrace wall as compared with the man on the ladder. Banaue. (Courtesy of the Bureau of Forestry.)

(lower right)—Rice terraces in central Bali. In the foreground a rice seedbed.

The shape and size of wet-rice fields may be determined by topography.

FIG. 53 (upper left)—Hoe-tilling on rice terraces in central Bali.

FIG. 54 (upper right)—Rice terraces in the Priangan Mountains, east of Garoet, Residency of Priangan, West Java.

FIG. 55 (center left)—Leveling a rice field with a carabao team, south of Buitenzorg, Residency of Buitenzorg, W

FIG. 56 (center right)—Rectangular fields surrounding a Toba-Batak *huta* (village), near Lake Toba, Residency of T Sumatra.

FIG. 57 (lower left)—Minangkabau women planting rice near Fort de Kock, Residency of Sumatra's West Coast.

FIG. 58 (lower right)—Weeding a rice field in the plains of southern Bali.

(upper left)—Peasant using a *landak* (a weeder developed in Japan) in a wet-rice field. East of Tjitjalengka, Residency of Priangan, West Java.

(upper right)—Javanese girls harvesting rice with the *ani-ani* (a small knife blade attached at a right angle to a short stick). The harvesters cut only enough of the stalk to permit the grain to be tied in bundles. South of Jogjakarta, central Java.

(center left)—Unterraced *tegalans*, seriously eroded and exhausted, on the slopes of Goenoeng Haroeman, north of Garoet, Residency of Priangan, West Java.

(center right)—Belated terracing of tegalans on outworn soil. Slope of Goenoeng-Kidoel, southeastern Jogjakarta, central Java.

(lower left)—Carefully terraced tegalans on slopes in Madura.

(lower right)—Peasant plowing his terraced tegalan, Madura.

These photographs illustrate aspects of dry-field terracing.

FIG. 65 (upper left)—Newly constructed terraces; the faces of the terraces not yet protected. Desa Rantjapoeroet, Res of Priangan, West Java. (Courtesy of the Department of Economic Affairs.)

FIG. 66 (upper right)—Four-year-old terraces, perfectly constructed. Notice *Leucaena glauca* on the edges of the three terraces. Desa Rantjapoeroet. (Courtesy of the Department of Economic Affairs.)

FIG. 67 (center left)—*Paspalum conjugatum* planted along the edge of a terrace. Desa Tjikonde, Residency of Priangan, Java. (Courtesy of the Department of Economic Affairs.)

FIG. 68 (center right)—Terraces with stone retaining walls give excellent protection against erosion to *tegalans* on Desa Tjikonde. (Courtesy of the Department of Economic Affairs.)

FIG. 69 (lower left)—Steep slope completely protected against erosion by well built terraces with stone retaining walls. Tjikonde. (Courtesy of the Department of Economic Affairs.)

FIG. 70 (lower right)—Terraced tegalans on slopes near Desa Leles, Residency of Priangan, West Java. (Courtesy Department of Economic Affairs.)

(upper left)—Terraced dry fields on a limestone ridge in western Madura.

(upper right)—Deforested and badly eroded section in central Madura.

(second row, left)—Unterraced dry fields on a limestone ridge, central Cebu.

(second row, right)—Gently rolling dry fields planted with maize and upland rice, central Cebu.

(third row, left)—A Cebuano immigrant has plowed a *cogon*-covered slope on the Bukidnon Plateau. Notice the signs of erosion. Terracing and contour plowing are needed. Bukidnon Province, Mindanao.

(third row, right)—Cebuano immigrants in Negros have planted recently deforested steep slopes with maize. Near Vallehermoso facing Tañon Strait between Negros and Cebu.

7 and 78 (lower left and right)—Deforested and eroded sections of central Cebu along the road from Toledo to Cebu City. Notice the newly planted patch in the center of Fig. 77.

Fig. 79—Looking toward the Roxas Range from a point west of Tupi, Middle Koronadal, Mindanao. (Courtesy of the National Land Settlement Administration.)

Fig. 80—Bilaan chief and his servant on their way to a hunt. Lagao, Southern Koronadal.

Fig. 81—Bilaan shifting cultivators on the way home from their *caiñgin*. The women and children carry the harvest. Notice the high *cogon* grass and scattered trees. Frequent fires damage the trees regularly and prevent the elimination of the grasses by the forest. A point 3 km. east of Marbel, Northern Koronadal.

(upper left)—Interisland steamer disembarking a new group of settlers on the beach at Dadjangas, Southern Koronadal, Mindanao. (Courtesy of the National Land Settlement Administration.)

(center left)—Group of newly arrived settlers going by truck from Dadjangas to Tupi, Middle Koronadal.

(lower left)—New homes being erected in front of temporary bunkhouse in *Barrio No. 3*, Marbel, North Koronadal.

(upper right)—Settler constructing his house with timber found on his lot, *Barrio No. 5*, Marbel.

(lower right)—Prize house, built of bamboo by a settler in Tupi.

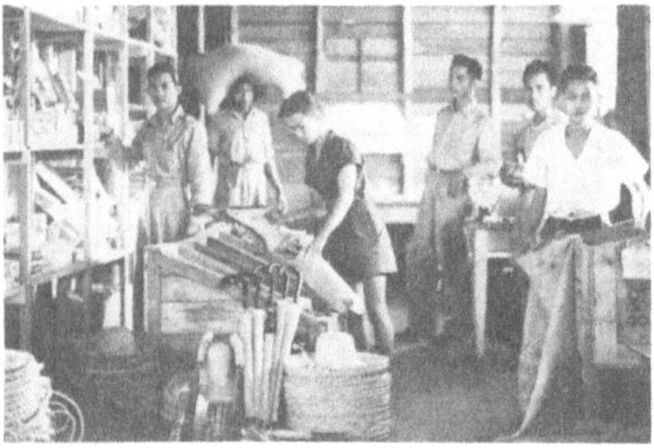

FIG. 87 (upper left)—Water jars to be delivered to settlers in the Koronadal Valley, Dadjangas beach, Mindanao.

FIG. 88 (upper right)—N.L.S.A. truck delivering rations to settlers living in a temporary bunkhouse in Polomolok, N Koronadal, Mindanao. (Courtesy of the National Land Settlement Administration.)

FIG. 89 (center left)—N.L.S.A. store at Marbel, North Koronadal. (Courtesy of the National Land Settlement Administr

FIG. 90 (center right)—Cement boundary markers at Mallig, Isabela Province, Luzon. Mr. Bernardo Torres, former sec treasurer of the N.L.S.A., on left; on right, Dr. Samonte, in charge of N.L.S.A. project at Mallig.

FIG. 91 (lower left)—Drawing of first farm lots in Lagao, Southern Koronadal. Seated at the right of the table is C Paulino Santos. First on the left is Prof. M. B. Raymundo, agricultural adviser of N.L.S.A. (Courtesy of the N Land Settlement Administration.)

FIG. 92 (lower right)—Settler delivering his first peanut crop at the N.L.S.A. warehouse in Lagao, Southern Kor (Courtesy of the National Land Settlement Administration.)

FIG. 93 (upper left)—Overseer meeting settlers of Polomolok to discuss the lay-out of a new *barrio*, Middle Koronadal, Mindanao. (Courtesy of the National Land Settlement Administration.)

FIG. 94 (upper right)—*Barrio No. 4*, Marbel. Some houses are already occupied; others are still under construction. Northern Koronadal. (Courtesy of the National Land Settlement Administration.)

FIG. 95 (lower left)—*Barrio No. 3*, Lagao. Mt. Matutum in the background. Notice the *buri* palms. Southern Koronadal. (Courtesy of the National Land Settlement Administration.)

FIG. 96 (lower right)—Building the Alunan Highway at the prospective site of Tupi, Middle Koronadal. (Courtesy of the National Land Settlement Administration.)

FIG. 97 (upper left)—Machine park of the N.L.S.A. in Lagao, Southern Koronadal, Mindanao. (Courtesy of the N Land Settlement Administration.)

FIG. 98 (center left)—Tractors on the march from Lagao to Tupi, in Middle Koronadal, and Marbel, in Northern Kor (Courtesy of the National Land Settlement Administration.)

FIG. 99 (lower left)—Bulldozer at work on the irrigation canal that was to tap the Siluay River, Lagao. (Courtesy National Land Settlement Administration.)

FIG. 100 (upper right)—Drilling a well in Polomolok, Middle Koronadal. (Courtesy of the National Land Settlement Ad tration.)

FIG. 101 (center right)—Second plowing of land that had been covered by *cogon*. McCormick-Deering Diesel pulling disk plow. Tupi. (Courtesy of the National Land Settlement Administration.)

FIG. 102 (lower right)—Uprooting a tree with a Cletrac Diesel. Koronadal Valley. (Courtesy of the National Land ment Administration.)

3 (upper left)—Women settlers in Marbel pulling rice seedlings from a seedbed. Northern Koronadal, Mindanao.

4 (upper right)—Harvesting maize in Tupi, Middle Koronadal. (Courtesy of the National Land Settlement Administration.)

5 (center left)—Planting germinated rubber seeds in the N.L.S.A. nursery near Kipalbig, Middle Koronadal. (Courtesy of the National Land Settlement Administration.)

6 (center right)—Cowpeas on the administration farm in Lagao, Southern Koronadal. (Courtesy of the National Land Settlement Administration.)

7 (lower left)—Budded rubber trees in the N.L.S.A. nursery near Kipalbig. (Courtesy of the National Land Settlement Administration.)

8 (lower right)—Watering Bermuda onions on the administration farm in Lagao. (Courtesy of the National Land Settlement Administration.)

FIG. 109 (upper left)—Administration headquarters of the N.L.S.A. in Lagao, Southern Koronadal, Mindanao. (Cour the National Land Settlement Administration.)

FIG. 110 (upper right)—Administration building in Marbel, Northern Koronadal. (Courtesy of the National Land Set Administration.)

FIG. 111 (center left)—School in Marbel. (Courtesy of the National Land Settlement Administration.)

FIG. 112 (center right)—Hospital in Marbel. (Courtesy of the National Land Settlement Administration.)

FIG. 113 (lower left)—N.L.S.A. Hog Farm in Lagao.

FIG. 114 (lower right)—Homemade carabao cart in Lagao. (Courtesy of the National Land Settlement Administration.

Chapter V

MINDANAO, THE FRONTIER

Mindanao, the second largest island of the Philippines, has been looked on by both the government and the public as the land of promise and unlimited opportunity, the frontier of the Philippines. Such a view is supported by the fact that Mindanao includes about 32 per cent of the country's territory but has today only 11 per cent of the population. With the exception of the two small coastal provinces in the north, its population densities are low (Maps, Figs. 48 and 115). The public and many government officials do not realize, however, that considerable areas of the island are unsuited for large-scale agricultural development because of the roughness of the topography. This unsuitability was brought out in a survey undertaken in 1939 by the Mindanao Exploration Commission on behalf of the Refugee Economic Corporation and is indicated on the accompanying map (Fig. 116).[1]

Government-Sponsored Agricultural Colonization Prior to the Establishment of the Commonwealth

Except in a narrow strip of Visayan settlements on the northern coastal plain of Misamis Oriental and Misamis Occidental, there were hardly any Christian Filipinos in Mindanao at the beginning of the American regime in the Philippines, in 1900. Mindanao's population consisted primarily of various indigenous Moslem groups—called *Moros* (Moors) by the Spaniards—such as the Magindanaos and the Lanaos, who live in the Cotabato Valley and around Lake Lanao, and of pagan mountain tribes, such as the Bagobos, Manobos, Bilaans, and Bukidnons, in the interior and in the eastern part of the island. These pagans were, and most of them still are, shifting cultivators, whereas the Moros practiced both shifting and sedentary agriculture. At that time Mindanao's population was not yet half a million, and the density was but five persons per square kilometer. Forest or grass covered most of the land, which amounts to 95,000 square kilometers. The grass areas were the result of shifting cultivation. There were no roads on the island so that travelers had either to go along the coast and up the rivers by boat or to walk along narrow trails. During the period of administration by the American Army in Mindanao a few roads were constructed along the coast or inland for short distances.[2]

Mindanao's wealth in timberland, as well as in undeveloped agricultural land, was recognized as soon as the island had been pacified by the Amer-

1 Pendleton: "Land Utilization and Agriculture of Mindanao," 1942, pp. 180–210.

2 When the administration of Mindanao was taken over by the Philippine Government in 1913, the island had a total of 339.6 kilometers of roads: of these 80 kilometers were first-class, 153.6 kilometers second-class, and 103 kilometers third-class (*Annual Report of the Governor of the Moro Province for the year ended June 30, 1913*, pp. 26–27).

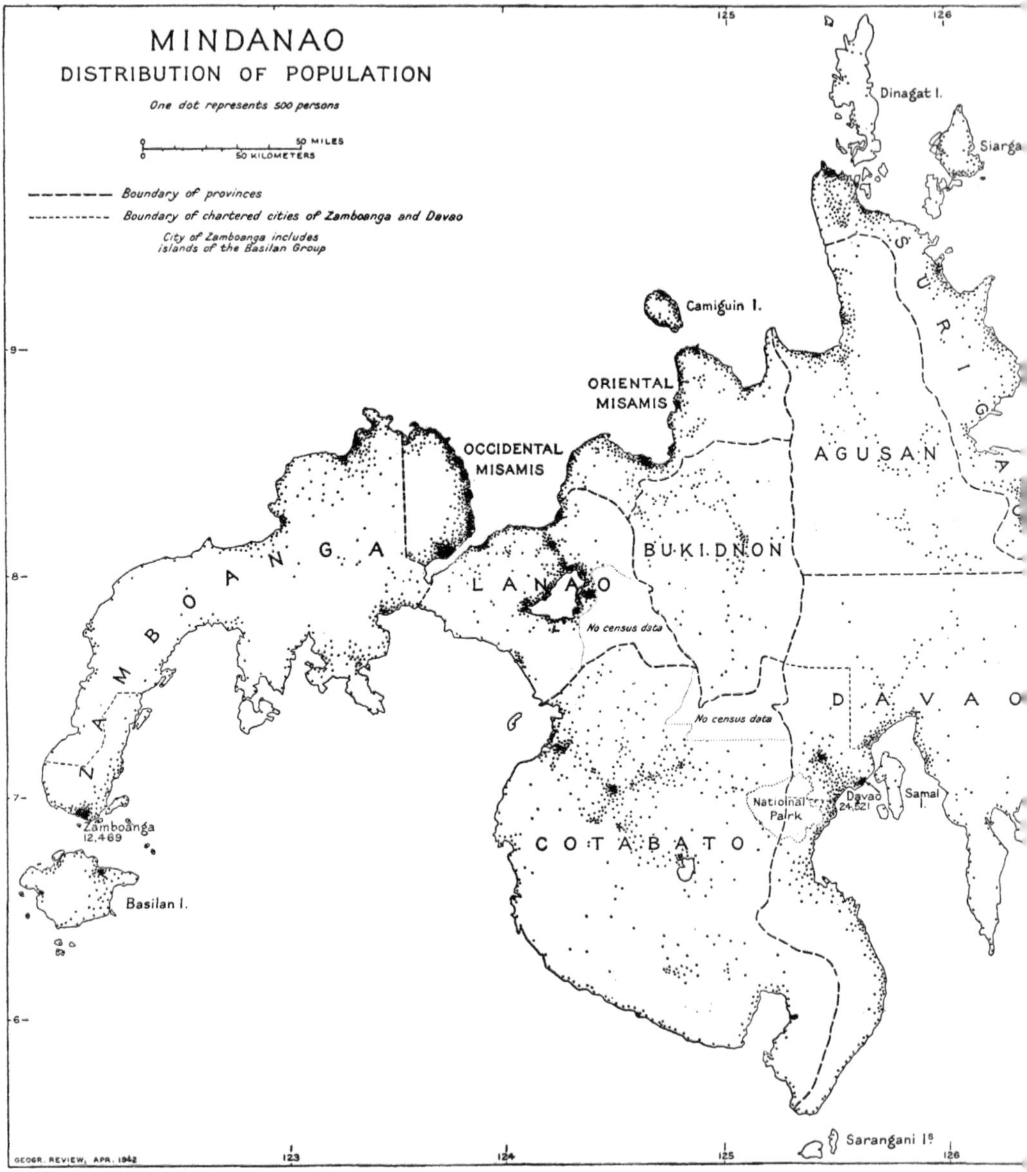

FIG. 115—This map is based on tabulations of the 1939 census. Each dot represents 500 people. The dots have be tributed with reference to locations of towns and *barrios*, cultivated lands, fisheries, etc., as shown on existing maps. (duced through the courtesy of the *Geographical Review*.)

ican Army and administrative control had been established. In pursuit of the land policy discussed in Chapter IV, the Filipino-controlled Insular Government was most anxious to prevent large-scale American investment and wished instead to direct homesteaders from the congested areas of the Visayas into the interior and the southern parts of the island. It also hoped that such a large-scale migration of Christian Filipino settlers from

TABLE X—PHILIPPINE ISLANDS: AGRICULTURAL COLONIES ESTABLISHED BY THE INSULAR GOVERNMENT, 1913–1917

COLONY	YEAR ESTABLISHED	AREA (HECTARES)
Pikit, Cotabato	1913	2,720
Silik, Cotabato	1913	2,708
Peidu-Pulangi, Cotabato	1913	1,380
Pagalungan, Cotabato	1913	4,475
Glan, Cotabato	1913	1,272
Momungan, Lanao	1914	2,344
Abulug, Cagayan	1914	2,677
Carmen, Bohol	1915	2,464
Talitay, Cotabato	1917	1,720

Source: Sitchon: Land Settlement Policies. These figures differ slightly from those given in Philippine Islands Bureau of Labor: The Activities of the Bureau of Labor, 1930, p. 75.

the north would lead to an amalgamation and a Filipinization of the Moros and pagans and thus remove the danger of a possible separation of Mindanao from the rest of the Philippines. Such a separation had been suggested repeatedly by Americans interested in plantation development who played up the differences and antagonism between Christian and non-Christian Filipinos.[3] In 1913 the government therefore adopted the policy of actively encouraging migration by establishing agricultural colonies.

Between 1913 and 1917 a total of seven agricultural colonies was founded in Mindanao, six in the province of Cotabato and one in the province of Lanao. Two additional colonies were established on other islands: one on Bohol and one in the Cagayan Valley in Luzon (Table X).

In all the colonies the settlers were grouped together in small towns instead of being compelled to live on their farmsteads. Each settler received his home lot of from 800 to 900 square meters in the town (plan of Peidu-Pulangi, Fig. 117) and, at some distance, his farm land, which was surveyed in blocks of 8 or 16 hectares.

In the Cotabato colonies the government intentionally mixed Christian and Mohammedan (Moro) settlers in an attempt to show that the two could work and live together peaceably.[4] This is why Moros were found

[3] As late as 1926, Representative Bacon introduced such a proposal in the U. S. Congress (*69th Congr., 1st Sess., H. R. 12772*, introduced on May 6, 1926, and *69th Congr., 2nd Sess., H. R. 15479*, introduced on December 20, 1926).

[4] *Report of the Governor General of the Philippine Islands to the Secretary of War 1917*, 1918, p. 39.

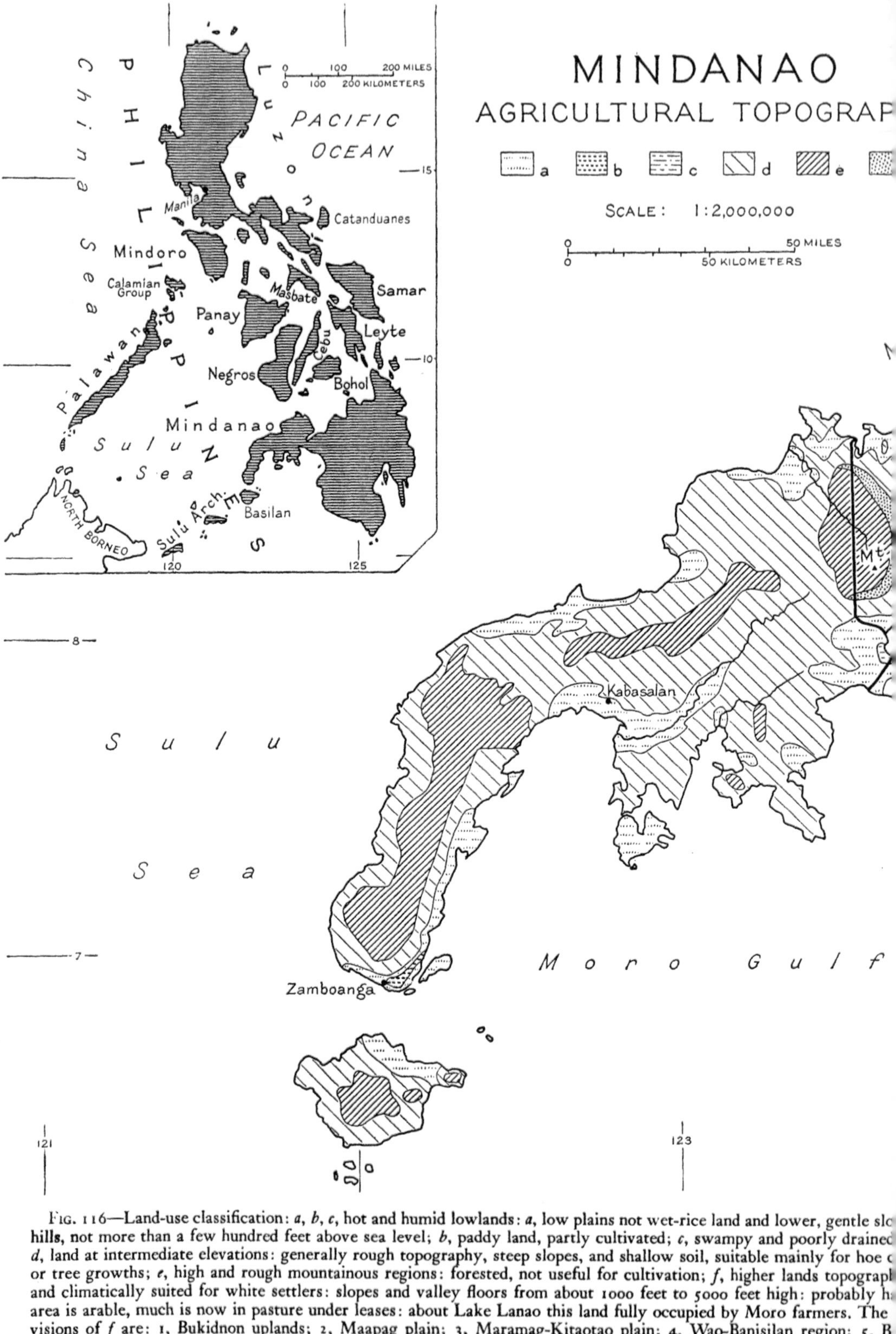

Fig. 116—Land-use classification: *a*, *b*, *c*, hot and humid lowlands: *a*, low plains not wet-rice land and lower, gentle slc hills, not more than a few hundred feet above sea level; *b*, paddy land, partly cultivated; *c*, swampy and poorly drainec *d*, land at intermediate elevations: generally rough topography, steep slopes, and shallow soil, suitable mainly for hoe c or tree growths; *e*, high and rough mountainous regions: forested, not useful for cultivation; *f*, higher lands topograpl and climatically suited for white settlers: slopes and valley floors from about 1000 feet to 5000 feet high: probably h area is arable, much is now in pasture under leases: about Lake Lanao this land fully occupied by Moro farmers. The visions of *f* are: 1, Bukidnon uplands; 2, Maapag plain; 3, Maramag-Kitaotao plain; 4, Wao-Banisilan region; 5, B Barira region; 6, Kidapawan region; 7, Upper Pulangi River Valley. (Reproduced through the courtesy of the *Geogr Review*.)

Compare with the population density map (Fig. 115), on which province names are given.

125
126
Surigao
S e a
a o
Diuata Mts.
Mt. Diuata
Butuan
Agusan R.
Macajalar B.
Cagayan
Diklom
Del Monte
1
Katanglad
Malaybalay
Dansalan
L. Lanao
Maging
Miarayon
Kalatungan
Valencia
2
Maapag
7
Musuan
Maramag
Mt. Piapayungan
Pangantocan
3
Kitaotao
Wao
4
Buldun
5
Barira
Banisilan
Parang
Malitobug R.
Maridagao R.
Pulangi R.
Cotabato
Kabakan
Pikit
Dulauan
Kidapawan
6
Mt. Apo
Davao R.
Davao
Kapalong
L. Buluan
Davao Gulf
Koronadal
Sarangani B.
9
8
7
6

in the colonies even when there was no need for them to seek land. The social goal of the colonization program, generally considered the primary objective, seems to have been attained. From the economic point of view, however, the colonies were a failure. Late in 1928 a survey showed that there were only about 8,000 colonists, including dependents, in the Cotabato colonies founded by the Insular Government between 1913 and 1917 inclusive. Half of the settlers were Moros; the other half were Christian Filipinos. They owed the government approximately 372,000 pesos, most of the sum being uncollectible.[5]

The migration to these colonies remained small for a number of reasons. A lack of funds was chiefly instrumental in limiting the number of settlers; the sites for the colonies were ill-chosen; and the settlers were unwisely selected, many of them having come, without farming experience, directly from the water fronts of Cebu City or Manila. The recruiting agents, in view of the reputation of the Cotabato Moro, had not been too particular. In the colonies a system of uniform administration and supervision was lacking, and little was done to solve the transportation and marketing problems, with the result that the settlers had to depend upon government support much longer than had been expected. Irrigation and drainage problems were neglected, and crops failed repeatedly because of droughts or floods.

In December 1917 the Insular Government, discouraged by the high costs and limited results of the first experiment discontinued its policy of founding agricultural colonies and of making loans to colonists who lacked funds and, in 1919, decided to encourage actively the migration of only those homeseekers who had funds sufficient to tide them over the first year.[6] Under no circumstances were they to receive loans. The Bureau of Labor, delegated to carry out this new policy,[7] established the Interisland Migration Division, which functioned from 1918 to 1939. The work of this division consisted of recruiting homeseekers and sending them usually to Mindanao but occasionally to thinly populated provinces in other islands, such as Mindoro. In order to qualify as a homeseeker entitled to the support of the Bureau of Labor, the prospective settler had to be the head of a family, or at least 18 years of age, and to possess the necessary agricultural implements, a carabao, and, most important of all, 250 pesos. A qualified homeseeker was advised about regions with settlement possibilities and was furnished free transportation from his home town to his

5 *Annual Report of the Governor General of the Philippine Islands 1928*, 1930, pp. 204–205.

6 The establishment of these colonies was authorized by Act No. 2806, Feb. 28, 1919. Between 1919 and 1926 a few provincial governments established agricultural colonies on their territories: the province of Zamboanga founded Lamitan on the island of Basilan; the province of Sulu founded Tawi-Tawi on Jolo; while in Mindanao Mailag was founded in Bukidnon province and Sarunayan and Magonoy were founded in Cotabato province. These, however, proved to be even less successful than those sponsored by the Insular Government.

7 Act No. 2727, Dec. 20, 1917.

destination. After this the government assumed no further obligation, and it was up to the settler to overcome the initial difficulties of his pioneering venture.

The Bureau of Labor directed a number of homeseekers to the agricultural colonies of Cotabato, where they took up lots that either had never been occupied or had been abandoned by the original colonists. As

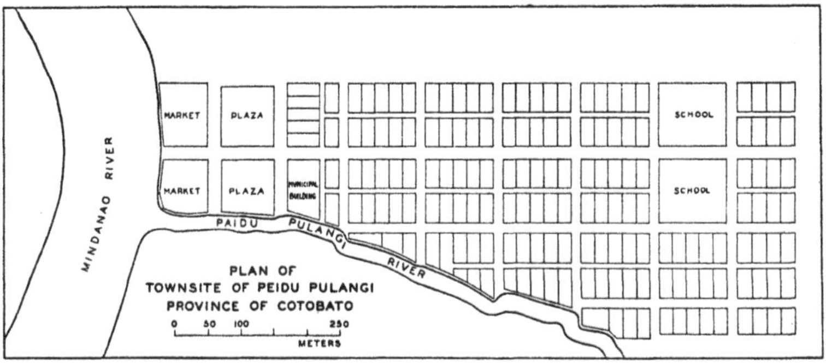

FIG. 117—Plan of the townsite of Peidu Pulangi, Cotabato: based on a blueprint plan of the Bureau of Public Lands, Philippine Commonwealth. See map, Fig. 118, for the location of Peidu Pulangi. For Cotobato, *read* Cotabato.

there was additional free land in the neighborhood of the colonies, these became the nuclei of slowly expanding settlements. Between 1918 and 1934 inclusive the Bureau of Labor supplied transportation to 6,080 homeseekers (31,408 persons), most of whom went to Mindanao. The average number of families a year to receive government support in their migration was thus only 357 (1,846 persons).[8]

Because a good deal of money seemed to have been spent inefficiently during the first years, when appropriations were relatively large, Governor General Wood cut the appropriations drastically during the years from 1923 to 1925. After 1925 the appropriations were slightly increased, but they never again reached the 1918–1922 level. Another restriction upon settlement was that the homeseekers, with the exception of those who were placed in the old agricultural colonies, came under the homestead system and so had to meet residence requirements in order to get titles to their land. This obligation was always considered burdensome. This was especially true in Mindanao because of the Christian Filipino's fear, hidden or open, of attack by the Moros and pagans. Delays both in road development and in the surveying and subdividing of the public land by the Bureau of Lands were another obstacle to a rapid influx of settlers. Also, many homeseekers became the victims of usurers and land-grabbers,

8 The migration of homeseekers was continued under the Commonwealth Government until June 1939. A total of 9,172 families (46,712 persons), or an annual average of 417 families (2,123 persons), migrated during the years from 1918 to 1939 inclusive.

who deprived them of the fruits of their labors.[9] Reports of such experiences traveled quickly to the home communities and deterred others from following these homeseekers, for it is everywhere true that nothing can retard migration more effectively than discouraging reports about economic and social conditions in the pioneer land.

Since the activities of the Bureau of Labor were limited to aiding persons able to qualify for support, as time passed, increasing numbers of settlers who could not qualify for government help went to Mindanao at their own expense. They became homesteaders, squatters, tenants, or agricultural laborers, depending upon the area to which they went and the opportunities which they found there. Unfortunately, no numerical data are available on this spontaneous, unsupported, and unguided migration; observations indicate, however, that it was greater than that of government-sponsored homeseekers.

So it was that up to the year 1935 the development of Mindanao had not been very great. Although from 30,000 to 35,000 persons had been sent there under the auspices of one government agency or another, in addition to those who had migrated on their own initiative, the island was still without adequate highways and much of the land was unsettled. Yet enough had been accomplished by the colonists and homeseekers to demonstrate that Filipinos are capable of pioneering even amidst a population not over-enthusiastic about their coming.

In the province of Davao, adjoining Cotabato, in the meantime, the Japanese had established a large colony engaged especially in the growing of abaca. When Japan became politically menacing, the dangerous potentialities of this colony brought the settlement of Mindanao by Filipinos once more to the forefront. Investigations revealed that Japanese agricultural companies were occupying considerably more land than they had a right to. The total area owned or controlled by Japanese in 1935 amounted to 57,350 hectares, of which 28,098 had been acquired legally; the other 29,252 hectares comprised public land, which had been alienated to Filipinos and Americans, who then turned it over to Japanese under lease or

9 Land-grabbers exploited the fact that "the bureau of lands based priority of claim upon priority of filing instead of priority of occupancy, as in the United States. This enables the shrewd late-comer to appropriate the land of . . . settlers by making a paper filing upon cleared and planted land" (*Annual Report of the Governor General . . . 1930*, Washington, 1932, p. 17). The members of the Mindanao Exploration Commission, which investigated refugee settlement possibilities in the Philippines, wrote in a letter dated Manila, July 7, 1939: "With the brisk demand for the subdivision of public lands now taking place all over the Islands, the Bureau of Lands is not always able to take care of all applicants as rapidly as might be expected or as the Bureau itself would like to do. They are greatly handicapped by the lack of sufficient experienced personnel. Unfortunately, as so frequently happens in such cases, it is the little fellow or the man without any influence, and who really can least afford it, who has to bear the brunt of such delays. The man of means or with good political affiliations always has access to the seats of the mighty. It, also, has often happened that in disputes over the occupancy of public land between the lowly and the politically powerful, the small man has in the end lost many years of hard work."

some other form of contract.[10] Although the government contended that most of these contracts were voidable since they violated public land laws, nothing was done to cancel the illegal contracts, probably because most of them would soon expire in any case, and the government wished to avoid a conflict with Japan. Instead, steps were taken to make further expansion of Japanese settlements impossible. Land surveys and road construction were undertaken to encourage the migration of Filipinos into Davao in order to block such encroachment.

The Commonwealth Opens Mindanao

In February 1935, just before the establishment of the Commonwealth of the Philippines, the legislature passed the Quirino-Recto Colonization Act,[11] which represented the first link in a new chain of legislative and executive actions to speed the development of Mindanao. The act called for elaborate undertakings designed to avoid past mistakes in the field of agricultural colonization; but before it could be put into operation the Commonwealth was established.

The Commonwealth Government, also recognizing earlier mistakes, decided to concentrate first on a program of extended road-building and public-land surveying before again attempting to establish government-supported agricultural colonies. In January 1936 it therefore diverted the funds that had been set aside for colonization to the construction of roads.[12] During the years from 1936 to 1941 the road-construction program was carried through rapidly (Map, Fig. 118). The important towns on the northern coast were connected with each other and with Cotabato and Davao on the southern coast, and Cotabato was linked with Davao. Homesteaders in unknown numbers,[13] coming at their own expense, took up the

10 Pelzer: Population and Land Utilization, 1941, p. 40.

11 Act No. 4197, Feb. 12, 1935.

12 Commonwealth Act No. 18, Jan. 2, 1936.

13 In the office of Survey Party 24-C in Buluan, Cotabato, I found a record that gives an indication of how rapidly the settlers were coming. Between 1938 and July 1940, 927 homeseekers and immigrants—coming at their own expense—registered with this office and applied for lots.

Dr. Stanton Youngberg, a member of the Mindanao Exploration Commission, wrote in a letter dated Davao, May 29, 1939: "My last previous visit to the island of Mindanao was almost six years ago. The changes that have taken place here during that period are really astounding. Then there were only disconnected strips of road in the different provinces, most of which started on the coast and ended nowhere in particular. Now one can travel by automobile from Surigao at the northeastern tip of the island through Agusan, Oriental Misamis, Lanao, Cotabato, and Davao to this rapidly growing town on the Gulf of Davao, a distance of seven hundred kilometers. In other words, road building has been rapid during the past five years but, fast as it has been, land settlement has been faster. People push on ahead and settle along the projected highways. The loud complaint of the Bureau of Lands personnel is over the fact that the people squat on the land and begin cultivation before they can get around to subdivide the areas that have been released by the Bureau of Forestry for agricultural purposes. Lands that were absolutely uninhabited two and three years ago are now occupied by settlers who are in the process of clearing the cut-over or non-commerical timber land and planting corn between the stumps. These people come from Luzon as well as from the heavily populated islands of the nearby Visayan group."

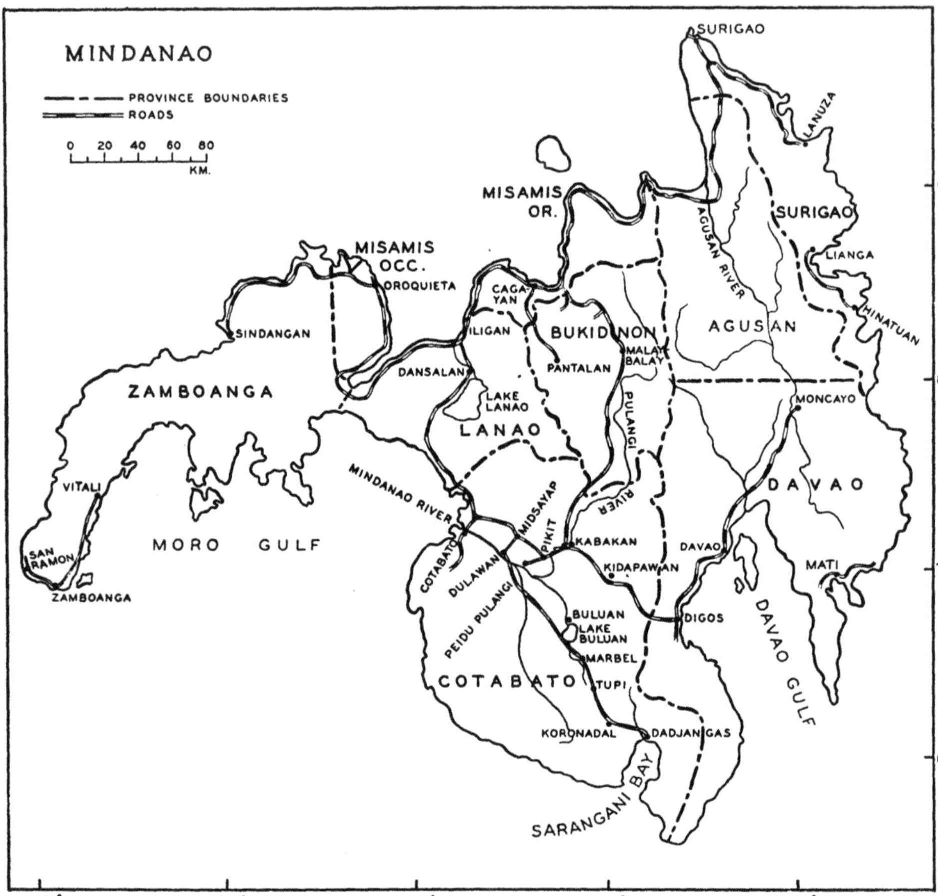

FIG. 118—Roads in Mindanao. The roads are of varying quality (none are cement highways). Most are usable by automobiles during the rainy season, but some only during the dry season. This map is based on official maps of the Bureau of Public Works and on commercial maps.

land along the new highways as soon as it was surveyed; sometimes they preceded the construction and survey crews, even though it was no longer possible to file applications for unsurveyed land.

While Mindanao was being opened up by the highway-building program, a number of government agencies and experts studied the problem of government-sponsored land settlement for those unable to migrate because of a lack of funds. The most important proposals to be submitted were the Silayan Plan,[14] developed by Hilarion S. Silayan, Director of the Bureau of Plant Industry, who had prepared it as the basis for the Quirino-Recto Colonization Act, and the plan advanced by Frederic Howe, economic adviser to President Quezon.[15] As a result of the various plans submitted, the National Land Settlement Administration was set up, answering the need for a better conceived and more completely integrated colonization procedure, particularly in Mindanao. (The activities of the N. L. S. A. in northeastern Luzon are discussed in Appendix A.)

The Filipinos themselves had been aware of the shortcomings of the government's earlier efforts. This is evidenced by the report by Manuel Roxas, Chairman of the National Economic Council, that was submitted to President Quezon on October 22, 1938, in which he stated:

> In no instance had there proceeded a careful study and selection of land to be settled nor of the people that were to be given State help to settle on those lands. No thought was apparently given to the important question of marketing the products of new settlers, neither was there any well-thought-out plan of assisting them to become efficient producers by providing for them demonstration and experimental stations or equipping them with modern tools and implements.[16]

Roxas felt that only communities with a measure of public order, sanitation, and education, and some means of transportation and communication could be expected to succeed. He had therefore proposed the organization of the National Land Resettlement Corporation, later renamed the National Land Settlement Administration, and listed four objectives,[17] the main parts of which were later adopted.

Upon receipt of the recommendations, President Quezon had appointed a special committee, composed of the Secretaries of the Interior, Finance, Agriculture and Commerce, to organize the National Land Settlement Administration. In order that work might be started immediately and that

---

14 Silayan: "Organized Land Settlement," 1938.

15 Howe: "Philippine Homestead Settlement Plans," 1938.

16 From an unpublished report.

17 Roxas' four objectives were: (1) the settlement of young reservists in unpeopled regions in order to strengthen the defense of the thinly settled parts of the islands; (2) the movement of tenant farmers in order to improve conditions in those provinces where agrarian disputes and unrest were most frequent; (3) the Filipinization of non-Christian Filipinos and their amalgamation with Christian Filipinos; (4) the development of new export crops in order to insure the country against economic reverses that might otherwise result from independence. (From an unpublished report).

the first crops might be planted in the spring, a working capital of 200,000 pesos was borrowed from the National Development Company on January 5, 1939, as the legislature had not yet passed the bill creating the N. L. S. A. On June 3 of that year, President Quezon signed the bill, which had meanwhile been passed by the National Assembly.[18]

The Organization of the National Land Settlement Administration

Objectives

The objectives of the N. L. S. A., following closely the lines that Roxas had suggested, were formulated by the law as follows:

(a) To facilitate the acquisition, settlement and cultivation of lands whether acquired from the Government or from private parties;

(b) To afford opportunity to own farms to tenant farmers and small farmers from congested areas, and to trainees who have completed the prescribed military training;

(c) To encourage migration to sparsely populated regions, and facilitate the amalgamation of the people in different sections of the Philippines;

(d) To develop new money crops to take the place of the present export crops, which may suffer from the loss of preferences which they enjoy in the American market.[19]

In order to attain these objectives the corporation was empowered to hold public agricultural land for 25 years, with the possibility of obtaining a renewal for another period of the same length. Furthermore, it could recommend to the President the reservation of public land suitable for settlement projects. Such land was to be surveyed, cleared, prepared for cultivation, and assigned to settlers in parcels not exceeding 24 hectares. The corporation was empowered to establish and operate electric-light and water-supply plants, irrigation systems, and trading stores, and to engage in manufacturing, milling, lumbering, and other enterprises considered necessary for the success of its activities.

Administrative Organization

The N. L. S. A. was a government corporation composed of five persons who acted as members of the Board of Directors charged with its management. The members were appointed by the President, with the consent of the National Assembly, for terms of 3 years.[20] The directors appointed Major General Paulino Santos, formerly Chief of Staff of the Philippine Army, as manager. Santos is a man of great energy, with experience in the

18 Commonwealth Act No. 441, June 3, 1939. For the full text of the act, see Appendix A.

19 *Ibid.*, Sec. 2.

20 The first Board of Directors was composed of Rafael R. Alunan, Secretary of the Interior, as chairman; Manuel Roxas, Secretary of Finance; Benigno S. Aquino, Secretary of Agriculture and Commerce; José Avelino, Secretary of Labor; and Ramon Roces, Manila publisher.

handling of problems related to agricultural colonization. Formerly, as the Director of Prisons, he had been especially successful in the development of agricultural penal colonies. As Chief of Staff of the Philippine Army, he had introduced vocational training, including modern farming technique for young recruits.

The manager in turn appointed the technical, clerical, and other personnel. All appointments to positions with salaries exceeding 2,400 pesos required the approval of the President—an illustration of the manner in which the long arm of the President of the Philippines reached down to control even minor appointments in the Commonwealth. After 2 years of operation the staff of the N. L. S. A. had grown to about 170 persons, including scientific agriculturalists, animal husbandmen, engineers, surveyors, mechanics, craftsmen, doctors, dentists, nurses, accountants, clerks, and recruiting officers. The central office was in Manila, and another large office was maintained in Lagao, the administrative center in Mindanao.

The capital of the N. L. S. A., amounting to 20 million pesos, was to be supplied from the coconut-oil excise tax fund in appropriations not exceeding 4 million pesos annually. The manager had to submit an annual report on the activities and progress of the N. L. S. A., as well as a balance sheet, to the President and the National Assembly.

Requirements for Settling

The act creating the N. L. S. A. laid down specific requirements for the people who were to be accepted as settlers. They must be Filipino or American citizens, of legal age, preferably not over 40, married and with children. They, and their dependents, should be healthy and fit for a pioneering life, should have had some agricultural experience, and should be of good character and reputation. The act further stated that settlers must be persons qualified to apply for public land under the Public Land Act.

However, in practice many of these requirements were disregarded. When the story of the work of the N. L. S. A. spread through the Philippines many families, both from other provinces of Mindanao and from other islands, moved to the settlement districts at their own expense. This made a careful examination of the settlers difficult if not impossible; yet many had to be accepted because of verbal instructions from the President that all persons presenting themselves as settlers in any of the settlement districts should be accepted if they seemed to fulfill the qualifications.

Contract Between the Settler and the Administration

As soon as a settler's application was accepted, he signed a contract with the N. L. S. A.[21] His main obligations were: to plant and cultivate the land given to him, using such plants as had been prescribed by the administration; to devote himself to this exclusively and not to engage in any

---

21 For the full text of the contract see Appendix A.

other trade or occupation without the approval of the administration; not to transfer his land in any way, except by bequest, to any other person for the first 10 years after receipt of the title to the land; to reimburse the administration within 20 years, at 4 per cent interest, for the transportation, building materials, food, and clothing that he might have received; to deposit all of his surplus products in administration warehouses for sale through the administration.

The administration's obligations were to advance the cost of transporting the settler, his family, and his belongings; to assign to him a parcel of 12 hectares of agricultural land; to supply him, on a credit basis, with building materials (and, if necessary, skilled assistance for the construction of his house), food until he should become self-sufficient, the use of machinery such as tractors and plows (Figs. 97 and 98), planting material and fertilizer, livestock, and agricultural implements; the administration was to sell, on his behalf, products deposited by him in the administration warehouses (Fig. 92) and to give him full title to the land after all obligations should have been met.

Payments were to begin after the third year and were not to exceed 30 per cent of the proceeds of sales made through the administration. The contract was favorable to the settler, relieving him of most of the usual hazards and burdens of pioneering except for labor—and even that was made much easier because of the availability of modern mechanical equipment beyond the means of the individual small-scale farmer in the Philippines.

Marketing Organization

The act under which the N. L. S. A. operated called for the establishment of "trading stores or cooperatives to engage in the buying or selling of commodities. . . ." A store was set up in each town to supply the settlers with daily necessities at cost price on either a cash or a credit basis (Fig. 89). It was expected that these N. L. S. A. stores would become cooperatives as soon as the administration should be ready to wind up its work in a district.

In order to help the settlers to market their products the N. L. S. A. established a central marketing organization. Both of these services proved of great value because they protected the settlers from becoming victims of the sharp business practices of predatory merchants and produce buyers.

Health, Education, and Religious Services

The planners of the N. L. S. A., recognizing the importance of health in a pioneering community, established a Health Division with three sections: medical, dental, and veterinary.

In the Koronadal Valley project (see p. 141, ff.) each one of the settlement districts received its own small hospital (Fig. 112) under the supervision of a physician assisted by two nurses. Each *barrio* had its first-aid station with a trained medical attendant. Among the tasks of the medical personnel were those of supplying vaccination against smallpox and inoculation against typhoid, cholera, and dysentery, and of supervising or undertaking suitable measures to control malaria. There was one dentist on the payroll of the project. The veterinary personnel was in charge of the hog and poultry farms of the N. L. S. A. and also cared for any diseases occurring among the settlers' livestock.

A school was built in each of the towns in the Koronadal Valley colony (Fig. 111). The religious needs of the settlers, all of whom were Roman Catholic, were met by American priests belonging to the order of the Oblates of Mary Immaculate (O. M. I.). The settlers built a church in each district out of material furnished by the administration.

Settlements in the Koronadal Valley

The Mindanao Reconnaissance Survey

In November 1938 General Santos had made a reconnaissance survey of Mindanao in order to select possible sites for settlement projects. He was accompanied by technical advisers in such fields as agriculture, forestry, livestock industry, soil science, and irrigation.[22] The party gathered in Zamboanga on November 12, 1938, and sailed to Dadjangas, at the head of Sarangani Bay in Cotabato, arriving there on November 14. Between the 14th and 19th the party moved rather quickly from Dadjangas through Koronadal Valley to Lake Buluan. Another 2 days were spent in the Kidapawan district along the Cotabato-Davao highway, and 3 or 4 days in the Compostela-Monkayo region on the upper Agusan River in northern Davao.

In a number of conferences held in Mindanao and on board ship returning to Manila, the possibilities of the various districts visited were discussed, and it was finally decided to recommend the reservation of Koronadal Valley and of the Compostela-Monkayo districts to the President. Koronadal Valley was given first choice because of its convenient location between Sarangani Bay and Lake Buluan. The topography would make road construction easy. Moreover, a large part of the valley's 97,000 hectares was open grassland that would not require costly and time-consuming clearing. These considerations seemed to outweigh doubts about

22 The party consisted of the following members: Major General Paulino Santos; Lieutenant Colonel Dionisio Gutierrez; Mr. Pablo T. Noroña, Assistant Director of Prisons; Mr. George M. Goodall, Adviser to the National Economic Council; Dr. N. B. Mendiola, Head of the Department of Agronomy, and Dr. A. L. Teodoro, Head of the Department of Agricultural Engineering, both from the College of Agriculture, University of the Philippines; Dr. M. M. Alicante, Chief of the Division of Soil Survey, Bureau of Science; Mr. D. Z. Rosell, Soil Scientist, Bureau of Science; Dr. Pedro S. Sales, of the Bureau of Animal Industry; Professor M. B. Raymundo, of the Bureau of Plant Industry; Forester P. San Buenaventura, of the Bureau of Forestry; Professor J. Elayda, Chief of the Vocational Training Division, Philippine Army; Captain R. T. Salacup, Medical Corps, Philippine Army; and Mr. M. Kasilag, Commissioner for Mindanao and Sulu.

the fertility, the climate, and the need for irrigation in the southern part of Koronadal.

On December 20, 1938, a week after the return to Manila, the survey party submitted a report on the areas of Mindanao selected as sites for settlement projects. On February 11, 1939, President Quezon signed proclamations nos. 383 and 384 reserving the two recommended districts for the N. L. S. A. Ten days later the first group of settlers, together with a number of employees, sailed from Manila to Koronadal under the leadership of General Santos.

The great haste in getting the project under way was due to the desire of the government to have a start made before the new planting season began. There were many questions that should have been studied before the first settlers were actually moved. Data on climatic conditions were lacking; only general information about the annual distribution of rainfall had been obtained through a questioning of the residents of Koronadal Valley. Specific data on suitable types of crops were scarce, although the experts had mentioned a number of cash crops, such as cotton and rubber, the cultivation of which might prove profitable in certain sections of the chosen sites. Notwithstanding the fact that the experts had pointed out the need for thorough surveys of the water supply available for irrigation and of the water-holding capacity of the light, sandy soils, such surveys were not undertaken until several months after groups of settlers had actually been established.

The survey party had found that Koronadal Valley was thinly populated, but nobody knew exactly how many Bilaans (a pagan tribe) (Figs. 80 and 81) and Moros were living in the region, or how much land should be reserved for them. These people were concerned about their land rights in the event of an establishment of agricultural colonies by the N.L.S.A. However, instead of making definite arrangements for the Bilaans and Moros, the members of the survey party merely gave these people oral assurance that their rights would be respected and told them to declare their land for tax purposes and to get tax receipts as evidence of ownership. Such advice was not very helpful, especially to the backward Bilaans.

It would have been better if the area could have been surveyed and sufficient land set aside for Bilaans and Moros. In fact, the lack of surveys slowed down the progress of the first settlers considerably, delaying for 15 months the allotting of their farm land.

The Valley: The Physical Environment

Koronadal Valley, in southeastern Cotabato, which was chosen by the Mindanao Survey Party as the first region of Mindanao to be developed by the N. L. S. A., is a narrow plain flanked by two mountain ranges (Map, Fig. 119). The higher eastern range has been named Quezon Range and the lower western one, Roxas Range (Fig. 79). The plain extends in a northwesterly direction from Sarangani Bay to Lake Buluan. About 80 kilometers long, and from 10 to 12 kilometers wide, it includes an area of about 97,000 hectares. The presence of elevated Pleistocene coral reefs, as in the vicinity of Fort Pikit, to the northwest of Lake Buluan,[23] indicates that the area southwest of a line from the mouth of the Mindanao River to Sarangani Bay, was once separated by water from the rest of Mindanao. Coralline limestone has also been found in several places in the hills a short distance north of Sarangani Bay. Geologically recent uplifts raised the shallow channel above sea level, and rivers began to deposit their sediment. The part of the old channel that connects Lake Buluan and Sarangani Bay formerly received considerable quantities of volcanic material erupted by Mt. Matutum (Fig. 95), a Pleistocene volcano in the Quezon Range. Koronadal is thus a raised plain that as recently as the Pleistocene epoch was a shallow sea channel.

The topography of Koronadal suggests a division into three parts: Southern Koronadal, Middle Koronadal and Northern Koronadal.

Southern Koronadal includes parts of the plains of Buayan and Makar and is bounded on the east by the Buayan River and on the west by the Klinan and Siluay rivers. It rises gently from the shore of Sarangani Bay northward to Napol Hill in front of the Quezon Range and northwestward toward Polomolok. The latter lies within Middle Koronadal at an elevation of about 300 meters above sea level. The soils of Southern Koronadal consist of ash-gray to nearly black, loose, volcanic sandy loam, very fine sandy loam, and coarse silt loam. They range in thickness from 25 to 30 centimeters.

Middle Koronadal is higher than either Southern or Northern Koronadal. As it lies near Mt. Matutum, whose deposits are of great depth, the general elevation (between 300 and 600 meters) is higher than that of Southern Koronadal. The soils have a greater depth, ranging from 30 to 40 centimeters, and consist of dark brown to very dark gray, loose, granular volcanic sandy loam mixed with a considerable amount of organic matter. The subsoil is formed of fine sandy loams and fine sand containing gravel of volcanic material.

Northern Koronadal slopes gently down toward Lake Buluan. Its soils, 30 to 40 centimeters deep, consist of very fine sandy loams and clay loams, dark gray to nearly black, with a clay loam forming the subsoil.

The word Koronadal is believed to be a corruption of two Magindanao words: *koron* or *kalon*, meaning cogon grass, and *nadal* or *dalal*, meaning plain.[24] As the name would seem to indicate, the dominant vegetation consists of grasses such as cogon (*Imperata cylindrica*), *talahib* (*Saccharum spontaneum*), and *silibon* (*Themeda triandra*). The last is especially widespread and makes good pasture when young. It needs less rainfall than

---

23 Smith: Geology and Mineral Resources of the Philippine Islands, 1924, p. 203.
24 Rosell: "The Koronadal Valley," 1939, p. 493.

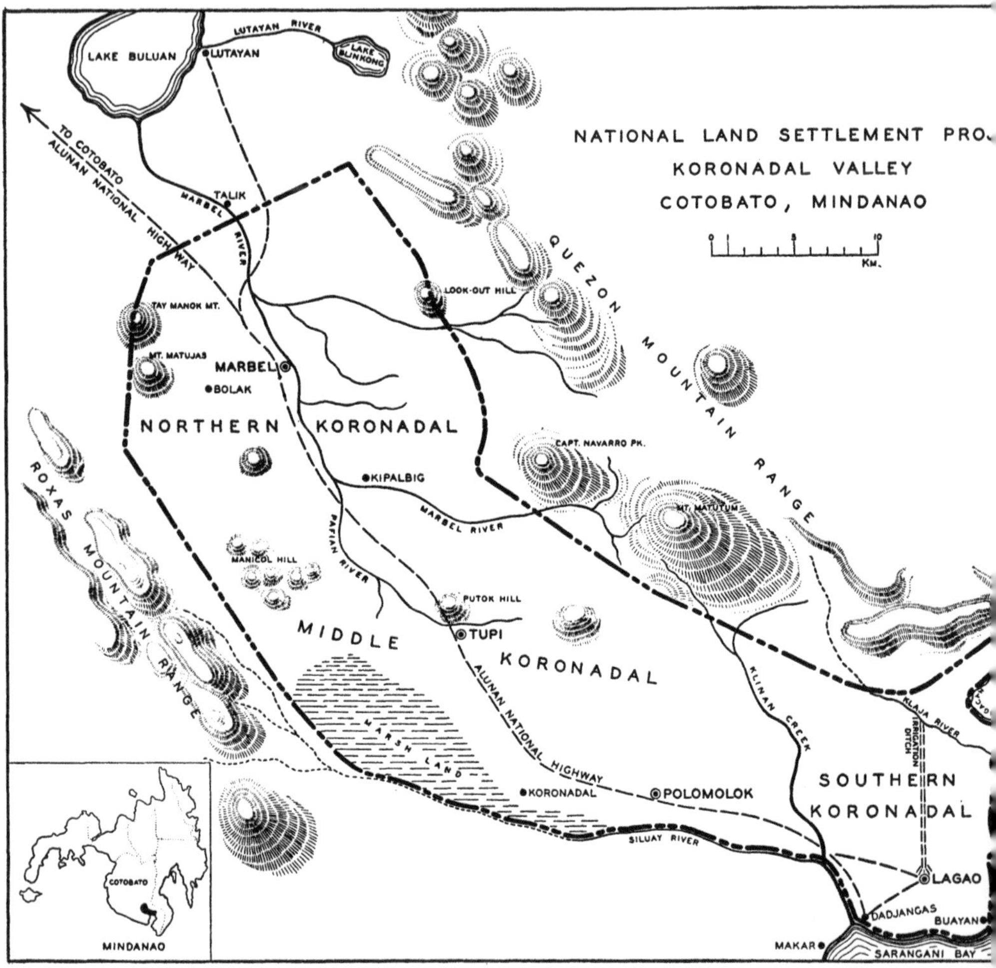

FIG. 119—Sketch map of the Koronadal National Land Settlement Project, based on a blueprint map of the N. The surveying of the area was incomplete at the time this map was drawn, and therefore the scale cannot be us precision in measuring distances upon it. For Cotobato, *read* Cotabato.

cogon or talahib and is frequently an indicator of a hardpan that prevents the free circulation of water.

Southern Koronadal is almost completely covered by grass, although *buri* palms are rather plentiful. It is not known whether this part of Koronadal ever had a forest cover extending beyond the narrow belts along the rivers. In Middle and Northern Koronadal there are occasional limited areas of primeval forest and more frequent patches of second-growth forest, especially along the foothills of the Quezon and Roxas ranges. Bamboos are also common in these parts and offer excellent and cheap building material. They are absent, however, in the southern part of the plain. It is probable that Middle and Northern Koronadal were once completely covered by forest vegetation that has since been largely destroyed by the shifting cultivation of the Moros and Bilaans. According to Dr. E. B. Copeland, Northern Koronadal had a considerable Moro population about 35 or 40 years ago, which would explain the extent of the present-day cogonales of Northern and Middle Koronadal. However, wars among various Magindanao *datos* have resulted in serious depopulation. After a siege, the inhabitants of the town of Talik were completely annihilated.

There are no statistical data on the climate of the Koronadal plain which, with the possible exception of its northern section, lies in that part of Mindanao that has no dry and no very pronounced rainy season. Rainfall is more or less evenly distributed throughout the year.

Compared with the rest of the plain, Southern Koronadal is considerably drier, as is indicated by the presence of silibon, mentioned above. In the middle of the day a strong, southerly wind blows across the plain of Southern Koronadal from the bay. Since it raises a great deal of dust in the fine loam and sandy soil, the planting of windbreaks would be advisable. In addition to their protective function, these would also supply firewood, of which there is a shortage. Because of its higher elevation, Middle Koronadal is somewhat cooler than the plains to the north or to the south. Here the rainfall is heavier, and rainy days are more numerous. The sky is frequently overcast, as Mt. Matutum gathers a daily cloud cover. It is said that Northern Koronadal has pronounced dry and wet seasons.

Koronadal can best be reached by ships approaching through Sarangani Bay. When the Commonwealth Government began its ambitious road-building program the Alunan National Highway was planned from the head of Sarangani Bay past Lake Buluan to Dulawan. By the time the N. L. S. A. began its activities the Bureau of Public Works had just begun construction of this road. In the days before the highway was constructed it was possible, during the dry season, to drive a car that had sufficient clearance over the grassy plains from Sarangani Bay to the shores of Lake Buluan. From there one could proceed by launch to Cotabato.

By 1941 there were two or three interisland steamers calling at Dadjangas (Fig. 82) weekly, while others called only at Glan, near the southern end of Sarangani Bay, whence Dadjangas could be reached by motor boat.

The Arrival of the Settlers

Upon their arrival in Koronadal, the settler and his family were assigned to one of the several settlement districts, to which they were transported by truck (Fig. 83). For the first few weeks they were housed in plain, long sheds or bunkhouses (Fig. 84) divided into small sections, each accommodating one family. The administration distributed rations once a week in quantities dependent upon the size of the families (Fig. 88). Rations were modeled on those of the Army. A check was made of the settlers' belongings, especially cooking utensils, tools, *bolos*, lanterns, axes, and farm implements (Fig. 87). Incomplete equipment was brought up to standard. Soon after their arrival all newcomers were examined by a doctor and inoculated against cholera and dysentery. Mosquito nets were issued in malarial districts. The officer in charge of the settlement explained in a public meeting the duties of the new settlers and assigned to each a home lot, usually 2,000 square meters in size.

The first task of the settler was to build a house so that his family could free the bunkhouse space for newer arrivals (Figs. 84–86). If there were no lumber available on the lot or in the neighborhood, the administration supplied the material necessary for a simple, Filipino-styled house, usually built of bamboo and roofed with cogon grass or the leaves of the nipa palm. The settler had full freedom in regard to the design of the house; and it was often possible for an observer familiar with Philippine house types to determine the home province of a settler by the style of his dwelling.

As rapidly as possible the home lot was prepared for cultivation and planted with vegetables and fruit trees. Seeds and seedlings were advanced by the warehouses and nurseries of the administration. It was suggested that a model home lot should have 36 banana trees, 12 papayas, a chico, a breadfruit, and a mango, in addition to such vegetables as eggplant, tomatoes, and squashes, and such tubers as sweet potatoes, yams, taro, and cassava. Most of the home lots contained an even larger variety of fruit trees and vegetables, depending upon the taste and industry of the occupants. Settlers were encouraged to keep small livestock, such as pigs and poultry. Breeding-stock came from the piggery and the poultry farm of the administration.

Within the course of a few months the settler was able to obtain from his garden all the vegetables and fruits he needed for his family, together with some eggs and occasionally a chicken. A well built house and a well developed home lot did a great deal to bind the settler to the new district and make him content.

The assignment of the farm lots was the next step (Fig. 91). At the beginning all farms measured 12 hectares, but in May 1940 it was decided to reduce those along the highway to 6 hectares, since the average Filipino family finds it difficult to cultivate 12 hectares.

In order to help the settler to begin the cultivation of the farm land as quickly as possible, the N. L. S. A. used such mechanical equipment as tractors, disk plows, and harrows to prepare at least 4 or 5 of the 12 hectares for the settler, who was then charged with the actual cost of their use (Fig. 101). Thus he was able to bring a large portion of his farm land under crops during the first planting season. The remainder of the land was broken by the settler with the help of his carabao and iron plow. In Koronadal much of the land was covered by cogon grass that could be burned away and tractor-drawn disk plows could then break this land easily. The cost of preparing cogon-covered land for cultivation was rather low as compared with that of preparing forested land.

In the 3 years of its operation before the war, the N. L. S. A. had no occasion to clear extensive primeval forest. In one or two settlement districts it was necessary to clear some first- and second-growth forest. In these instances mechanical equipment (Fig. 102) was used to pull out the stumps.[25]

It was the plan of the administration that about one-half of the land should be devoted to subsistence crops, such as rice, maize, sweet potatoes, and cassava, and the other half to cash crops, such as rubber, coffee, cacao, cotton, soy beans, peanuts, cowpeas, Bermuda onions, and others for which there was a market in the Philippines. The type of cash crops to be recommended would depend upon conditions of soil and climate in the different settlement districts. At the time of the outbreak of the war the choice of cash crops was still of great concern to the administration which was opposed to the usual cash crops of the Philippines—tobacco, abaca, coconuts, and sugar—as these were grown almost exclusively for the foreign and especially the American markets and were faced by the prospective loss of the latter after 1946. Since the cash crops should enable the settler to pay off his debt to the N. L. S. A., the right choice of these crops was of the utmost importance not only to the administration but also to the settler.

---

25 In the Compostela-Monkayo area, however, forest clearing would have been necessary on a large scale. The survey party estimated that there was an average of 2,500 trees per hectare in this region. It was decided to have settlers work together in groups of 24 under a foreman. The costs of clearing were to be prorated by farms. It was thought that, once the trees were cut, one hectare could be cleared within a week by three stump-pullers if two carabaos were used on each puller. This estimate seems too optimistic, for it means that each pulling team would have had to remove 17 stumps per hour if it operated for six 8-hour days. The Compostela-Monkayo project had not been started, however, by the time the war broke out, as the road that was to open the region was still unfinished.

A married settler received a credit of 1,000 pesos and a bachelor one of not more than 600 pesos. This credit was gradually established as the settler gave evidence of making progress. Theoretically the debt was covered by the house and improvements on the land as collateral, so that if the original settler should leave, another could take over the major part of his obligations. The administration could stop issuing credit to a settler whose

TABLE XI—KORONADAL: ACCOUNTS OF TWO SETTLERS WITH THE N. L. S. A., AS OF JULY 26, 1940

ITEM	CREDIT EXTENDED BY THE N. L. S. A. TO— Settler A*	Settler B*
Food	₱174.76	₱142.76
Clothing	27.01	32.59
Construction material and help	197.58	205.79
Planting material	34.32	85.51
Passage to Mindanao	81.60	112.20
Transportation from harbor to settlement region	———	6.13
Farm implements	12.41	19.68
Work animals	73.00	142.34
Miscellaneous	56.05	46.89
Total	₱656.73	₱793.89
Credits (labor)	₱74.70	₱78.60

* Settler A arrived on June 19, 1939, and Settler B on June 12, 1939.
Source: Unpublished records of the N.L.S.A.

account grew faster than his collateral and so could oblige him to work for a daily wage from 5 o'clock in the morning until noon. This left him time to work for several hours for himself but prevented his debt from increasing. He received his wage in the form of rations and not in cash.

In recognition of the absolute necessity for keeping the accounts as low as possible, an attempt was made to calculate the minimum requirement of food, clothing, implements, and other items. Settlers were directed to present written requests for food and other supplies so that these could be checked and approved, or disapproved. The stores sold only inexpensive merchandise to the general public. Canned milk, for example, was kept only for babies, or for medical patients whose need of milk had been certified by the doctor.

The accounts of the first groups of settlers were not carefully supervised and grew relatively large. By June 30, 1940, 30 settlers each had debts exceeding 700 pesos. One had already passed the thousand-peso limit, and two were approaching it. An indication that the settlers' requests for supplies, at least during the first year, were not handled systematically or supervised strictly, may be seen in Table XI, which shows the accounts of two typical settlers who had been in Mindanao for just one year. Although Settler A had only two dependents, whereas Settler B had six, the food item was smaller for the larger family.

The Settlement Districts

The N. L. S. A. divided Koronadal into four settlement districts: Lagao, Polomolok, Tupi, and Marbel. Each of these has one town, for which the district has been named, representing the administrative and economic center around which the villages, or barrios, are located. The first settlers to be sent to a newly opened district were placed within the area set aside for the town. After all the home lots in the town were occupied, the new arrivals were directed to the locations chosen as sites for the barrios. The latter had a smaller number of lots than had the towns. This division into town and barrios was necessary in order to spread out the population so that the individual farmer should not be too far away from his farm land.

Each district was under an overseer and a staff of assistants. The overseer lived in the town, where he had a small office. The store, hospital, church, school, administration warehouse, tool and machinery sheds, and whatever other buildings were needed either by the community or by the N. L. S. A. were also in the town. Each barrio was under the charge of a barrio lieutenant and an assistant barrio lieutenant, who were chosen by the overseer from the ranks of the settlers on the basis of leadership, intelligence, and industry.

Lagao. This district takes in Southern Koronadal and has a total area of approximately 30,000 hectares, of which only about 10,000 were available for distribution to settlers. The town of Lagao, established on March 3, 1939, lies 3½ kilometers northeast of Dadjangas, the port of Koronadal, and was the administration and distribution center of the whole Koronadal project of the N. L. S. A. Hence General Santos had his office in Lagao (Fig. 109); and a number of other houses there were occupied by employees of the N. L. S. A. As a supply center, it had an especially large warehouse, store, and machine shop. In this shop farm machinery (Fig. 97), trucks, and motor cars were overhauled and repaired.[26] A power plant supplied power for the machine shop and for the maize and rice mills, as well as light for the community. The town consisted of three barrios, closely grouped together (Fig. 95).

Lagao was chosen as the supply, distribution, and administration center for a number of reasons. It is near Dadjangas which, though not included in the settlement district, served as the landing point for settlers and goods and was the terminal of the national highway between Sarangani Bay and Dulawan on the lower Cotabato River. It is free from malaria—which cannot be said of the settlements further inland.

The settlement at Lagao was started by the N. L. S. A. with 62 settlers and 14 employees. But the settlers had been recruited in Manila too quickly. Many of them were unmarried and had had little agricultural experience, having been taxi-drivers, craftsmen, professional workers, or laborers. A few of them soon gave up, discouraged by the dryness of the light soil of Lagao. As has been mentioned, the members of the survey party

[26] At the beginning of 1941 Koronadal had the following mechanical equipment or machinery: 49 tractors, 18 trucks, 9 "pick-up" trucks, 3 passenger cars, 1 motorcycle, 2 station wagons, 25 plows (4, 5 or 6 disks), 16 harrows, 9 planters, 1 thresher, 1 haybaler, 1 cotton gin, 1 maize mill, 1 rice mill, 1 battery charger, 1 road grader, 2 peanut shellers, 2 light plants and 3 windmills.

had pointed out the absolute necessity for an irrigation system if the plain of Lagao was to produce crops, but at first this advice was disregarded, so that by June 1939 drought was endangering the gardens started by the settlers. Pessimistic reports reached Manila, and newspapers began to speak of the failure of the Koronadal colony.

General Santos, however, ordered the immediate construction of an irrigation canal large enough to water at least the home lots of Lagao. This canal, which tapped the small Klaja River 9 kilometers north of Lagao, was started on July 3 and finished on August 19, 1939. It had a discharge of 130 liters per second. This small system not only saved the gardens of Lagao but also gave new courage to the settlers, as well as to the administration, and checked the outburst of criticism directed against the N. L. S. A. The Klaja was, however, too small to supply irrigation water for the farm land. Early in 1940 engineers began to survey an irrigation system for distributing the waters of the Siluay River and its main tributary, the Klinan, over this farm land (Fig. 99). The Siluay, according to the records of the engineers, has an average flow of 6,300 liters per second; the maximum observed was 11,250 and the minimum, 4,400. The construction of the main canal was started in July 1940. The system was to begin operation in July or August 1941 and was expected to supply about 4,000 hectares of land with sufficient water to ensure favorable conditions for the growth of cotton, Bermuda onions, soy beans, peanuts, and cowpeas on a large scale.

The experts consulted by the N. L. S. A. considered Lagao especially well suited to the mechanized cultivation of cotton. Experimental plantings of a variety called Batangas White[27] were made from June to December 1939 in order to determine the best time for planting, which was found to be September and early October. During 1940 the cotton area amounted to only 80 hectares, but in 1941 the administration planned to plant as much as 500 hectares of cotton on its own farm and to suggest that each settler in Lagao plant one or two hectares as well.

Agricultural experts of the N. L. S. A. were convinced that Lagao farm lots would bring the highest financial returns of all the farms in Koronadal when the irrigation problem had been solved. Nevertheless, many of the incoming settlers preferred to move to one of the districts farther inland as they were anxious to get land that was suitable for the customary staple crops, such as rice and maize, for which the Lagao soil, although it can support these crops, is not particularly well adapted. In February 1941 Lagao's settlers had increased from the initial 62, who, with 6 dependents, had landed on February 27, 1939, at Dadjangas, to 582, with 1,709 dependents, who were distributed over eight barrios clustered around the new town of Lagao. In addition, 85 employees, with 135 dependents, lived in the town.

Polomolok. The townsite of Polomolok lies 21 kilometers northwest of Lagao. This district was not opened until late in 1940, because the presence of malaria and the lack of easily accessible drinking water made special preparations necessary (Figs. 93 and 100). It includes 18,000 hectares, of which about 13,500 were considered for distribution. According to agricultural experts, Polomolok is especially well suited to upland rice and maize as subsistence crops, and to rubber, cacao, and soy beans as cash crops. By the end of February 1941, 5 months after its opening, the district had 347 settlers with 718 dependents, living in four barrios.

Tupi. About 19 kilometers northwest of Polomolok lies the townsite of Tupi, extending for 3 kilometers on both sides of the new highway. It is the center of the third settlement district. This district comprises 23,000 hectares, of which 14,000 were marked for distribution. The site chosen for Tupi was covered by dense second-growth forest. The first settlers were sent here on July 3, 1939; clearing was begun in September. By the end of 1939 the employees of the Bureau of Public Works, who were constructing the highway, reached Tupi and cleared an additional strip 15 meters wide (Fig. 96).

In July 1940, when I passed through the town, the clearing done by the settlers had made considerable progress. The home lots already contained houses and gardens. The farm lots, however, had not even been surveyed, although settlers had been there a year; for the administration did not have enough surveyors to take care of all the settlers at the rate at which they were coming in. The older settlers of Tupi, whose home

---

[27] A variety introduced by the Spaniards from Mexico that is similar to the Acala cotton grown in California.

lots were already developed, were therefore given two hectares of plowed land on a temporary basis so that they might begin field cultivation and become self-sufficient. A report dated February 1941 showed, however, that many of the Tupi settlers were still waiting for their farm land.

The main subsistence crops of the district are upland rice and maize, the latter yielding two or three crops (Fig. 104). The most promising money crops are rubber, coffee, cacao, black pepper, soy beans, and cowpeas.

Near the barrio of Kipalbig, established in May 1940 about 9 kilometers northwest of Tupi, the administration set aside 1,000 hectares for a rubber plantation under the management of a young Filipino who had formerly been in the employ of a rubber plantation in Borneo. It was planned, furthermore, that each Tupi settler should plant several hectares with rubber as soon as the nursery could supply him with budded seedlings.

Early in 1941 Tupi had 582 settlers, with 1,891 dependents, and 39 employees of the N. L. S. A., with 58 dependents, living either in the town, in one of the four outlying barrios, or on the rubber plantation.

Marbel. The town of Marbel, founded on January 10, 1940, lies on the highway 27 kilometers northwest of Tupi and about 70 kilometers northwest of Lagao (Fig. 110). The district includes 21,000 hectares, 14,500 of which were to be given to settlers. Because of Marbel's topography and climate, N. L. S. A. officials believed that it would become the rice and maize granary of Koronadal. The district is well suited to both upland and lowland rice (Fig. 103) as well as to maize. At the time of my visit to Marbel the question of additional perennial cash crops had not yet been decided.

This district's population increased much more rapidly than any other in Koronadal; after one year it had 956 settlers with 2,930 dependents and 13 employees with 12 dependents, who were living either in Marbel or in one of the seven barrios located around the town (Fig. 94).

Administration Farms

The administration farms played an important role. They were experiment and demonstration stations as well as seed and livestock farms. Each of the settlement districts was eventually to have one farm on which to grow all crops suited to the particular area, with especial attention given to cash crops (Figs. 106 and 108). The planners hoped that in time the profits from these farms would contribute a large part if not all, of the administrative expenses. In the early years these farms offered employment to settlers whose credit had to be limited. The main function of the farms was to produce the seeds needed in large quantities by new settlers. Here agriculturists experimented with new varieties of standard crops and with the introduction of new crops before settlers were encouraged to plant them. A promising start was made with a thousand-hectare rubber farm just a year before the Philippines were attacked (Figs 105 and 107).

The largest hog farm in the Philippines was managed by the N. L. S. A. (Fig. 113). This project, near Lagao, under the supervision of Dr. Gil Pabillo, a veterinarian, was begun early in 1939 with 18 sows and 2 boars bought from the Bureau of Animal Industry and the San Ramon Penal Colony for 1,177 pesos. The objectives of this venture were first to provide each settler with a pair of good breeding pigs and thereafter to produce pigs for the market at the rate of 10,000 a year. After 2 years of operation the farm had 600 pigs of all sizes, excluding those that had been sold to settlers, and the value of the stock had risen from the original 1,177 to about 25,000 pesos. A five-year program was intended to increase the number of pigs to 18,000, worth half a million pesos. It was estimated that the profits of the hog farm alone would be great enough to pay the salaries of the employees of the N. L. S. A. Dr. Pabillo pointed to a number of factors that promised success: absence of animal diseases, suitability of climate, low costs of fodder and building materials, and a good market.

TABLE XII—GROWTH OF POPULATION IN THE KORONADAL VALLEY PROJECT FEBRUARY 1939 TO FEBRUARY 1941

		NUMBER OF RESIDENTS*					
YEAR	MONTH	Employees	Dependents	Settlers	Dependents	Total	Monthly Increase
1939	February	14	...	62	6	82	....
	March	14	....	65	8	87	5
	April	18	17	79	61	175	88
	May	21	31	102	80	234	59
	June	21	35	150	148	354	120
	July	21	44	184	203	452	98
	August	23	39	195	227	484	32
	September	26	39	200	246	511	27
	October	28	39	209	259	535	24
	November	29	39	217	282	567	32
	December	29	39	289	576	933	366
1940	January	43	55	419	1,169	1,686	753
	February	43	55	513	1,272	1,883	197
	March	43	55	568	1,473	2,139	256
	April	52	62	607	1,610	2,331	192
	May	62	61	970	3,033	4,126	1,795
	June	72	76	1,221	3,640	5,009	883
	July	72	76	1,574	4,346	6,068	1,059
	August	75	83	1,924	5,046	7,128	1,060
	September	79	142	2,044	6,332	8,597	1,469
	October	82	145	2,134	6,458	8,819	222
	November	88	148	2,151	6,648	9,035	216
	December	118	164	2,410	7,231	9,923	888
1941	January	145	185	2,467	7,248	10,045	122
	February	150	190	2,513	8,163	11,016	971

* This table does not include the following groups:

Hired laborers	239
Dependents	62
Probationary settlers for Ala Valley	1,207
Dependents	389
Total	1,897

The probationary settlers came into the valley in December 1940 and January and February 1941, at their own expense.

Source: Unpublished records of the N. L. S. A.

The poultry farm of the N. L. S. A. was not so successful as the hog farm because of diseases. There were further plans calling for a cattle and carabao ranch on which to raise work animals for the settlers. A horse-breeding station, for which high-bred Australian mares and stallions were imported, had already been started. Then there were plans for producing large quantities of vegetables, such as tomatoes and cucumbers, for a cannery to be established and operated by the National Development Company.

All of these projects were entirely sound, since the N. L. S. A. was concentrating chiefly on those crops and products that were still being imported into the Philippines.

Growth of the Koronadal Settlements

The existence of the N. L. S. A. was endangered in July and August of 1939 when critics of the government were speaking of the "failure" of the new colonization project; it weathered the storm, however. As the first year progressed, the staff of the N. L. S. A. learned from its mistakes and worked out a better technique in locating settlers and getting them started. Recruiting officers became more skilled in selecting the right type of settlers. Within the first 12 months the number of settlers grew from 62 to 513. During the second year exactly 2,000 settlers were added. Table XII shows the monthly growth in the number of settlers as well as of employees.

If the original plan of giving each family 12 hectares had been carried out, the 2,513 settlers present at the end of February 1941 would have occupied about 30,156 hectares of the available 52,500. In May 1940, however, the directors of the N. L. S. A. decided to limit to 6 hectares the land allotted to any settlers whose farms were located along the national highway. It was felt that these farmers had an advantage over those whose farms were less conveniently located. There are no data available on how many additional farms might be located along the highway as a result of this ruling, for the surveys were far from complete when the war interrupted all activities. The total could not, however, have been more than 350 farms, assuming that each highway farm measured 200 by 300 meters, instead of 400 by 300, as formerly. If all farms had been kept at 12 hectares, Koronadal would have had room on its second anniversary for only an additional 1,900 settlers, or less than the number accepted during its second year of operation.

At the end of February 1941 an additional 1,207 settler-applicants had accumulated in the four districts of Koronadal, especially in Marbel. They had come at their own expense and were clamoring for admittance. The N. L. S. A. directors, faced with an urgent need for new settlement areas, decided to expand by taking in the Ala Valley, which runs parallel to Koronadal and is separated from the latter by the Roxas and Matujas ranges. Like Koronadal, the Ala Valley is largely covered with grasses. According to reports, its potentialities are comparable to those of Tupi and Marbel. It has about 50,000 hectares that could be subdivided.

During the first months of 1941 the N. L. S. A. employed the settler-applicants who were waiting in Northern Koronadal to build a road from

Marbel into the Ala Valley and promised to locate them there under the same conditions as those enjoyed by the settlers of Koronadal.

Home Provinces of the Koronadal Settlers

The realization of one of the main purposes of the N. L. S. A., the reduction of pressure in overpopulated provinces, was hindered by political considerations. Assemblymen who felt that their constituents should receive an equal share in the benefits of the N. L. S. A. wrote the following provision into the law:

> The applicants shall be recruited *from all provinces in proportion to their respective population* [italics are mine], and in case a province shall not be able to fill the quota assigned, the unfilled portion of the quota may be covered from other provinces having greater number of applicants.[28]

Because of this provision and because of political pressure, the recruiting officers at one time concentrated their efforts for several weeks on the province of Tayabas, although Tayabas is an area of immigration rather than of emigration and there is no need for the peasants of this province to turn to Mindanao in search of land. On the whole, however, the administration did not pay too much attention to this provision, knowing it had been inserted purely for vote-getting purposes. Although 41 out of a total of 50 provinces in the Philippines were represented by the settlers in Koronadal, the Visayan provinces of Iloilo, Leyte, Cebu, Capiz, and Bohol furnished nearly 70 per cent of the settlers. The Luzon provinces of Pampanga, Pangasinan, and Tarlac came next in importance. Table XXI[29] shows the home provinces of the settlers in the various settlement districts.

Expenditures of the National Land Settlement Administration

By December 31, 1940 the government had released 4 million pesos out of the 20 million that had been set aside for the N. L. S. A. According to a statement of the comptroller, the N. L. S. A., by December 31, 1940, had received an income of 89,531.06 pesos from interest charges, electric-light fees, charges for transportation and for preparation of settler farm lots, sales of seeds and seedlings, and various other sources. This increased to 4,089,531.06 pesos the total amount that had been at any time available for expenditure.

Out of this sum, 1,836,616.02 pesos had been expended by December 31, 1940.[30] The main items were: salaries, 202,819.87 pesos; wages, traveling expenses, freight and handling charges, rentals of buildings and offices, postal service, and other miscellaneous expenses, 233,079.90 pesos;

28 Commonwealth Act No. 441, Sec. 3c.

29 Table XXI is found in Appendix C, p. 253.

30 By March 15, 1941, this amount had increased to 2,233,250.00 pesos. These and the accompanying figures in the text are from an unpublished report, drawn up by the accountant of the N. L. S. A.

costs of operating the various administration projects such as the piggery, poultry farm, and seed farms, 71,222.16 pesos; buildings, farm machinery, transportation equipment, office furniture, technical and scientific equipment, hospital and dispensary instruments, 534,928.17 pesos; advances to settlers, 794,565.92 pesos.[31]

On December 31, 1940, the administration had 2,262,049.37 pesos, either in banks or on hand in its offices. To this should be added 138,044.41 pesos, representing accounts receivable and inventories in stores and warehouses. The accounts payable at the same time amounted to 147,178.74 pesos.

For the 17-month period from January 1939 to June 30, 1940, the directors of the N. L. S. A. approved a budget of 1,187,311.00 pesos. The regular budget of the N. L. S. A. for the fiscal year, July 1, 1940 to June 30, 1941, allowed for an expenditure of 813,870.00 pesos; and supplements approved on July 30, September 12, November 16, and December 24, 1940, increased this sum to 2,031,714.28 pesos. The full amount of the appropriation for the 1940–1941 fiscal year is not known, as further additions for projects such as the Siluay irrigation system were necessary. The original estimate of the costs of this irrigation system called for 131,926.00 pesos; but in the spring of 1941 the administration asked for an additional 129,555.74 pesos. This was but one instance of the directors' having to approve an increase in the budget. In the budget for 1940–1941, as drawn up on December 24, 1940, 41.8 per cent was allocated for advances to settlers; 31.3 per cent for the purchase of equipment, construction of buildings, and similar outlays; 12.6 per cent for administrative salaries and wages; and the remainder for operating and general expenses. This distribution of expenditure did not depart greatly from that of the first 23 months.

It is probable that by the outbreak of the war the N. L. S. A. had expended between 4 and 5 million pesos, a great part of which had been advanced to settlers or had been used for machinery, buildings, wells and irrigation systems, and to give the settlers health protection.

The time had, of course, been too short for the N. L. S. A. to receive a return on its investment in the form of payments from settlers on their debts or in the form of large sales of cash crops from administration farms.

Summary and Appraisal of the Work of the National Land Settlement Administration

The National Land Settlement Administration was one of the many activities of the Commonwealth of the Philippines designed to strengthen the economy and social structure of the Philippine nation. The government was making concentrated efforts to develop Mindanao's resources. An energetic road-building program was begun immediately after the establishment of the Commonwealth. The Bureau of Lands at the same time at-

31 By March 15, 1941, advances had increased to 967,551.81 pesos.

tempted to speed up surveys and divide the newly opened territory. This led to increased migration, especially of Visayans. Not content with the progress of settlement in Mindanao, and harassed by agrarian disorder in central Luzon, the government established the National Land Settlement Administration to encourage pioneer colonization and to help solve social and economic problems in central Luzon and the Visayan Islands.

Final judgment and critical evaluation of the activities of the N. L. S. A. are impossible because the work was interrupted by war before the objectives had been reached. It would be unfair to compare the number of settled families with the total amount of money expended, since large sums were spent on heavy equipment and machinery which was to be used in project after project. The N. L. S. A. had initial difficulties—what pioneer undertaking does not experience birth pains? It had, however, acquired valuable experience, developed a working routine, and assembled a staff which had become familiar with the problems of opening up and colonizing pioneer country, when the Japanese invasion disrupted everything.

The venture of the N. L. S. A. was a democratic experiment in a very new country. It profited by the enthusiasm and optimism of a young people enjoying for the first time the privilege of working out their own destiny, but it sacrificed some of the efficiency achieved by the Netherlands Indies for the sake of establishing democratic procedures. Its position in the limelight and its dependence upon the good will of the country's legislators made quick, spectacular results imperative.

The greater part of the N. L. S. A.'s achievements can be credited to the choice of General Santos as manager. This energetic and resourceful personality was a constant inspiration to his all-Filipino staff. Together they devoted all of their energies to making the N. L. S. A. a success.

The goal was to create a class of agriculturists who would be more than subsistence farmers working to cover their barest needs, as is the case with most rural Filipino families. It was to achieve this goal, that the N. L. S. A. allowed each settler a liberal amount of land to enable him to produce, in addition to his own food, cash crops such as cotton, rubber, coffee, peanuts, soy beans, and onions. Recognizing the great need for diversification in agriculture, the administration discouraged the cultivation of the usual Philippine cash crops (abaca, tobacco, sugar, and coconuts), encouraging instead the introduction of products in which the country was deficient.

Centralized marketing was a very important feature in the scheme of the N. L. S. A. This assured the settler much better prices than he could get from the middlemen. Liberal credit, low interest rates, and a long period of amortization effectively protected the settler against moneylenders and land-grabbers. The activities of these last two groups were further discouraged by the stipulation that a settler could not turn over his land to anyone but his descendants for a period of 10 years after he had been granted title. Past experience had shown that many pioneer homesteaders lost their land and years of work through their ignorance of land laws and the machinations of the usurers and land-grabbers.

Of cardinal importance were the measures taken by the N. L. S. A. to protect the health of its settlers. Success or failure depended largely upon the ability of the medical staff of the administration to fight disease, above all malaria, and to keep the death rate at a normal level.

The bringing together into agricultural villages of people from all parts of the islands served to break down provincial barriers and to create a feeling of nationhood. It was also instrumental in spreading Tagalog as the national language (Tagalog has been declared the national language and will be the language of instruction in schools after independence is achieved).

Many of the difficulties in Koronadal developed from the haste of the N. L. S. A. in rushing the first settlers to Mindanao before the recommendations made by the experts in the survey party that had selected Koronadal could be properly studied. Moreover, no surveyors or road constructors had preceded these settlers. Indeed, when the administrative staff first arrived in Dadjangas it was accompanied by the initial group of colonists.

Glowing accounts appeared in the press about the bright future of the pioneers, and the impatient public in Manila expected dramatic results within a few months. When the reports from Koronadal showed that pioneering in Mindanao was not a bed of roses, pessimists referred to the N. L. S. A. as a failure, before giving General Santos and his men time to become familiar with their new task in Koronadal and with the obstacles that faced them there.

Experts had stressed the absolute need of irrigation in Southern Koronadal; youthful optimism ignored the warning. The first settlers were placed in Lagao and saw their freshly planted gardens suffer severely from drought. At that moment General Santos and his office staff joined the settlers in the field and helped to dig a small irrigation system. Both the settlers and the staff of the N. L. S. A. showed true pioneer spirit! But it might well have been wiser to have placed the first settlers in Middle and Northern Koronadal, where irrigation systems were not required. That would have given an opportunity for an auspicious start there and at the same time in the Lagao district for surveying and for constructing an irrigation system before the first settlers were brought in.

Settlers had to wait so many months for their farm land that they were dependent upon support from the administration much longer than would have been the case had the land been surveyed and subdivided before their arrival. This defect was soon recognized, however, so that in the Mallig project in Luzon (see Appendix A) colonists were not to be admitted until all property lines had been definitely fixed and suitably marked.

The settlers at Lagao were encouraged to build houses larger and more expensive than the usual Filipino ones. In most pioneer communities the first houses are not much more than shelters, to be replaced by substantial dwellings once the pioneers have weathered the early years. The administration, however, had to impress visitors, and above all visiting legislators upon whose vote so much depended. Nothing strikes the casual observer so favorably as well built houses. He does not see the account of the settler in the files of the administration! It would, of course, have been more economical to insist upon smaller and less costly houses at the beginning and to leave it up to the settler to make enlargements and improvements when his financial resources would allow him to do so; but in Lagao, the point beyond which visitors seldom venture, political and psychological considerations forced a certain amount of display. In Middle and Northern Koronadal the houses were simpler in style and considerably cheaper.

In Mallig, in Luzon—another spot fairly accessible to visitors—I noticed the same tendency towards showmanship. "We will save on the houses in the barrios off the main highway" was the answer of the administrator there when I asked him if he could allow every settler as much credit for his house as he allowed those with home lots in the direct neighborhood of the administration building.

The American has inculcated in the Filipino a great admiration for modern technology. In striving for quick results the N. L. S. A. relied heavily on mechanical equipment, all of which had to be imported from abroad. At times it seemed to me that the administration had been the victim of high-pressure salesmanship by agents for American farm machinery. In some instances machinery had been bought which was not even usable in Koronadal. Because of the generally prevailing high prices for imported American machinery, the low prices for Filipino agricultural products, and the low agricultural wages, modern farm machinery was not extensively used elsewhere in the Philippines. In 1941 the N. L. S. A. was without doubt the largest user of modern mechanical farming equipment in the islands. Since the average settler was inexperienced in handling the machines, the administration had to hire outsiders rather than settlers as mechanics and tractor drivers, thus adding to the cost of operation. The geographical isolation of Koronadal also made the machinery division expensive to operate. An unusually large stock of spare parts had to be kept on hand in order to prevent long delays in making repairs and replacements. Indeed, this tendency of relying heavily on imported equipment that could be operated only by specially trained men made Koronadal anything but self-sufficient as far as labor was concerned.

According to the plans of the N. L. S. A., the machinery was to be used only during the initial stage of a settlement project in order to help the peasants bring their land under cultivation. After that the equipment was to be transferred to a new project and the settlers were to rely on the old standby of the Filipino farmer—the carabao. In 1940 and 1941 there was a great demand for carabaos in Koronadal, and it seemed more difficult to purchase good work animals than tractors.

Although the administration tried to remedy defects in its work wherever it recognized them, in some instances a lack of skilled technical personnel caused bottlenecks that slowed progress. Yet, on the whole, the few defects appear unimportant beside the commendable features of the N. L. S. A.'s activities.

Chapter VI

OVERCROWDED JAVA

Java and Madura suffer from acute population pressure. The latest census of the Netherlands Indies, held October 7th and 8th, 1930, showed 40,-891,000 Indonesians[1] on an area of 131,973 square kilometers in those islands. The average density of population was nearly 310 persons per square kilometer, if only Indonesians are taken into account, or 316 persons if non-Indonesian groups are included. A census had been planned for 1940, but it was cancelled because of the war. Having no indication that the rate of growth during the period from 1930 to 1940 would have differed from that of the previous decade, officials estimated that Java's population had probably increased to 48 million by 1940. If this rate of growth had been uninterrupted by war we might assume that there would have been a population of nearly 50 million in 1945, or a density of about 380 persons per square kilometer.

The Growth of Population

During the eighteenth century the population of Java seems to have been kept more or less stationary by internal wars, epidemics, the severe oppression of the people by their own princes as well as by the Dutch East India Company, and the general insecurity.[2] In 1795 the total population was estimated at 3,500,000;[3] in 1815 it had grown to 4,500,000;[4] during the next 30 years it more than doubled and by 1845 had reached 9,400,000. In 1880 it had again more than doubled, approximating 19,540,000, and the estimated total in 1940 is slightly more than twice that of 1880. Table XXII summarizes the population growth of Java and Madura and gives the average rate of yearly increase for the various periods.

This growth of population in Java since the beginning of the nineteenth century has been truly remarkable. Today Indonesians are ten to eleven times as numerous there as they were under the governorship of Sir Stamford Raffles during the British interregnum from 1811 to 1816. The growth set in when some of the repressive factors had been removed and all of Java had been brought under a strong and efficient central government. Boeke points out that the increase in the population of Java under the culture system began after "the last Java war [1825–1830] had been con-

Note: Tables XXI–XLVII are in Appendix C, p. 253 ff.

1 The Indonesians of Java and Madura are composed of three main groups in the following proportions as of 1930: Sundanese (20.7 per cent) principally in West Java, Javanese proper (65.6 per cent) principally in central Java, and Madurese (10.5 per cent) in East Java and Madura. All are often loosely referred to as Javanese.

2 Raffles: The History of Java, 1817, Vol. 1, pp. 66–73.

3 Bleeker: Nieuwe bijdragen, 1870, p. 7. In 1802 Engelhardt also estimated Java's population at 3,500,000.

4 Raffles: *op. cit.*, Vol. 1, p. 65.

cluded; the Western administration penetrated the interior; the division of the duties in respect to the culture service once more stimulated the desire for larger families."[5] This growth has slowed down somewhat in the present century. The exceptionally low rate of increase in the interval between 1905 and 1920 was partly a result of the bubonic plague but more especially of the influenza epidemic in 1918 and 1919; the latter took the lives of more than a million persons.

A similar increase in population in Western Europe and in Japan arose from changes in the economic and social structure; there the Industrial Revolution led to a growing urbanization and drew an ever-increasing proportion of the population into non-agricultural pursuits. In Java, however, the growth was not accompanied by such changes. Instead, the great mass of the people remained dependent upon agriculture and age-old village industries for subsistence and when the importation of manufactured articles caused a decay of even the native crafts and home industries, the families previously dependent on these skills were forced into agriculture. In recent years, however, Indonesian native crafts and home industries have seen something of a revival.

In 1816 Raffles estimated "the quantity of land in cultivation to be to the land still in a state of nature as one to seven."[6] Since then the forest has shrunk steadily under the axes of increasing numbers of peasants in need of more and more agricultural land. This is revealed in the statistics of the land tax. In 1882 the government collected land taxes on 2,608,054 hectares, in 1922 on 5,967,427, in 1930 on 6,623,638, and in 1938 on 6,942,137 hectares.[7] Yet the fact that the number of landowners who had to pay land taxes increased at a faster rate than the area subject to this tax indicates that the average amount of land per landowner was declining. In 1922 the average amount of land per taxpaying landowner stood at 1.15 hectares, but in 1938 it had shrunk to 0.86 hectare.

During the ten-year period from 1928 to 1938 the total area cultivated by Indonesians was enlarged by 4 per cent, yet between 1930 and 1940 the Indonesian population of Java is estimated to have grown by 20 per cent. In one case, the population of a district, Kebonongan, in the regency of Bantoel, Jogjakarta, is reported to have increased by 15 per cent between October 1930 and January 1940, but the area cultivated by the peasants expanded by only 4.2 per cent. In 1930 Kebonongan had a density of 840 persons per square kilometer and in 1940, of 961 persons.[8] These

---

5 Boeke: The Structure of Netherlands Indian Economy, 1942, p. 159.

6 Raffles: *op. cit.*, Vol. 1, p. 71.

7 These data do not include the principalities of central Java. The publications giving the statistics for 1930 and 1938 also report the area subject to the land tax in the principalities, where the totals for 1930 and 1938 are 7,253,469 hectares and 7,612,188, respectively. The data for 1882 and 1922 are from *Statistisch jaaroverzicht van Nederlandsch-Indië 1929*, 1930, p. 241. The figure for 1930 is from *Indisch verslag 1931*, 1932, Vol. 2, p. 237. That for 1938 is from *Indisch verslag 1939*, 1940, Vol. 2, p. 248.

8 Jonkers: Monographie van het landrentedistrict Kebonongan (ms. dated 1940), p. 22.

figures show that the increase in population is only partly compensated for by an expansion in the agricultural area at the disposal of the Indonesian peasant. The disparity in growth was especially great in the days of the culture system, under which the peasants were obliged to cultivate certain crops for the government on one-fifth of their land. At that time the people had no interest whatsoever in enlarging the area under cultivation, since the newly cleared land would also have come under the culture system and would only have increased their burden.

There is a limit beyond which the cultivated area cannot be enlarged at the expense of the forest cover without serious consequences. This limit has already been passed in Java, where the natural vegetation consists of forest. Foresters have estimated that 30 per cent of the land should be forest-covered in order to protect the water supply, so vital to the *sawahs* of the island. Instead, forests occupy only 23 per cent of the land. The Forestry Service has found it necessary in a number of instances to take land out of the control of Indonesian communities and under great difficulties to reforest it because it had become subject to serious erosion and a menace to adjacent land.[9]

Densities of Population

Java has thus become one of the most densely populated agricultural areas in the world, rivaled in this respect only by some districts in China, parts of the Ganges Valley, and the Nile Valley. There is considerable unevenness in the distribution of population, the densities ranging all the way from 20 to more than 1,600 persons per square kilometer. Table XIII gives the densities in 1930 in the residencies of the three provinces and the two principalities of Java. Most of West Java and East Java is less densely settled than the middle part of the island.

The Netherlands colonial administration divided Java into some 400 districts, for which the census of 1930 gives separate figures of density. These show that there is a close relationship between the type of soil (as well as the climate and the modes of lànd use) and the density of population. The richer the soil, the higher the yields, the percentage of cultivated land, and the density. Java's most fertile soils are found around young volcanoes: on the slopes, at the bases, and in the basins of the rivers that get water and silt from these volcanoes[10] (Map, Fig. 1A). Volcanic ash soils are so extraordinarily rich that the peasants risk the danger of new eruptions in order to cultivate the slopes as high up as possible. The higher parts of the slopes are less densely settled than the plains below only because of the difficulties in bringing steep slopes under cultivation.

9 A commission investigating the land rights of Indo-Europeans in 1937 estimated that there were about 300,000 hectares of arable land still available in Java, sufficient to take care of two-fifths of an annual increase in population of 500,000 for a period of 7 years, or of a total increase at this annual rate for 3½ years.

10 Mohr: "The Relation between Soil and Population Density," 1938, p. 479.

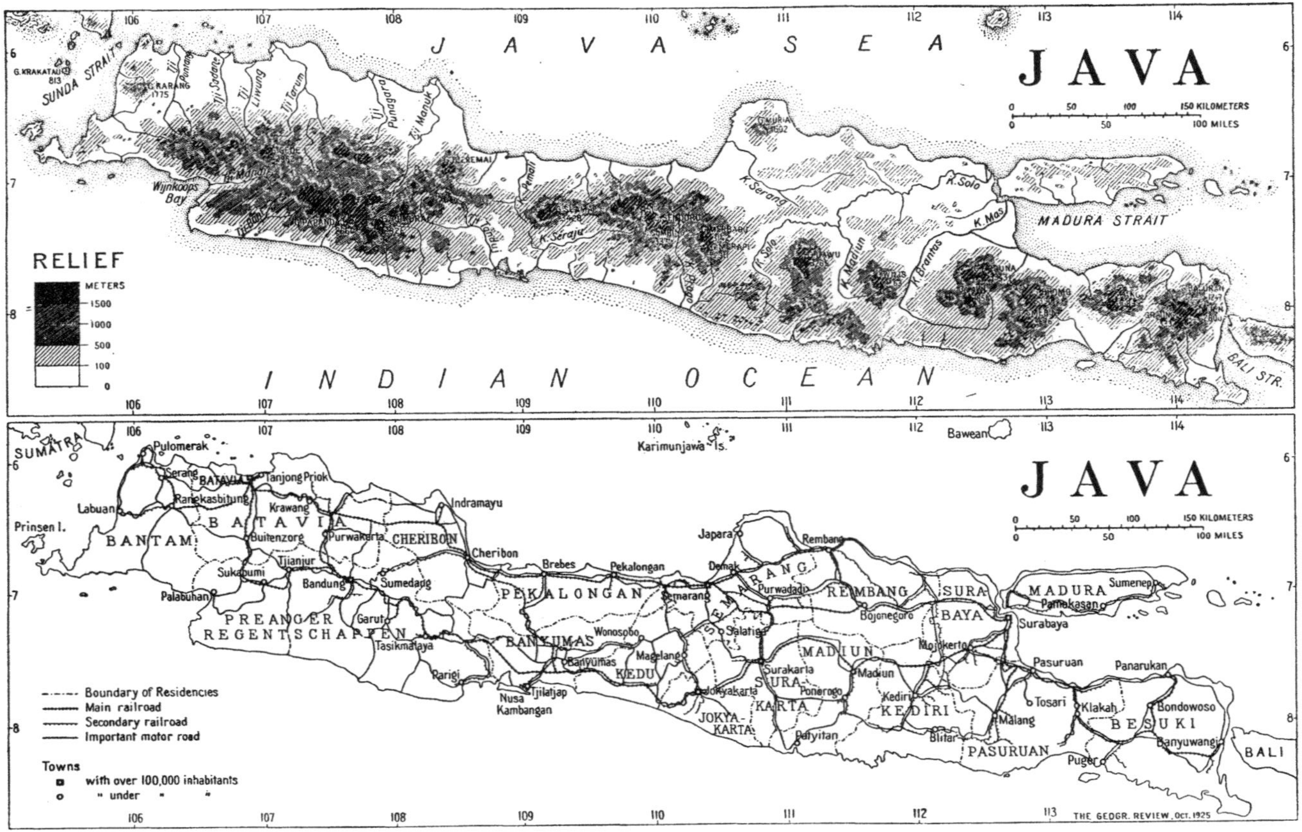

Figs. 120 and 121—Map of Java showing relief and map showing the residencies and principalities, the important towns, and the chief roads and railroads as of 1925. (Reproduced through the courtesy of the *Geographical Review*.)

As we pointed out in the discussion of the density of population in the Philippines, the ratios between population and cultivated area are of greater significance than are the figures of average density based on the total area. In Java and Madura there was an average of 535 persons per square kilometer of cultivated land in 1930 (Table XIII). With the excep-

TABLE XIII—JAVA AND MADURA: DENSITIES OF INDONESIAN POPULATION, BY RESIDENCIES, 1930

RESIDENCY	DENSITIES Per sq. km. of total area	Per sq. km. of cultivated land*	Per sq. km. of *sawah*
West Java	*244*	*496*	*1,038*
Bantam	130	383	886
Batavia	329	452	784
Buitenzorg	191	551	1,205
Priangan	253	506	1,544
Cheribon	381	569	851
Middle Java	*396*	*570*	*1,261*
Banjoemas	386	562	1,635
Pekalongan	468	759	1,310
Kedoe	457	548	1,485
Semarang	372	537	1,047
Djapara-Rembang	311	466	961
Jogjakarta	*492*	*615*	*2,351*
Soerakarta	*425*	*554*	*1,328*
Klaten	535	769	1,569
Soerakarta	379	475	1,217
East Java	*314*	*531*	*1,364*
Bodjonegoro	291	469	805
Madioen	314	459	1,198
Kediri	351	605	1,597
Soerabaja	540	762	1,198
Malang	310	551	1,769
Besoeki	206	531	1,341
Madura	359	443	2,647
JAVA AND MADURA	*316*	*535*	*1,248*

*This category is described in the source as "Native Lands."
Source: Data for West Java from *Volkstelling 1930*, Vol. 1, Table 8; data for Middle Java and the principalities from *Ibid.*, Vol. 2, Table 9; data for East Java from *Ibid.*, Vol. 3, Table 8.

tion of Bantam, all the residencies in Java had a much larger number of people per unit of cultivated land in 1930 than had the Philippines as a whole in 1939 (405 persons per square kilometer of cultivated land), although no residency in Java had so high a density as the province of Ilocos Sur in Luzon, which claimed 804 persons per square kilometer of cultivated land. Klaten, with a density of 769 persons, came nearest to this in

Java. All but one of the residencies of Middle Java exceeded the average for Java as a whole; in West Java only two out of five, and in East Java only three out of seven, residencies surpassed Java's over-all average density.

The ratio of persons to cultivated land varies greatly from district to district; the highest densities in 1930 were found where a large proportion of the area is irrigable and therefore suitable for the cultivation of wet rice and where European plantations regularly leased irrigated land from the peasants for the cultivation of sugar cane and tobacco. Those districts having little irrigable land are usually only thinly settled. Table XXIII lists six rural districts in Java, three of which had the highest, and three the lowest, densities in 1930. A little more than 60 per cent of the cultivated area in the districts of Adiwerno, Ploembon, and Wiradesa was irrigated for the cultivation of wet rice. As we have seen, this type of cultivation requires a great deal of labor and produces comparatively high yields. On the other hand, in Poenoeng, although nearly 93 per cent of the soil was in use, only 5 per cent was irrigated; hence the density was low. In Java the density of population in a district depends to a large extent upon the mode of land use rather than upon the amount of land in use.

Effects of Overpopulation

The extraordinary increase in population was neutralized only in part by various countermeasures: extensive forest clearing and cultivation of virgin land (the *tegalan* area alone increased from 355,000 hectares in 1875 to 4,503,000 hectares in 1938); intensification in land use (i.e., the conversion of tegalans into sawahs wherever possible); improvement of Indonesian-built irrigation systems and the construction of new ones under government auspices; improvement in yields through better methods of cultivation and the use of selected seeds; and employment opportunities offered by Western plantations. In spite of all of these measures, the growth of population resulted in an increased parcelling of the land into dwarf holdings, in the accumulation of large holdings wherever land-tenure practices allowed it, in the formation of a class of landless agricultural laborers, in widespread indebtedness, in the growth of tenancy, in the decline of rural incomes, and in poverty, malnutrition, and food shortage; briefly, in overpopulation with all its frightful implications.

Since Indonesian society in Java is essentially agrarian, the possession of land, especially of sawah, is of the greatest importance; without land the Javanese villager has no social standing. This emphasis is reflected in the structure of the Javanese *desa*, which includes a number of classes based on landownership. First comes the group of villagers who own both fields and garden land plus a house on the latter. They are known as *gogols*. The gogol is a "full peasant" in the sense that he has all the rights that a member of the desa may have and also shares in all the burdens. In the next lower group are the *stengah gogols*, or "half peasants," who own garden land and houses, but no fields. The *menumpangs* are villagers who own houses, which, however, stand on the garden land of others. Still lower in rank are the villagers who live in others' houses and own neither house nor land of any kind. These are called *kumpulans* or *nusups*.[11] The numerical proportion of these various groups is a good indication of the crowded condition of a village. If a desa or a whole district is divided into small holdings and has a large number of stengah gogols, menumpangs, and nusups, we may assume that the population exceeds the carrying capacity of the land. The excess population must rely in part or wholly upon wages for their livelihood.

The Parcelling of the Land

Peasant holdings in Java are small. In 1938 the average area was only 0.86 hectare per taxpaying landowner. This included both sawah (0.36 hectare), and tegalan and garden land (0.50 hectare). The average size of the holdings, as we have seen, has apparently been steadily declining.

Unfortunately we lack recent detailed data on the size and ownership of peasant holdings in Java as a whole and must rely on general observations and on case studies of specific districts and desas. Investigations held in 1903,[12] however, show that in Java, of all those holding land in individual hereditary possession or having fixed shares of communal land, 70.9 per cent had less than 0.7 hectare, another 18.2 per cent from 0.7 to 1.4 hectares, 7 per cent from 1.4 to 2.8 hectares, and only 3.9 per cent had more than 2.8 hectares. In Bantam and Semarang more than 80 per cent of all peasant holdings were less than 0.7 hectare in size. Table XXIV shows that even in 1903 Java was a country of dwarf holdings. According to a Javanese popular saying, a family requires one *bouw* (0.709 hectare) of sawah or two bouws of tegalan in order to make a fair living without having to look for supplementary income.[13] As early as 1903 a large percentage of the peasants in Java did not have this amount of land, and a sizable landless class already existed. According to a study by Burger comparing conditions in 1868 and 1928, in the desa of Pekalongan, regency of Pati, residency of Djapara-Rembang, the average peasant-proprietor had from about 0.7 to 1.1 hectares of sawah in 1868 but only 0.5 hectare in 1928.[14]

Burger gives the following distribution of sawah area, for 1929, in the desa of Ngablak, also in the regency of Pati, where 63 per cent of all

---

[11] The terminology is not uniform throughout Java. The gogol, for example, is also known by such terms as *baku*, *sikep*, *kulie*, *kulie kentjeng*, and *wong kentjeng*.

[12] Hasselman: Algemeen overzicht van de uitkomsten, 1914, Appendix R. These figures do not include the principalities of central Java. The Javanese have the following types of land tenure: (a) registered property (*agrarisch eigendom*) (b) individual hereditary possession (*erfelijk individueel bezit*); (c) communal ownership with fixed shares and the same partakers (*communaal bezit met vaste aandeelen en steeds dezelfde deelgerechtigden*); (d) communal ownership with fixed shares but changing partakers (*communaal bezit met onveranderlijke aandeelen maar wisselende deelgerechtigden*); (e) communal ownership with periodic redistribution (*communaal bezit met periodieke verdeeling*); (f) land set aside for village officials (*ambtsvelden*).

[13] *Ibid.*, p. 296.

[14] Burger: "Rapport over de desa Pekalongan," [1929], p. 8.

sawah owners possessed less than about 0.7 hectare and 37 per cent less than about 0.3 hectare:[15]

Less than 0.142 hectare	22	owners
0.142 or more but less than 0.284 hectare	76	"
0.284 " " " " " 0.426 "	27	"
0.426 " " " " " 0.568 "	19	"
0.568 " " " " " 0.709 "	23	"
0.709 " " " " " 1.42 hectares	46	"
1.42 " " " " " 2.13 "	18	"
2.13 " " " " " 2.84 "	13	"
2.84 " " " " " 4.26 "	11	"
4.26 " " " " " 7.09 "	7	"
7.09 or more "	1	"
	263	"

In the district of Kebonongan in the principality of Jogjakarta each sawah owner had an average of about 0.3 hectare in 1939.[16] Many villagers possessed garden land only. In 1939 Kebonongan had 17,702 adult men, of whom 10,901 owned land. Of these latter, 6,862 were gogols, 3,706 were stengah gogols, and the remaining 333 had plow land but lived on others' garden land. The respective numbers of menumpangs and nusups were not reported.[17]

Table XIV gives the ratios between gogols and stengah gogols in the districts of the regency of Toeloengagoeng, Kediri Residency, in the province of East Java.

Landless Agricultural Laborers

The census of 1930 gives no data that throw light on the size of the landless population. Hasselman, however, reports the number of landowners and of adult men without land in 1903 (Table XXV). The ratio of landowners to landless men was especially unfavorable in the Preanger highlands, Banjoemas, Pekalongan, Kedoe, and Madioen. Since then the number of villagers who own no land has grown considerably. In the district of Adiwerno, which in 1930 had the highest population density in Java, only 21.8 per cent of the working population was engaged in native agriculture, 25.5 per cent in industries, and another 28.9 per cent was engaged in "insufficiently described occupation," which in most cases means that the individuals were unskilled laborers who worked wherever they could find employment, usually upon sugar plantations. This part of the rural population has been seriously affected by the decline of Java's sugar industry since 1931. The landowners who could no longer lease their land

15 Idem: "De desa Ngablak," 1933, p. 232.

16 Jonkers: *op. cit.*, p. 22.

17 *Ibid.*, p. 22. In Kebonongan the gogol is known as kulie kentjeng, the stengah gogol as *kulie karangkopek*, and the owner of plow land without garden land as *kulie gundul.*

to sugar mills have been partly compensated by having all their land at their own disposal instead of having only two-thirds or three-fourths as before; but those Indonesians without land have had great difficulty in finding other employment. Moreover, the wages paid by the sugar industry decreased markedly from 1929 to 1937. In 1928 the industry paid an average wage bill of 545.14 guilders per hectare of sugar-cane land, but in

TABLE XIV—EAST JAVA: NUMBER OF GOGOLS AND STENGAH GOGOLS IN THE REGENCY OF TOELOENGAGOENG, KEDIRI RESIDENCY, BY DISTRICTS, 1939

DISTRICT	GOGOLS (a)	STENGAH GOGOLS (b)	RATIO OF (a) TO (b)
Toeloengagoeng	6,910	8,322	0.8: 1.0
Ngoenoet	12,811	7,138	1.8: 1.0
Tjampoerdarat	10,099	4,334	2.3: 1.0
Kalangbret	8,210	5,550	1.5: 1.0

Source: Soenario: "Onderzoek naar de praktische mogelijkheden van schuldbevrijding," 1940, p. 97. Ratios computed by the author.

1937 a bill of only 178.49 guilders. In 1928 the Indonesians received slightly more than 106 million guilders in wages and in 1936 only 7½ million guilders. In 1937 economic recovery brought a sudden extension of the area planted in sugar cane and doubled the total sum of wages paid, although the wage rates were lower than in 1936. Table XXVI shows payments by the sugar industry to the Indonesian population of Java from 1920 through 1938.

TENANCY

In the Philippines the population problem was aggravated by a marked concentration of landownership in the hands of a few and by widespread tenancy. Java differs greatly from the Philippines as regards both the proportion of the total area owned by large landed proprietors and the size of the areas held by them individually. There are no large contiguous landed estates owned by Indonesians and worked with wage laborers that would correspond to the large *haciendas* in the Philippines. However, in Java, there are persons who, by Indonesian standards, own considerable quantities of land in scattered parcels. Land is sought as an investment by Indonesians of means, especially government officials and those who live in the urban centers of Java.[18] In villages where communal landownership in one form or another still prevails, there is no possibility of building up large holdings. The same is true of villages that allow the sale of land to outsiders only when no one in the village will buy it. A concentration of land in the hands of wealthy individuals is found in West Java and especially in the former Preanger regencies, where as early as 1903 11.3 per cent of all landowners had more than 2.84 hectares. Of the 150 Indonesians owning more than 25 hectares in 1903, 71 were living in the Preanger regen-

18 Boeke: The Structure of Netherlands Indian Economy, 1942, p. 42.

cies. Meyer Ranneft and Huender have compared the number of Indonesian landowners possessing more than 17.7 hectares (25 bouws) in 1905 and 1925 (Table XXVII) and found that there was a great increase, especially in the residencies of West Java and in Besoeki in East Java.

The most effective barrier against a concentration of landholding was the Agrarian Decree of 1870, which forbade the purchase of agricultural land by any but Indonesians. This prevented the Chinese and Europeans from becoming landed proprietors living on the labor of Indonesian tenants. The Indonesian, who is always in need of cash, would have sold his land long ago for a mere pittance had it not been for this agrarian legislation. Some non-Indonesians, however, have acquired land in the name of their Indonesian spouses, though such methods are illegal.

As it is, impecunious Indonesians may lose their land to well-to-do Indonesians who make loans against mortgages on sawahs. Interest rates are high, so that frequently the borrowers are unable to repay the loans; then the land becomes the property of the mortgage-holder.

Like most large landowners of Southeastern Asia, those of Java lease their land in small parcels to tenants on a share basis. We have no recent data on the extent of tenancy in Java. In 1905, 5.3 per cent of all persons engaged in native agriculture owned no land but, instead, were working the land as tenants. Table XXVIII gives data on the frequency of tenancy at that time in the various residencies. It would appear to have been relatively frequent in West Java and Jogjakarta, but less frequent elsewhere. A survey made in 1926 shows that in the residencies of Bantam, Cheribon, Semarang, Kedoe, Soerabaja, and Madura most of the peasants owned the land they were tilling; in the residencies of Pekalongan, Madioen, Kediri, and Pasoeroean tenancy existed to a moderate degree, and in the residencies of Batavia, Priangan, Banjoemas, and Besoeki it was relatively widespread.[19]

The terms of tenant contracts vary greatly and depend upon such local conditions as the pressure of population, the fertility of the soil, and especially upon custom. Frequently the tenant gets one-third of the rice crop if the land produces high yields, one-half if the fertility is only average, and two-thirds if the fertility is low. The Javanese term for the half-and-half division is *maro*, or *maron*, and for the two others, *mertelu*.[20] Contracts vary also in regard to the division of expenses and tax-payments between landowner and tenant. Sometimes the landlord must supply his tenant with seeds, a work animal, and tools, as well as with some cash or foodstuff to be paid back with high interest at the time of the harvest. In such cases the tenant is really a sharecropper. His share is smaller than is that of a tenant who has no such need of backing from the landlord.

In those sections of Java where the sawahs are owned communally one finds a group of tenants under a different form of contract. These work

19 Scheltema: Deelbouw in Nederlandsch-Indië, 1931, p. 273.

20 For a detailed discussion of tenant contracts see Scheltema's excellent study.

their own land on a share basis for a limited period. The small peasant is habitually in need of cash to pay his taxes or to meet extraordinary expenses. Those who own their land may mortgage it, but those who hold a share of communal land do not have that right. Their only way of raising cash is to lease their land to someone in the village or to an outsider by the

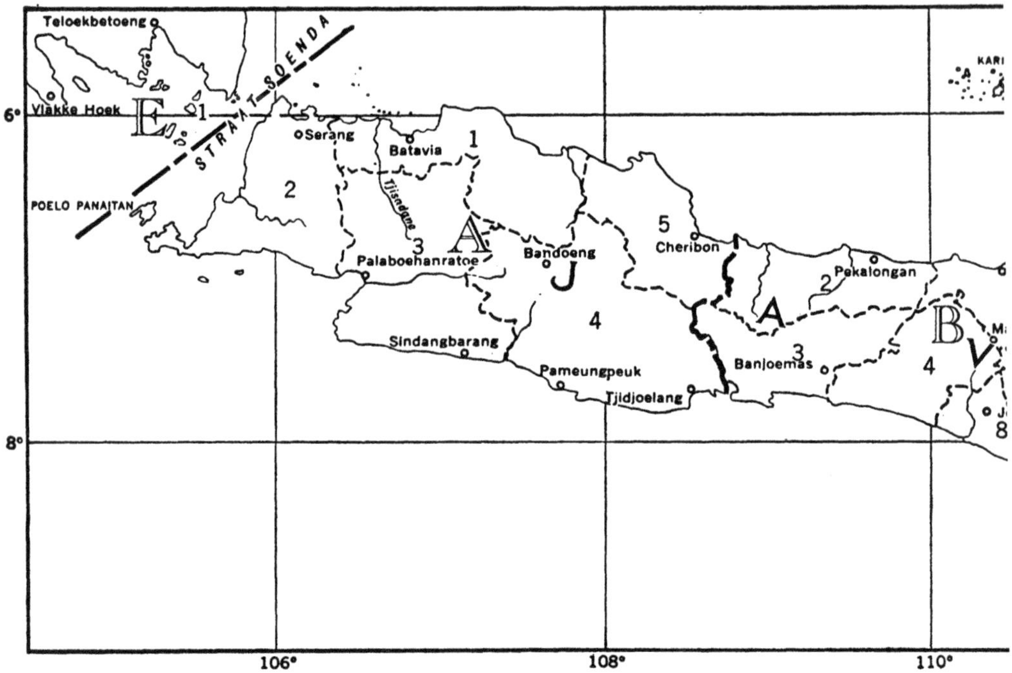

Fig. 122—Map showing the chief administrative divisions of Java and Madura, Bali and Lombok, and southern Sumatra. The divisions designated by the letters and figures are as follows:

A—Province of West Java
1. Residency of Batavia
2. " " Bantam
3. " " Buitenzorg
4. " " Priangan
5. " " Cheribon

B—Province of Middle Java
1. Residency of Semarang
2. " " Pekalongan
3. " " Banjoemas
4. " " Kedoe
5. " " Djapara-Rembang
6. " " Soerakarta
7. " " Klaten
8. " " Jogjakarta

(*Continued on the opposite page*).

season or the year for a cash rent paid in advance. The payment is sometimes made one or two years before the land is actually put at the disposal of the lessee. The latter in most cases does not intend to cultivate the land himself but employs the lessor as tenant, or, rather, subtenant—that is, he subleases the land back to the owner on a share basis and takes one-half, or even five-sevenths, of the crop. The lessees are usually rice merchants, who by this method assure themselves of a cheap supply of rice, or occasionally individual investors, who in this way obtain high interest rates. The longer in advance the rent is paid, the lower it is.

By this roundabout arrangement the landowner, or to be more correct

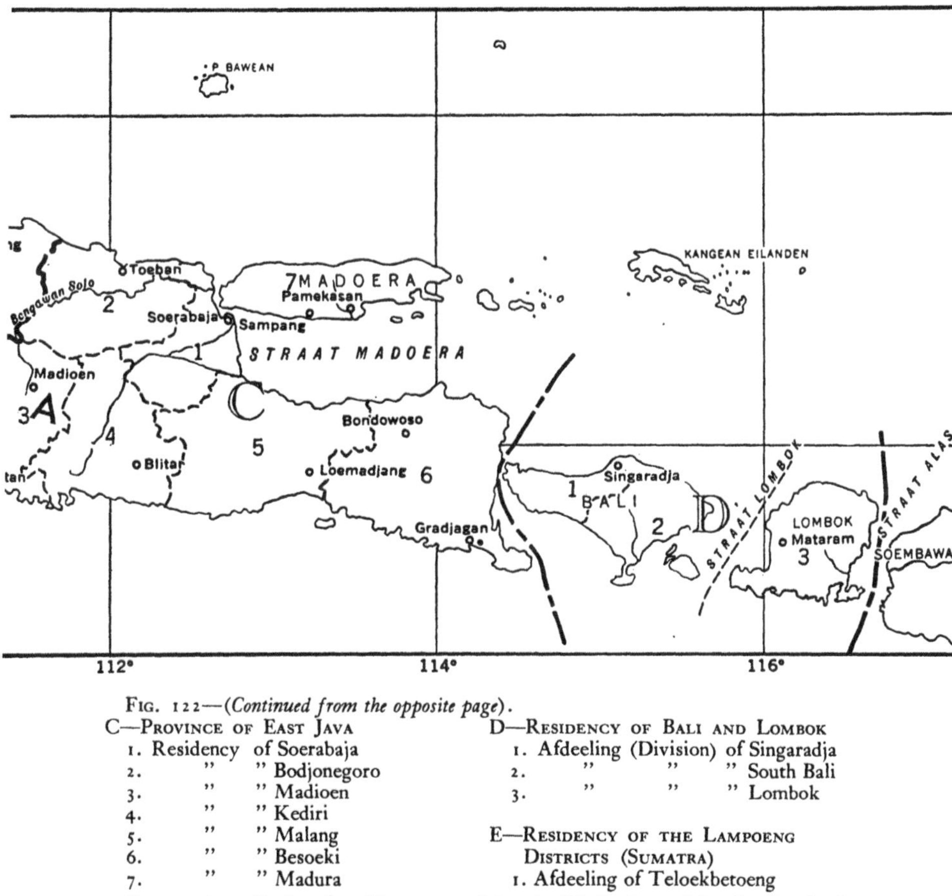

FIG. 122—(*Continued from the opposite page*).

C—PROVINCE OF EAST JAVA
1. Residency of Soerabaja
2. " " Bodjonegoro
3. " " Madioen
4. " " Kediri
5. " " Malang
6. " " Besoeki
7. " " Madura

D—RESIDENCY OF BALI AND LOMBOK
1. Afdeeling (Division) of Singaradja
2. " " " South Bali
3. " " " Lombok

E—RESIDENCY OF THE LAMPOENG DISTRICTS (SUMATRA)
1. Afdeeling of Teloekbetoeng

Source: Area Designation Map prepared by the U. S. Army Map Service for its Gazetteer to Maps of Java, Madura, and Bali (June 1944). (Reproduced through the courtesy of the Army Map Service.)

the landholder, becomes a subtenant on his own land for a limited period, whereas the lessee—since he leases the land back to the owner—functions as landlord. Like an ordinary landlord, he shares in the risk of a low yield or a crop failure just as much as in the profit of a bumper crop. This system represents a buying in advance of a share of a future crop of which the yield is unknown. Because of the risk involved, the rent is rather low, yet the landholder is forced to accept it because of his need for cash.[21]

Contracts of this kind are usually unregistered and therefore escape the knowledge of the authorities unless special surveys are made to investigate the degree of indebtedness of the peasants. Such a survey was made at the beginning of 1939 in the regency of Toeloengagoeng, residency of Kediri, in the province of East Java (Table XXIX); it brought to light the fact that a high percentage of the peasants were in debt. They had leased their land and were working it as subtenants, except when the lessees were tilling it themselves. Out of a total of 2,740 landholding peasants, 74.2 per cent had leased some of their land in order to obtain cash and as many as 33.9 per cent had been forced to lease all of their land. In 3 out of the 21 villages included in the survey every landholder had leased either a part, or all, of his land, and in 9 of the 21 villages 90 per cent or more of all landholders had gone into debt in this fashion.

The Leasing of Land to Sugar Plantations

Similarly in those sections of Java where Western-owned sugar and tobacco plantations operated, the peasants became accustomed to leasing land to them against cash rent payable in advance. In contrast, however, to the Indonesian lessee just described, the plantations themselves made use of the land, although they were willing to hire the peasants as workers. In spite of the fact that the average peasant in Java dislikes to work as a laborer, he was often forced to do so, for the rent that he received a year and a half in advance had usually been spent by the time the plantation began to cultivate his land.

Thus, because of his financial weakness and proverbial need for cash, the peasant was hardly in a position to bargain with the plantation, and as the plantation was ready to pay the rent well in advance of occupation the peasant was confronted with an irresistible temptation. Often the individual peasant had no direct contact with the plantation that leased his land. This was the case in all those villages that still had communal landownership with periodic redistribution of sawahs. In 1932, 8.8 per cent of all sawahs were held in this form of tenure, and it was still found in the sugar districts of Java in 1941, although the government had favored replacing it by communal ownership with fixed shares, or, preferably, by individual hereditary possession; for when the land is redivided every year or every 3 years, as is frequently done, the peasant takes no interest in permanent improvements, since someone else would reap the benefit. But the sugar

21 The cash rent for 0.7 hectare of sawah paid by the lessee for one wet season averages from 10 to 12 guilders if paid 2 years in advance of the season to which the lease applies, from 16 to 18 guilders if paid a year in advance, and from 20 to 25 guilders if paid at the beginning of the wet season. The average share paid to the lessee may be put at 20 piculs of paddy which has a value of 32 guilders; hence, a lessee who paid a cash rent of 20 guilders at the beginning of the season would receive an increment of 12 guilders, or 60 per cent of his investment, within 6 months.

plantations preferred to negotiate land leases with one *lurah*,[22] or village head, rather than with several hundred individual peasants, for it saved a great deal of time and, moreover, the cost of such leases was less than that of many individual ones. The peasant who held his land in hereditary possession was often unwilling to lease it at the rate at which a lurah was ready to hand over the village land; the lurah was frequently swayed by the remuneration received from the plantation as a commission for the transaction. In desas with communal landownership the lurahs had great power, abuses of which were not uncommon. It was an easy matter in such desas for a lurah to crush a gogol who in the village assembly opposed the signing of a lease with a sugar plantation. To overcome these disadvantages to the peasant, many Netherlands officials struggled for a change in the system of tenure, but their efforts were openly, or more often secretly, opposed by the administrators of the sugar plantations and their faithful servants, the lurahs, for whom a change would have meant the loss of a good source of income.

In spite of this opposition, the Netherlands colonial administration, knowing the respective weakness and strength of the two sides, attempted to protect the peasant against abuses by keeping a watchful eye on all rental transactions. In each residency, in conformity with a government ordinance, the resident from time to time established minimum rent-rates for each class of sawahs. There were usually seven or eight classes, differentiated according to their fertility. Advance payments were subject to restrictive conditions, and the length of the lease was limited. Short leases of sawahs ran for 3½ years, long leases for 21½ years. The lessee could occupy only one-third of the village land at any one time; the other two-thirds had to be at the disposition of the peasants.

This relationship between the peasants and the great capitalistic enterprises was the subject of long and heated controversy among Dutch students of colonial affairs, for, although the plantation industry brought many advantages to Indonesian communities, there were also marked disadvantages associated with it. The government tried to hold the balance by protecting the Indonesian from the Western entrepreneur and at the same time taking measures to give the latter security in his investment. The fact remained, however, that the Indonesian who leased his sawah to the plantation was disadvantaged, since in most cases he could have earned more if he had worked it himself. Why, then, did he lease it? The simple answer has already been indicated: by leasing, he received the rent far in advance of the time when he would have been able to sell his own crop.[23]

The landless agricultural laborers in the sugar districts, on the other hand, benefited when the sugar plantations were working, since the wages

22 The lurah and the other village officials, such as the *tjarik* or scribe, the *modin* or religious official, and the *kapetengan* or policeman, are elected, usually by the landowning villagers alone.

23 Gonggrijp: Over de invloed van het westerse grootbedrijf, 1930, p. 14.

paid by the plantations were higher than those paid by Indonesian employers.

Efforts to Relieve Population Pressure

At least 60 years ago Dutch students of colonial affairs began to express their fear of the specter of overpopulation in Java and Madura. It was believed by some even then that Java's soil was producing all that it possibly could. Since those days more intensive methods of land use, large-scale construction of modern technical irrigation systems, and extensions of the agricultural area have greatly enlarged the food production and hence the carrying capacity of the islands. But now the situation has become really serious. There is no longer room for large new irrigation projects that would make further intensification of cultivation possible, nor is there any longer any land to speak of that could be added to the area under cultivation. Double cropping is already being practiced to a remarkable degree: wherever possible, the Indonesian peasant today plants a dry-season crop in addition to the wet-season crop. In 1926 Java's crop area amounted to 131 per cent of the cultivated area (that is, sawahs and tegalans only, with garden land excluded); by 1936, it had grown to 142.4 per cent.[24]

In spite of strenuous efforts by the Agricultural Advisory Service and the Irrigation Service to increase the food production at the same rate at which the population increased, recent investigations reveal that the diet of the people of Java has definitely deteriorated in food value, if not in actual volume, in the course of the past few decades, and one may add that "the Javanese at no time had a diet which allowed of much curtailment."[25] Table XXX shows an irregular decline in the consumption of rice since 1913, and a great increase in that of cassava, which is markedly deficient in nutritive value.

The balance between population and means of support has become so unstable that it is in constant danger of being upset if the islands should experience a bad harvest or other catastrophe. Hart was of the opinion that overpopulation had existed for years in Java, but that at least until the invasion of 1942 the frugality of the people, the intensive and multiple efforts of the government, and good harvests had preserved the precarious balance.

No wonder that in recent years the Government of the Indies, as well as individual observers, has devoted a great deal of attention to the population problem. Whereas about forty years ago Van Deventer recommended "education, irrigation, and emigration," the remedies suggested today are "intensification of agriculture, industrialization, and emigration." The Agricultural Advisory Service is of the opinion that there is still room for some intensification of agriculture in Java with the help of better selection of seed, the use of green manure, the control of soil erosion, and other means discussed in Chapter III. The possibilities offered by industrializa-

24 Hart: "Volkstal en bestaansmiddelen," 1937, p. 575.

25 Idem: "Recent Development in the Netherlands Indies," 1942, p. 94.

tion lie beyond the scope of this study and have been treated in recent works.[26] In this and the following chapter our principal concern is with the attempts of the Government of the Indies to relieve the pressure of population by the encouragement of migration from the overcrowded to the less densely populated districts within the island, and, particularly, of migration to the Outer Islands.

Although the peasants of Java are strongly attached to the soil, they have always migrated when the pressure became too great and when they knew of another area that seemed to offer relief. Raffles spoke of such internal migrations from Bantam, Batavia, Cheribon, Demak, and Banjoewangi to the Javanese native states:

> It was fortunate . . . that the native governments were less oppressive than the sway of their European conquerors, and that their states afforded a retreat from a more desolating tyranny. It has been ascertained, that, on the first establishment of the Dutch in the eastern part of the island, the inhabitants of whole districts at once migrated into the Native Provinces. Every new act of rigour, every unexpected exaction, occasioned a further migration, and cultivation was transferred to tracts which had previously scarçely a family on them. . . . During the administration of Marshall Daendels, in the years 1808, 1809, and 1810, nearly all the inhabitants of the province of *Demák*, one of the richest in the eastern districts, fled into the Native Provinces; and when an order was given for the rigid enforcement of the coffee monopoly, every district suffered in its population, in proportion to the extent of service levied upon it.[27]

We know that during the time of the culture system many peasants tried to escape the burden of the forced cultivation of money crops by fleeing into less accessible forest regions.[28] People from Bantam moved across the Sunda Strait into southern Sumatra.

The census of 1930 investigated for the first time the extent of movements of population in the Netherlands Indies and revealed considerable migration within Java as well as movements of people from Java to the Outer Islands. In contrast to the days of the East India Company when people fled from the coastal regions to the interior, recently migration within Java has been from the crowded interior to commercial centers on the coast. The cities of Batavia, Semarang, and Soerabaja in the residencies of the same names attracted large numbers of people because of the employment offered. Among the residencies of Java, Besoeki occupies a unique position, as nearly one-fourth of its population was born elsewhere and migrated to this residency, especially to the regencies of Djember and Banjoewangi. The construction of the railroad between Kalisat and Banjoewangi (about 1901) encouraged the migration from central Java. The

26 Broek: Economic Development of the Netherlands Indies, 1942; Mitchell: Industrialization of the Western Pacific, 1942; Shepherd: Industry in Southeast Asia, 1942; Sitsen: Industrial Development of the Netherlands Indies, 1943.

27 Raffles: The History of Java, 1817, Vol. 1, p. 65.

28 De Vries: Landbouw en welvaart, 1931, Vol. 1, p. 100.

coffee plantations also attracted a large number of people, who settled permanently in this residency. However, it is still thinly settled as compared with central Java.

The census returns of 1930 brought out the fact that the number of emigrants who had left the heavily populated residencies of Kedoe, Madioen, Kediri, and Madura and the principality of Jogjakarta and had migrated either to other residencies in Java or to the Outer Islands represented in each case 10 per cent or more of the population living in 1930 in those residencies and the principality (Table XXXI). The Madurese migration across the Strait of Madura from the island of Madura to East Java had been going on for such a long time that by 1930 only 45 per cent of the persons of Madurese culture and language were found in their homeland, whereas most of the others lived in the eastern corner of Java.[29] In 1930 Madurese composed 97.8 per cent of the population in the regency of Panaroekan, 98.3 per cent in the regency of Bondowoso, both in the residency of Besoeki, and 94.2 per cent in the regency of Kraksaan in the residency of Malang, all these regencies being located on or near the Strait of Madura. In the residency of Besoeki, as a whole, Madurese composed about two-thirds of the population. The people of Madura, however, have not yet begun to migrate on any large scale to the Outer Islands.

29 Lekkerkerker: "Java's overbevolking," 1937, pp. 875–880.

tire Javanese *desa* migrates to Soekadana colony, Residency of the Lampoeng Districts, Sumatra.

23 (upper left)—Villagers, old and young, awaiting transportation at a railroad station in Java. (Courtesy of the Department of Economic Affairs.)

24 (upper right)—Night on board ship between Java and Sumatra. (Courtesy of the Department of Economic Affairs.)

25 (center left)—Approaching Teloekbetoeng at the head of Lampoeng Bay in southern Sumatra. (Courtesy of the Department of Economic Affairs.)

26 (center right)—Leaving the South Sumatra Railroad. (Courtesy of the Department of Economic Affairs.)

27 (lower left)—Boarding buses that take the migrants to the colony. (Courtesy of the Department of Economic Affairs.)

28 (lower right)—Each settler is issued kitchen utensils and a bamboo signal drum. (Courtesy of the Department of Economic Affairs.)

FIG. 129 (upper left)—Rice bundles, earned as *bawon* by newly arrived Javanese colonists, lined up before the house o perous *tani* whom the newcomers assisted. Gedongtataän, Residency of the Lampoeng Districts, Sumatra. (C of the Central Commission for Migration and Colonization of Natives.)

FIG. 130 (center left)—Cooking pots to be issued to new settlers. (Courtesy of the Department of Economic Affairs.)

FIG. 131 (lower left)—Distribution of seed rice to new settlers. Soekadana colony, Residency of the Lampoeng D Sumatra.

FIG. 132 (upper right)—New settler clearing his farm lot. Belitang colony, Residency of Palembang, Sumatra.

FIG. 133 (lower right)—Weighing seed rice for new settlers. Soekadana colony, Residency of the Lampoeng Districts, S

4 (upper left)—*Bedengs* (bunkhouses) awaiting new settlers. Soekadana colony, Residency of the Lampoeng Districts, Sumatra.

5 (upper right)—Three weeks after arriving in the colony, a settler has made a small clearing and started a simple shelter. Belitang colony, Residency of Palembang, Sumatra.

6 (center left)—Settler of one year's residence has started to expand his primitive shelter into a house. Toegoemoeljo colony, near Loeboeklinggau, Residency of Palembang, Sumatra.

7 (center right)—One-year-old village, showing gardens surrounding houses; the adjoining forest has not yet been completely cleared for fields. Soekadana colony.

8 (lower left)—Settler of two years' residence no longer needs to plant vegetables right up to his doorstep, as his field is now entirely cleared. Notice the bicycle, a luxury which the settler did not enjoy in Java. Toegoemoeljo colony.

9 (lower right)—Well built house in old Javanese style with tiled roof, owned by a colonist of 20 years' residence. Gedongtataän colony, Residency of the Lampoeng Districts, Sumatra.

FIG. 140 (upper left)—Site of a future village. The forest on both sides of the road must be cleared by settlers befc can build their houses and plant their gardens. Soekadana colony, Residency of the Lampoeng Districts, Su

FIG. 141 (upper right)—Three-year-old village. Settlers have cleared the forest and developed flourishing gardens. Toego colony, near Loeboeklinggau, Residency of Palembang, Sumatra.

FIG. 142 (center left)—Clearing a heavily forested area for the *sawah* of a newly arrived Soekadana settler.

FIG. 143 (center right)—Sawahs, villages, and coconut groves have replaced the heavy jungle in the colony of Gedon Residency of the Lampoeng Districts.

FIG. 144 (lower left)—House of the agricultural adviser in Belitang colony, Residency of Palembang, Sumatra.

FIG. 145 (lower right)—Clinic of the Belitang colony.

6 (upper left)—Distribution works at Trimoerdjo; the bed of the primary canal in the foreground. The inlets will regulate the flow in the secondary canals. Soekadana colony, Residency of the Lampoeng Districts, Sumatra.

7 (upper right)—Argogoeroeh Weir, which dams the waters of the Wai Sekampoeng, the source of irrigation water for the Soekadana colony.

8 (center left)—Floating dredge digging the main irrigation canal. The hull is made of cement. Belitang colony, Residency of Palembang, Sumatra.

9 (center right)—Constructing an irrigation canal. Soekadana colony.

o (lower left)—Constructing a weir to dam the Soengai Klingi, the source of irrigation water for the Toegoemoeljo colony, near Loeboeklinggau, Residency of Palembang, Sumatra.

1 (lower right)—Settlers of the Soekadana colony digging a secondary irrigation canal.

FIG. 152—Javanese colonists have been called together to witness the payment of the *marga* taxes. Toegoemoeljo colony, near Loeboeklinggau, Residency of Palembang, Sumatra.

FIG. 153—*Pasirah* of marga Proatin Lima and Javanese *lurahs* of Toegoemoeljo colony.

FIG. 154—*Desa* officials. The differently colored bands indicate the offices the men hold. The one in front is the lurah. Gedongtataän colony, Residency of the Lampoeng Districts, Sumatra.

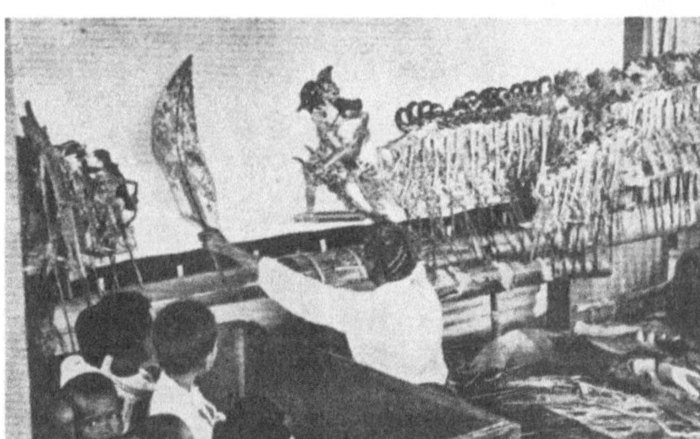

This triad makes the Javanese *seneng* (contented).

FIG. 155—Rice stacked in the yard of a rice mill waiting to be threshed and milled. Gedongtataän colony, Residency of the Lampoeng Districts, Sumatra.

FIG. 156—Part of the *gamelan* orchestra presented to the Gedongtataän colony by His Highness Paku Buwono X, Susuhunan of Soerakarta.

FIG. 157—A *wajang kulit* (shadow play) performance at Gedongtataän colony. The *dalan* (puppeteer) is a colonist.

FIG. 158 (upper left)—"Colonization land is fertile" is the text of this postcard given to colonists who wish to send messages to relatives or friends in Java.

FIG. 159 (upper right)—"Migrate to Tanah Sabrang" is the title of a booklet for *lurahs*, published in 1938 in an edition of 200,000 copies by the Central Commission. The figure depicts one of the beloved heroes of the Javanese *wajang kulit* (shadow play) dressed as a Javanese official.

FIG. 160 (lower left)—"Javanese People are already in Colonization Areas of Celebes" is the title of a booklet published in 1940 by Balé Poestaka, Batavia, for the Central Commission in an edition of 20,000 copies.

FIG. 161 (lower right)—"Go to the Colonies" is the title of a school booklet published in 1940 by Balé Poestaka for the Central Commission in an edition of 100,000 copies.

Chapter VII

THE LAND BEYOND

The Outer Islands as Outlets for Java's Surplus Population

The overcrowded islands of Java and Madura are surrounded by the sparsely inhabited Outer Islands,[1] and nothing would seem more natural than to attempt to solve Java's problem of overpopulation with large-scale migration.

Demographically the Outer Islands present a sharp contrast to Java. They comprise 1,772,000 square kilometers, or 93.1 per cent of the total area of the Netherlands Indies; but in 1930 they contained only 18,247,000 Indonesians, or 30.8 per cent of the total Indonesian population of the Netherlands Indies. They had an average population density of 10.3 inhabitants per square kilometer, if only Indonesians are taken into account, and of 10.7 inhabitants if non-Indonesians are included.

Lacking data, we cannot trace the growth of population in the Outer Islands as far back as that of Java. Most of these areas came under full Dutch control in comparatively recent times. They were covered for the first time by the census of 1905, which reported a population of 7 million; by 1920 this had grown to about 14 million, and in 1930 to 18,247,000 inhabitants. The enumerations of 1905 and 1920 are not very reliable, however, and for the Outer Islands even the census of 1930 is incomplete, so that it is still too early to make any precise statement about the growth of population in these islands. We do know that between 1920 and 1930 the population there increased faster than that of Java, or at a rate of about 2.5 per cent a year. In 1941 the population was estimated to be 22 million; thus it is likely that the total had more than trebled since 1905.

Yet large sections of the Outer Islands are still thinly settled. Only in a few areas in northeastern Sumatra, Sumatra's West Coast, northern and southern Celebes, Bali, and Lombok is the density of population relatively high, but in none can it be compared with that in Java. About 77 per cent of the area of the Outer Islands has less than the average density of 10 persons per square kilometer. The differences in density of settlement among some of the larger political divisions are shown in Table XXXII and on the map (Fig. 163).

What is the reason for this enormous contrast between Java and the Outer Islands? Probably it may be attributed to the same close relationship that was stressed in our discussion of Java in Chapter VI between

Note: Tables XXI–XLVII are in Appendix C, p. 253 ff.

[1] The Outer Islands are divided administratively into the three "governments" of Sumatra, Netherlands Borneo, and the Groote Oost (Great East), each under a governor. The Groote Oost includes Celebes, the Lesser Sunda Islands, the Moluccas, and Netherlands New Guinea.

density of population on the one hand and on the other, modes of land use and soil fertility as determined by climate and the type of soil.

In Sumatra only those sections that have soils derived from young volcanoes show a considerable concentration of people, such as the highlands on the border between the residencies of Palembang and Benkoelen, the Padang Highlands, and the Batak country in the neighborhood of Lake

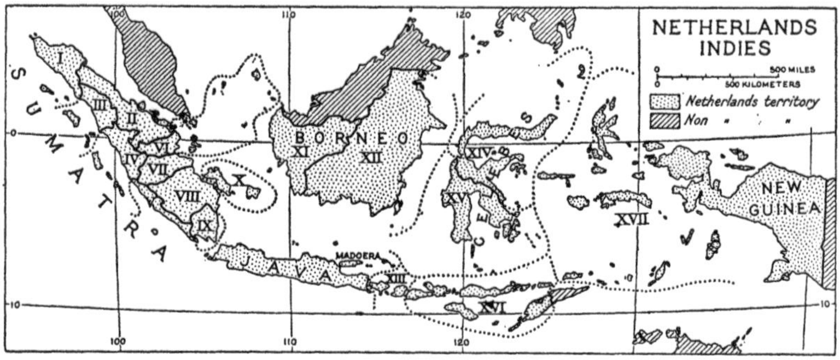

FIG. 162—Location map showing the chief administrative divisions of the Outer Islands of the Netherlands Indies. For those of Java, see Fig. 122. The residencies designated by Roman numerals are as follows:

I—Atjeh
II—The East Coast of Sumatra
III—Tapanoeli
IV—Sumatra's West Coast
V—Benkoelen
VI—Riouw and Dependencies
VII—Djambi
VIII—Palembang
IX—The Lampoeng Districts
X—Bangka and Billiton
XI—West Borneo
XII—South and East Borneo
XIII—Bali and Lombok
XIV—Manado
XV—Celebes and Dependencies
XVI—Timor and Dependencies
XVII—The Moluccas

(Reproduced through the courtesy of the *Geographical Review*.)

Toba (Toba Meer). A similar situation is found in the island of Celebes where densities rise above 100 inhabitants per square kilometer in the young volcanic areas in the southwest, in the Minahasa region in the northeast, and in the country of the Toradjas in the central part of the island; whereas in the less fertile, non-volcanic sections of the island the density is less than 5 inhabitants per square kilometer. Neither Borneo nor New Guinea has a single young and active volcano (Fig. 1A), a fact that is reflected in the low density of population. Bali and Lombok, where there are a number of volcanoes, are the most densely inhabited islands outside of Java and serve as excellent illustrations of the relation between young volcanoes, with the resulting fertility, and concentrations of people.

The amount and distribution of rainfall also greatly affect the fertility of tropical soils, and hence population densities. Heavy rainfall at all times of the year, combined with high temperature, leaches the soil. Such rainfalls occur in large parts of Sumatra, Borneo, and New Guinea and help to account for the inferiority of their soils as compared with those of Java, where, especially in the middle and eastern sections, the rainfall is concentrated in the months of the west monsoon.

Since in large parts of the Outer Islands the land is not irrigable, it is unsuited for the cultivation of wet rice; yet it is generally the intensive

TABLE XV—THE OUTER ISLANDS: TOTAL AREA AND FOREST AREA, BY GOVERNMENTS

GOVERNMENT	TOTAL AREA (SQ. KM.)	FOREST AREA (SQ. KM.)	PER CENT OF FOREST AREA IN TOTAL AREA
Sumatra	474,000	292,000	62
Borneo	539,000	416,000	77
Groote Oost	759,000	499,000	66
Outer Islands	1,772,000	1,207,000	68

Source: Schokkenkamp: "De Boschexploitatie in Nederlandsch-Indië," 1941, p. 93.

cultivation of wet rice that supports dense rural populations in this part of the world.

Hence we may conclude that, because of their comparative infertility and limitations in modes of land use, the Outer Islands, except for a few places, will probably never be able to support so large a population per unit of area as Java.

Nevertheless, the Outer Islands have considerable possibilities of agricultural expansion, as much more land is still forest-covered than is necessary for the maintenance of the water supply and the control of soil erosion. The situation is not, however, so favorable as Table XV would seem to indicate. The "forest area" of 1,207,000 square kilometers includes not only primary forest but also extensive areas of *alang-alang*, ferns, second-growth forest, and bush vegetation—which are largely the result of the *ladang* culture discussed in Chapter II. Probably more than 90 per cent of the people inhabiting the Outer Islands are shifting cultivators, who require considerably more land per individual than do the peasants in Java who cultivate the same pieces of land year after year. Shifting cultivation is justifiable as long as the population remains small in numbers, land is plentiful, and the old ladangs have a chance to revert to forest and thus recuperate fully before they are used again for the production of a crop; but with a rapid growth of population and, above all, the introduction of money crops such as rubber, coffee, and coconuts into the economy of the ladang cultivators, the need for land greatly increases. Because of the need for land, the fallow period of abandoned ladangs in many districts has been shortened to such a degree that, notably in the Lampoengs and Palembang in southern Sumatra, exhaustion and erosion of the soil and a rapid spread of alang-alang are constantly reducing the area suitable for shifting culti-

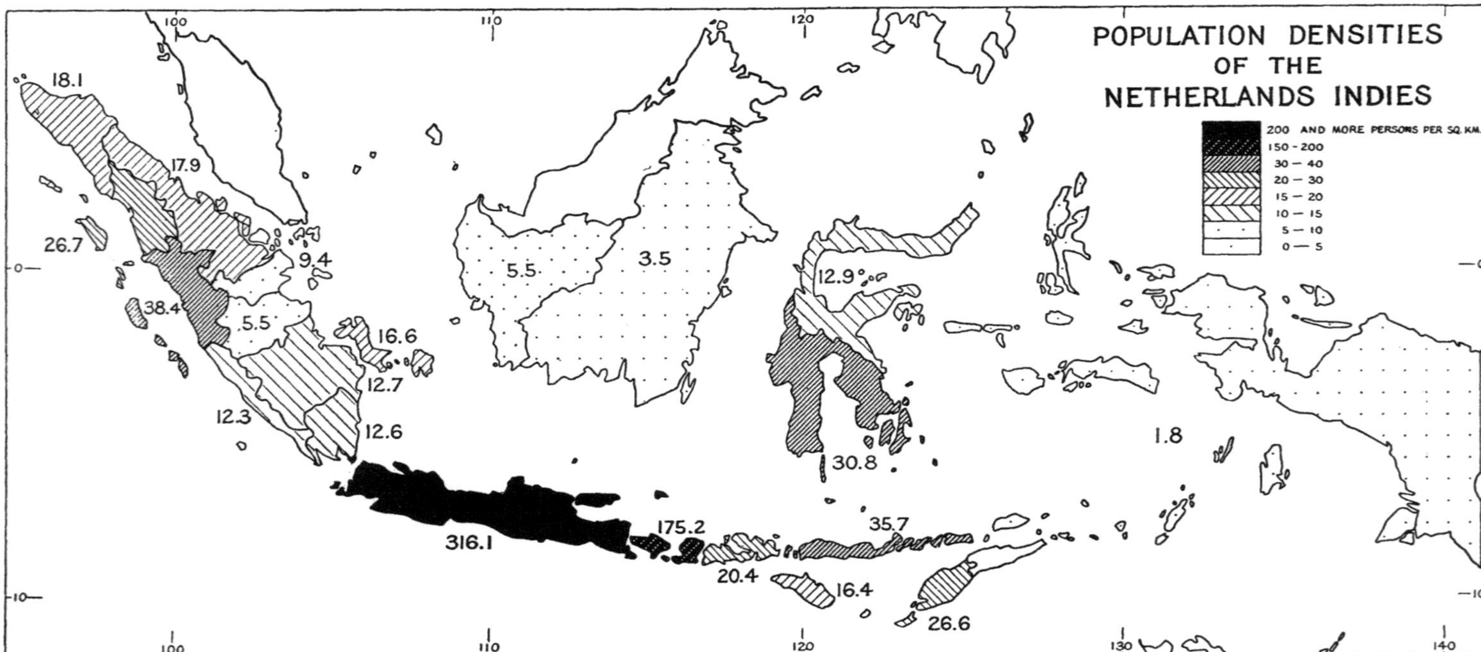

Fig. 163—Density of population in the Netherlands Indies, by large administrative divisions, as recorded by the Census of October 7 and 8, 1930. Both Indonesians and non-Indonesians are included.

vation.[2] The time seems to have come for the shifting cultivators of Sumatra, Celebes, and some of the Lesser Sunda Islands to abandon their time-honored form of land use in favor of a more intensive form. This change becomes all the more urgent if the Outer Islands are to continue to receive a constant stream of peasants from Java.

Persons of Java Ancestry in the Outer Islands as Enumerated by the Census of 1930

In 1816 Governor Raffles predicted an overseas migration from Java to the Outer Islands when he wrote,

> Afterwards the immense tracts of unoccupied or thinly peopled territories on Sumatra, Borneo, and the numerous islands scattered over the Archipelago, may be ready to receive colonies, arts and civilization from the metropolis of the Indian seas.[3]

Since the days of Raffles this migration has been considerable. The full extent of the movement is unknown, as the only recorded migration aside from the migrations discussed in this chapter, was that of laborers going from Java to the East Coast of Sumatra or to other parts of the Outer Islands after 1911.[4]

The first attempt to obtain data on the extent of interisland migration was made in the census of 1930, which enumerated 825,000 Java-born Indonesians in the Outer Islands. Many of these, especially the wage laborers, were without doubt sojourners who planned to return to the island of their birth. That others had settled down in *Tanah Sabrang*, or The Land Beyond, as the Javanese call the islands across the sea, becomes apparent when one compares the number of Java-born (825,000) with the total number of persons in the Outer Islands for whom Java was the land of ancestry (1,151,000). In 1930, 967,000 Javanese, 129,000 Sundanese, 16,000 Madurese, and 39,000 other Indonesians of Java ancestry were found in the Outer Islands.

In the past Sumatra attracted the bulk of the migrants leaving Java. Of the 825,000 Java-born Indonesians enumerated in the Outer Islands in 1930, 767,570 persons, or 92.2 per cent, lived in Sumatra (Table XXXIII). Of these latter, 462,360 persons, or about 60 per cent, were concentrated in the residency of the East Coast, where they formed 31.4 per cent of the population. So extensive had been the migration from Java to Sumatra that by 1930, among the Indonesian population of Sumatra, the total number of persons for whom Java was the land of ancestry was 1,046,000, or 13.5 per cent. In fact, the Javanese from central Java had become the third largest ethnic group in Sumatra, comprising 11.4 per cent of the total

2 Danhof: "Bijdrage," 1940; Vonk: "Systematisch beheer van het ladangareaal," 1937.

3 Raffles: The History of Java, 1817, Vol. 1, p. 71.

4 Pelzer: Die Arbeiterwanderungen in Südostasien, 1935, pp. 97–108.

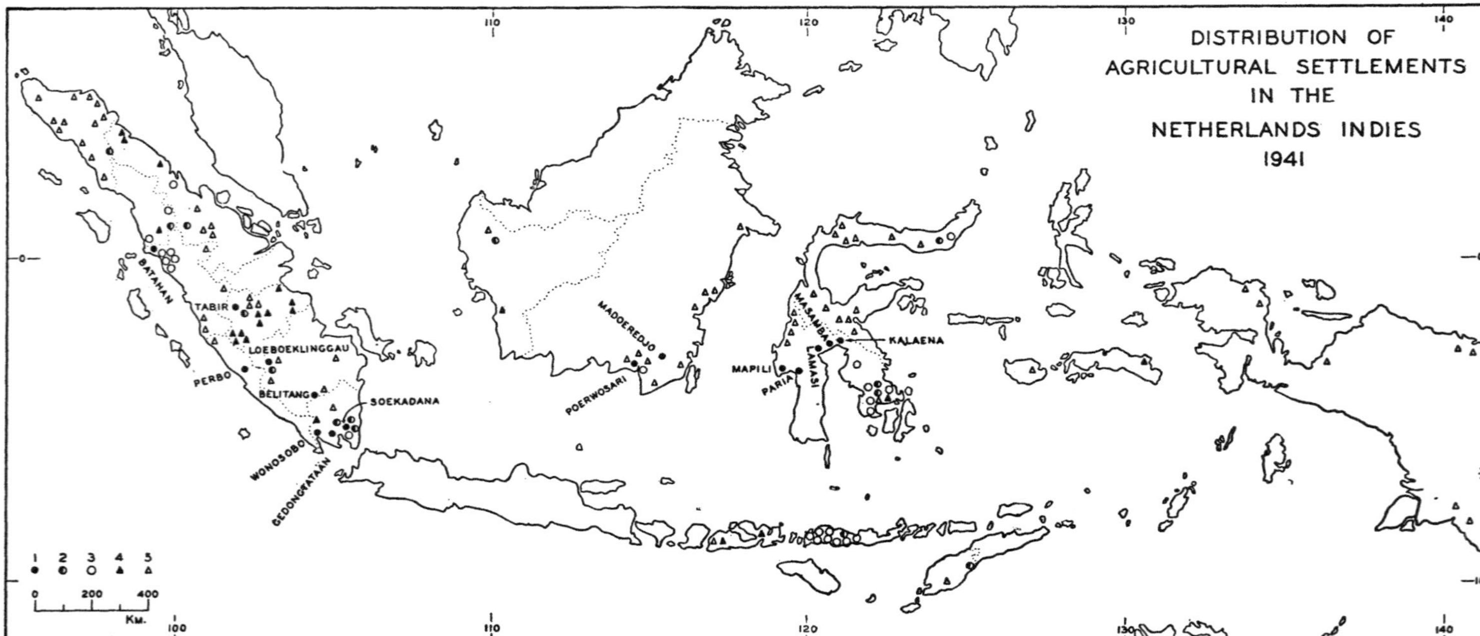

Fig. 164—Agricultural colonies and areas under consideration for settlement in the Netherlands Indies, 1941. The numbered symbols indicate: 1—Agricultural colonies actually existing in 1941; 2—Areas approved by preliminary soil and irrigation surveys; 3—Areas rejected; 4—Areas visited by reconnaissance parties but not yet scheduled for irrigation survey; 5—Areas considered, but not yet surveyed by reconnaissance parties. (Based on data supplied by the Central Commission for the Migration and Colonization of Natives.)

Indonesian population of that island; they were exceeded in number only by the Minangkabaus (25.6 per cent) and the Bataks (15.8 per cent). On the other hand, the migration of Indonesians from Java to such islands as Borneo, Celebes, or New Guinea, had hardly begun by 1930.

Agricultural Colonization During the Experimental Period 1905–1931

The Founding and Early Development of Gedongtataän, 1905–1911

Although individuals had long been pointing out the need for an alleviation of Java's population problem, the situation did not receive official recognition until the beginning of the present century. After it had been admitted in the Speech from the Throne in 1901 that the condition of the native population in Java was unsatisfactory, the Minister of Colonies requested C. Th. van Deventer, whose famous article, "A Debt of Honor,"[5] had called attention to Java's misery, to propose ways of improvement, and it was then that Van Deventer put forward his formula of "education, irrigation, and emigration."

At the same time the Government of the Indies, alarmed by reports of overpopulation and poverty in the residency of Kedoe, authorized Assistant-Resident H. G. Heyting to study the question of the migration of Javanese to the Outer Islands. Heyting began his survey in Kedoe and then, taking two assistant *wedanas*[6] and six villagers from Kedoe with him, he went to Sumatra. For 3 months they traveled in the residencies of Palembang, Benkoelen, Sumatra's West Coast, and the Lampoeng Districts, searching for suitable colonization terrain. In December 1903 Heyting submitted a report in which he proposed five colonization projects in Java and six in the Lampoengs.[7] According to his plan, 22,940 families were to migrate within 10 years from Kedoe to the new colonies at an estimated cost of 7 million guilders. Heyting wanted to start with a nucleus of 500 families in each project. These families would receive full support from the government during the first pioneer year and would be followed in the second year by 100 families and in the third year by 200 families, the settling of which should not be quite so expensive as that of the nuclear families, since the earlier settlers could provide the newcomers with temporary shelter. Heyting also foresaw that each family brought to Sumatra at the expense of the government would attract related families and thus set in motion a spontaneous colonization.

The government considered Heyting's plans too ambitious and decided to experiment on a much smaller scale in one area only. The selected locality was Gedongtataän (Map, Fig. 164), a terrain in the Lampoengs

5 Van Deventer: "Een eereschuld," 1899.

6 Wedanas and assistant wedanas are Indonesian officials of the Department of the Interior.

7 "Schema van een emigratie-plan," unpublished.

lying south of the Wai (river) Sekampoeng in the vicinity of a highway running from Teloekbetoeng to Kotaägoeng. The area seemed to have ample water and therefore to be suitable for the making of *sawahs*. With an assistant wedana and two *mantris*[8] of the Irrigation Service, Heyting arrived in Gedongtataän in May 1905 to prepare for the first settlers. He had a small clearing made and sheds built to house them until they could construct their own homes.

Late in 1905 Heyting transferred the first group of colonists, consisting of 155 families, from Kedoe to Gedongtataän, where they founded the *desa* of Bagelen. The establishment of Gadingredjo, a second desa of 555 families, was authorized in March 1906. During 1907 and 1908 no new colonists came over from Java, but a third settlement was established in 1909, a fourth in 1910, and a fifth in 1911. Then new colonists were added to one or another of these five settlements.

The pioneers who were recruited between 1905 and 1911 received a premium of 20 guilders a family before leaving Java and free transportation, which cost the government about 50 guilders a family, from their homes to the settlement. While clearing the land and preparing the irrigation systems they were paid 0.40 guilders a day for subsistence. In 1909 it was estimated that it cost the government 150 guilders to provide food for one family until it became self-supporting. In addition, the settlers were supplied with building and roofing material for their houses, worth 65 guilders, and tools, worth 13.50 guilders. The financial aid granted by the government thus amounted to approximately 300 guilders a family. Each family received about a hectare of land, made up of 0.7 hectare of sawah and up to 0.3 hectare of garden and dwelling-land in the village. If local circumstances permitted, they were also given some *tegalan*. In his report for 1903 Heyting wrote:

> Everyone who is familiar with Javanese conditions understands that colonization on a large scale has a chance of success only if the Javanese receive an area where, under their own leaders and administrative systems, they devote themselves quietly to agriculture as the mainstay of their livelihood.

He therefore persuaded village officials also to migrate from Kedoe to Sumatra, promising them that all their Javanese institutions would be preserved and that the sawah area would be held in communal ownership without periodic redistributions. Communal possession was considered by Heyting especially advantageous since it would make cooperation and group work necessary. As remuneration for his work, each village official was given a good-sized piece of land (*ambtsveld*) in addition to the normal amount that each settler received, and until the colonists were able to cultivate the ambtsvelden belonging to their desa officials, the latter were paid a small salary.

The colonization experiment was supervised by Heyting until the end

8 A mantri is an Indonesian technical assistant.

of October 1906, when his successor, Van Dissel, continued along the lines that he had worked out.

By March 31, 1911, there were 4,818 Javanese in the four desas of Bagelen, Gadingredjo, Poerworedjo, and Wonodadi. According to Van der Zwaal these settlers had cost the government 750 guilders per family of five (or 600 guilders for a family of four). This statistical average cost per permanently settled family is twice that of the government outlay of 300 guilders to each migrating family; death and desertion and the cost of irrigation systems and administration account for the difference. The death rate was very high because of malaria and dysentery. Furthermore, many colonists had deserted to seek employment on plantations in the neighborhood.

Settlement Problems During the Phase of the Lampoengs Credit Bank 1911–1928

During the first 6 years of the colonization experiment the government continued its policy of paying a premium, the costs of transportation and of building material, and living expenses covering 2 years, while the settlers were constructing their homes and clearing the land. In March 1911 the government decided to revise its system of partly supporting settlers, by substituting loans for outright gifts. It established the People's Credit Bank of the Lampoeng Districts, the chief function of which was to finance Javanese colonization. Under the new plan the settlers received a premium of 22.50 guilders a family and free transportation. For the rest, they were entitled to borrow a maximum of 200 guilders a family from the People's Credit Bank at an interest rate of 9 per cent. The borrowed money was to be paid back within 10 years, and repayment was to begin in the third year. This sum was expected to take care of their subsistence during the first 2 years. After they had brought their land under cultivation they could borrow additional sums for the purchase of livestock, seeds, and tools, and for home improvements. This financial policy in the colonization work was continued until 1927.

One may question whether this departure from Heyting's plan improved matters. Both schemes were too generous with financial support. Moreover, the fact that the first settlers were not required to repay anything may have influenced later settlers not to take their loans too seriously. The Javanese villager was unaccustomed to working with bank credit and must have considered the government a very patient and lenient moneylender. In the first years, especially, the bank gave credit too easily, and the settlers failed to make the best use of it. According to reports, they lived and dressed better than ever before, but the consumption of alcohol spread among them, and 50 of them in 1916 were opium addicts. Unfortunately, no investigation was made as to the number who had been habitual consumers of alcohol, or opium, before their migration. Further reports of those years complain that many colonists had not been *tanis* in Java; they lacked initiative and required constant supervision.

While the settlers were partly to blame for the slow progress of the settlement, the administration, too, had its shortcomings. Although Heyting had stressed the importance of irrigation in the early stages, no experienced irrigation engineers had been employed but instead only mantris of the Irrigation Service. Heyting had mistakenly believed that the Javanese were sufficiently skilled and familiar with the methods of irrigation to handle the problem without the help of trained engineers. We may venture the explanation that in the overpopulated areas of Java most of the sawahs had been laid out long ago; hence, although the people of the present generation in such districts may be excellent cultivators of sawah, they have had little or no experience in turning tegalans into sawahs. Because of this lack of expert knowledge, the irrigation system was far from satisfactory. Schalkwijk, the manager of Gedongtataän, reported in 1917 that many fields had too much, others, too little water. Furthermore, several villages had been laid out on low-lying land with the fields on higher ground. As a result, the villages were flooded and unhealthy during the wet season, while the rice fields suffered from a lack of water. It therefore became necessary to change the locations of villages and fields, as was done in the desas of Karanganjer, Tambahredjo, Poerworedjo, and Baledono. Unfortunately, such mistakes in the original planning were costly and retarding.

The failure to establish an efficient health service in the early years was also a serious and costly matter. Until the end of 1914 there was only a small, primitive hospital in the colony, and the death rate was high, because of widespread malaria, hookworm, and dysentery. Only late in 1914 did the government appoint a doctor for the colony. He found that the deaths exceeded the births and that the growth of the colony depended entirely upon immigration. As a result of the health measures insisted upon by the doctor,[9] the death rate declined in 1915 and 1916, but during the influenza epidemic of 1918 and 1919 it reached new highs (Table XVI).

In the early years the administration tried to persuade other colonists

[9] Schalkwijk: "De kolonisatieproeven," 1918, pp. 423–425. He recommended (1) keeping a systematic death register; (2) arrangements with the Central Hospital at Tandjoengkarang for the hospitalization of all serious cases; (3) vaccination of all new colonists against smallpox, treatment of hookworm, and examination for intestinal diseases, malaria, anemia, tuberculosis, and leprosy—all this to be done before colonists were to be admitted to the settlement; (4) conference with health officials before a decision should be made on the location of a new desa; (5) supervision of the sale of rice (settlers frequently did not keep enough for their own needs); (6) improvement and extension of irrigation; (7) continuation of the anti-hookworm campaign; (8) fight against infectious intestinal diseases by improving the drinking water and insuring a proper disposal of feces: (a) construction of at least one well in each village—up to that time the colonists had obtained drinking water wherever rain water collected in depressions, as it had been the opinion of the European officials that the population did not feel the need of an elaborate and expensive drinking-water supply system and would not use it; (b) transfer to the hospital of all patients suffering from intestinal diseases, in order to prevent epidemics; (9) study of the problem of malaria and the distribution of quinine free of charge through the desa officials; (10) hospitalization of all persons infected with yaws.

to take over the land and debts of deceased settlers. But this proved to be nothing more than a juggling of the items in the balance sheet, since very

TABLE XVI—GEDONGTATAÄN COLONY: POPULATION AT THE END OF THE YEAR, AND BIRTH RATES AND DEATH RATES PER THOUSAND, 1910–1938

YEAR	TOTAL POPULATION	BIRTH RATE‡	DEATH RATE‡
1910	4,818	*	*
1911	5,099	*	*
1912	6,251	*	*
1913	6,701	*	35.0
1914	6,930	*	43.7
1915	7,327	*	28.7
1916	7,677	*	26.5
1917	7,888	*	37.2
1918	8,903	*	54.3
1919	11,997	*	74.3
1920	16,349	*	42.6
1921	19,251	37.5	41.7
1922	19,554	42.2	31.9
1923	19,823	40.8	40.3
1924	20,144	36.0	27.2
1925	20,650	44.8	38.5
1926	21,234	36.9	25.7
1927	21,827	43.4	26.2
1928	25,090	41.2	23.9
1929	25,889	37.7	27.9
1930	28,345	36.8	19.7
1931	29,863	31.2	23.4
1932	32,465	30.0	37.9†
1933	31,965	30.0	26.5
1934	31,935	33.8	46.1†
1935	32,534	30.6	24.3
1936	33,374	34.9	24.8
1937	34,358	41.1	23.4
1938	35,350	39.3	21.0

* Data unavailable.
† These high death rates were caused by malaria epidemics among newly arrived colonists.
‡ Rates computed by the author.
Source: Data for 1910 through 1932 from Van der Zwaal: "De Javanenkolonies Gedong Tataän en Wonosobo," 1936, pp. 205, 206, and 208. Data for 1933 through 1938 from the second volume of the *Indisch Verslag* for each of the years from 1934 through 1939.

few of the settlers were able to pay off two accounts, and in the end the debts had to be written off.

The recruiting system, which was in charge of Indonesian civil service officials in Java working through village heads, or *lurahs*, had its faults, especially in regard to the type of colonists selected. Many lurahs used this occasion to rid their villages of undesirable members, and of sick and aged people, who of course make poor pioneers. Believing that better settlers could be obtained by officers who had the interest of the colonies at heart, the manager of the Gedongtataän colony, accompanied by an as-

TABLE XVII—WONOSOBO COLONY: POPULATION AT THE END OF THE YEAR, AND BIRTH RATES AND DEATH RATES PER THOUSAND, 1922–1938

YEAR	TOTAL POPULATION	BIRTH RATE†	DEATH RATE†
1922	2,720	14.7	41.0
1923	4,288	14.0	72.0
1924	5,927	11.5	61.0
1925	5,558	16.6	41.0
1926	3,356	30.1	70.9
1927	3,337	35.4	56.6
1928	3,200	*	33.4
1929	3,310	10.0	22.4
1930	3,414	17.6	23.5
1931	3,438	17.6	25.0
1932	3,623	23.3	13.0
1933	3,644	30.7	10.1
1934	3,748	32.5	16.5
1935	4,048	34.3	26.2
1936	4,579	12.6	22.3
1937	4,807	30.2	10.4
1938	5,280	29.3	12.1

* Data unavailable.
† Rates computed by the author.
Source: Data for 1922 through 1932 from Van der Zwaal: *op. cit.*, p. 214. Data for 1933 through 1938 from the second volume of the *Indisch Verslag* for each of the years from 1934 through 1939.

sistant and several lurahs, went to central Java in 1913 to select colonists. This method of selection was an improvement; the more carefully selected settlers had better gardens, better crops, and smaller debts. The new procedure was apparently not pursued for long, however, for in 1920 Schalkwijk, the manager, reported the unsatisfactory state of recruiting. Thereupon he went to Java and selected 1,500 families within 3 months. He left a mantri, who had worked in the colony of Gedongtataän and knew how to select good, sturdy settlers, to be stationed in Koetoardjo, Kedoe Residency, Middle Java, and to continue recruiting in that province as well as in the principalities.

From 1905 to 1929 the colonization experiment had to compete with the recruiting agencies of great plantation organizations in the Outer Islands, which were almost entirely dependent upon Java for their labor supply. Their recruiting agents spread many rumors in the desas of Java to discredit Gedongtataän, so that, were it not for the premium, it would probably have been impossible to persuade any Javanese to go to a government colony in the Lampoengs. For this reason it was necessary to continue paying the premiums until the depression forced the plantations in the Outer Islands to reduce their labor force and thus temporarily to stop recruiting. Even though the plantations in the Lampoeng Districts, especially the coffee plantations, had found a welcome labor market in Gedongtataän, since the colonists were glad to increase their income by earning supplementary wages, most of the plantations looked on the colonization experiment with displeasure. The government itself showed a vacillating attitude. During periods of great labor demand in Java and the Outer Islands it lost interest and drastically cut the colonization budget. In 1914, for example, the representative of the Ministry of Colonies declared in the discussion of the budget before the States-General in the Hague that the necessity for colonization had been eliminated by the increasing demand for wage laborers. Then between 1918 and 1923 the government again increased the funds so that Gedongtataän grew at a rapid pace; but from 1924 until the depression little was done to further agricultural colonization. During the second phase of the colonization, between 1911 and the end of 1927, the population of Gedongtataän increased from 5,100 to approximately 21,800 persons, or by a yearly average of 1,000 persons. This increase was the difference between the sum of immigration and births and that of desertions and deaths (Table XVI).

In 1921, at a moment of renewed interest, the government authorized the opening of a colonization terrain at the head of Semangka Bay, west of the town of Kotaägoeng. The project, which was originally named after this town, is now known as the Wonosobo colony (Map, Fig. 164). It comprises about 3,500 hectares. The first group of 870 families, including 2,720 persons, arrived in 1922. Most of the men, having once been contract laborers on plantations in the Outer Islands, were not prepared for pioneer life, since as laborers they had been accustomed to taking orders instead of working on their own initiative. Largely because of this early experience with former contract laborers at Wonosobo, the recruiting regulations in later years closed agricultural colonies to plantation workers. In 1924 the population of the Wonosobo project had increased to 5,927 persons, an all-time record, at least until 1938, the last year for which data are available. The death rate was shockingly high between 1923 and 1927, because of malaria (Table XVII). Frightened by the many deaths and the unexpected difficulties in connection with irrigation, a great number of the Wonosobo colonists fled, leaving their debts.

The Wonosobo project emphasizes the danger not only of inadequate preparation but also of poor management. The financial policy of the Lampoengs Credit Bank was unfortunate. Credits were given without supervision of the use to which they were put. The first group of settlers had borrowed about 36,000 guilders from the bank during their first year. Later groups went into debt at a similar rate. This expense, together with the high losses incurred from death and desertion, made the costs of the project very great. It has been estimated that by 1928 the Wonosobo colony had absorbed 630,000 guilders; for each permanently settled colonist the government had spent 200 guilders, or 800 guilders for a family of four persons.[10]

In addition to this dubious policy in granting credits, as a result of which the greater part of the debts had to be written off as uncollectible, the Lampoengs Credit Bank had been seriously mismanaged. Fraud, embezzlement, and faulty bookkeeping, which were discovered in 1926, had created chaos. Finally, in 1928, the government decided to write off the entire deficit and ended this second phase of the colonization experiment with the liquidation of the Credit Bank.

The Final Phase of the Period of Experimentation, 1928–1931

During the years from 1928 through 1931, an interval that may be designated as the third phase of the period of experimentation, the government seems to have had no clear policy regarding agricultural colonization. Officials were discouraged by the fiasco of the Lampoengs Credit Bank and the bad reputation that colonization had earned.

In spite of the general discouragement one important innovation was introduced. Someone noticed that in 1927, 914 persons, including dependents, had migrated of their own accord to the colonies in southern Sumatra. These spontaneous migrants, relatives or friends of colonists, had had no need of the inducements of premiums, free transportation, and other aid from the government, but had been attracted solely by the prospect of getting free land and of being able to support themselves until their first harvest by working for established settlers. The settlers established in Gedongtataän were anxious to see a growth of their desas since each new homestead lightened the burden of upkeep of the roads and other public work that had to be taken care of, and in general strengthened the village. The desa officials, too, favored a rapid growth of the village population for personal financial reasons. Despite Heyting's emphasis on the importance of reducing colonization costs by encouraging spontaneous family migration and incorporating it systematically into the colonization scheme, the government had heretofore failed to do so. Instead, the concept had become fixed that each settling family would require the same amount of financial support. Now, however, for the first time the government took cognizance of the possibility of mutual aid between established and incoming colonists and reduced its financial support to the payment of the costs of transportation alone. That the new experiment was successful is shown by the fact that in 1928 as many as 1,693 persons were sent to

10 Van der Zwaal: "De Javanenkolonies," 1936, p. 214.

southern Sumatra at a total governmental expenditure of about 25,000 guilders, or not quite 15 guilders a person, for transportation. In the same year an additional 687 persons went to the colonies in southern Sumatra entirely at their own expense.

But the government did not follow up this important innovation immediately. In 1929 and 1930 the colonies in southern Sumatra received only such colonists as migrated independently of any financial aid from the government. Altogether a total of 3,500 Javanese migrated to these colonies without depending upon public funds during the years from 1927 through 1931. Thus, finally, after 25 years of priming, a small but steady stream of migrants had begun to flow. These new pioneers did not have to be enticed by the 22.50 guilder premium and were definitely of a better type than the earlier colonists. For the premium in the opinion of a Javanese reduced pioneering to the level of selling one's self as a contract laborer to a plantation. These settlers, on the contrary, did not belong to the poorest strata but had sold their meager possessions in Java in order to profit from the greater opportunities awaiting them in the Outer Islands.

While from 1905 to 1931 the main stream of colonists flowed into Gedongtataän and Wonosobo, there was a small trickle directed to the neighboring residency of Benkoelen, where it separated into numerous tiny colonies. During this period two other experiments in colonization were made, in Celebes (1906) and in Borneo (1920); both were failures.

The Costs of Colonization during the Experimental Period

It is difficult to obtain a clear picture of the achievements and costs of colonization because of insufficient data—the government has not opened all of its records—and because of the confusion in the records available, due to the mismanagement of the Credit Bank. According to the annual budgets from 1905 to 1927, the government spent 2,500,000 guilders. To this sum C. C. J. Maassen[11] adds the 1,150,000 guilders that were lost by the bank, which would bring the cost of colonization to 3,650,000 guilders, not including the sums invested in irrigation systems.[12] Subtracting some 140,000 guilders, which were spent on limited experiments in Benkoelen, Borneo, and central Celebes, Maassen estimates that Gedongtataän and Wonosobo together had cost about 3,500,000 guilders. He divides this sum by 24,300, representing the number of persons present in the colonies in 1928, and concludes that each person had cost the government 144 guilders, or 576 guilders a family.

This conclusion of Maassen's would seem to require qualification. In the first place, the number of families who had been transferred to Sumatra

---

[11] Maassen: De Javaansche landbouwkolonisatie, 1937, pp. 10–11. Maassen, an official of the Department of the Interior, was in charge of agricultural colonization in the later 1930's.

[12] The irrigation project of Gedongtataän cost 1,486,000 guilders and that of Kotaägoeng, 166,700 guilders.

and thus had profited by the expenditure was considerably larger than the number living in Gedongtataän and Wonosobo in 1928. Many of the former had deserted the colonies and found employment on plantations or with local pepper growers, or had settled in the Lampoengers' villages; in other words, since they had left Java for good, they might well have been included in the reckoning and thus have reduced the estimated cost per individual. In the second place, it seems unsound to include the sum lost by the bank through fraud and mismanagement. If this sum is excluded, Maassen's reckoning would show that each permanent settler represented an expenditure of 100 guilders (400 guilders for an average-sized family).

After 1931 agricultural colonization was greatly expanded, so this year may be said to mark the end of the experimental stage. A great deal of experience had been gained by the men actually working among the Javanese in the Outer Islands as social engineers during the years after 1905. The recommendations[13] contained in the annual reports of these col-

[13] The annual reports of Heyting and his successors contain the following recommendations summarized from the annual reports covering the years 1907 through 1915 published in the colonial reports of those years, and also from Schwalkwijk: "De kolonisatieproeven," 1918, and Idem: "De kolonisatie met van Java afkomstige gezinnen," 1921: (1) Select terrain for agricultural colonization that has dry land for the desas as well as extensive areas suitable for the laying out of wet-rice fields. Only such fields allow for the intensive cultivation of rice year after year. Only colonies with ample wet-rice land give a guarantee of success. (2) Make careful surveys of the soil and especially of the possibilities for irrigation. Do not start any new colony without such surveys. (3) Select only forested land. Unless work animals and steel plows are furnished, grasslands are more difficult to cultivate than forest soils. Forest soils, because of their acidity, require 3 to 4 years of use before they can be irrigated. This gives time for the building of irrigation systems. (4) Employ efficient irrigation specialists to construct satisfactory irrigation systems. Attempts to save here lead only to losses. (5) Select large areas that offer space for many villages so that the first settlers, once established, can become a nucleus for new arrivals. The establishment of a nucleus is always expensive; but nuclei cut down the costs of settlement of later colonists who settle in the neighborhood. Older colonists need additional help, above all during harvest time, and such help cannot be found locally. (6) Send new colonists to well established colonies just before the harvest season so that they can earn sufficient rice to carry them over to the time when they can harvest their own first crop. (7) Establish your first villages—the future nuclei—in the neighborhood of existing kampongs. The latter will offer a market for the Javanese and will give them employment at a time when they need money badly. Women, for example, can find employment as rice huskers. (8) Select the colonists with the greatest possible care. Take only strong, young people who are really tanis. Do not take persons who have been laborers on Western plantations; they have lost a great deal of initiative and capacity for working on their own and, above all, have become used to regular weekly or monthly payments of wages, whereas the tani must wait for months before he gets a return for his labor. Take only married people; single men are likely to desert as soon as they are confronted by a difficulty. (9) Do not try to attract settlers by offering them a premium as inducement. This may have been necessary when colonization was in its first stages and had to compete with recruiting for plantations, but is no longer so. (10) Do not offer relatively large money credits to colonists, for they are not accustomed to these, and experience has shown that the credits are frequently used unwisely. (11) Protect the health of the settlers as much as possible during the pioneer stage when conditions are most primitive and the work most exhausting. Efforts to economize here will only cause losses.

onization officials were based on their practical experience and later formed the gist of the "ten commandments of colonization" drawn up by Maassen (see pp. 209,210).

Large-Scale Colonization 1932–1941

The pessimism that prevailed in the Government of the Indies over the results of the colonization experiments at the end of the 1920's was based primarily on their costliness. In 1926 it was estimated that future settlements would require an expenditure of 300 guilders per family, apart from the construction of irrigation works, so that for the next few years it was the general opinion that the cost of settling Javanese in the Outer Islands was too high for the country to undertake large colonization projects. As we have seen, this opinion was encouraged by powerful plantation interests hostile to the colonizing efforts of the government, because of the latter's competition with their own recruiting activities.

Then the depression came: the markets for export crops shrank rapidly, and the prices obtainable for the produce declined even faster. The Javanese lost most of their opportunities for employment in Western enterprises in Java at the same time that the plantations of the Outer Islands were sending Javanese laborers home by the thousands.[14] Conditions became especially strained in the districts of Java that had suffered from congestion even during the prosperous years. Relief was badly needed, all the possibilities had to be surveyed, and Gedongtataän was examined once more.

So it came about that the officials in Batavia for the first time saw the possibilities of greatly increasing the scale of the spontaneous migration already taking place, by utilizing the need of the Javanese colonists for

14 At various times, especially during the early years of the depression of the 1930's, long negotiations were carried on between the government and the planters about the possibility of establishing permanent Javanese colonies on, or in the neighborhood of, the plantations of the East Coast of Sumatra, in the hope that this would assure the planters a reliable labor force and also prevent the return of thousands of laborers when Java, too, was suffering from a decline in employment opportunities. These negotiations led nowhere. At the time of the granting of agricultural concessions no space had been left even for a growing indigenous population; so that establishment of large-scale Javanese colonies in the immediate neighborhood of the plantations but outside of the concession area was impossible. On the other hand, the planters refused to agree to the founding of colonies on concession land for several reasons, such as that these would require too much land, and that the planters anticipated that their laborers would no longer be at their beck and call if given sufficient land to provide a livelihood. The planters were therefore willing to furnish each laborer of good standing with only a tiny garden plot of one-tenth of a hectare—large enough collectively to provide the atmosphere of a desa but too small to grow enough foodstuffs for the families. At the same time the planters reserved the right to evict a laborer if he did not work regularly. This arrangement did not, of course, merit the name "colonization" and could not forestall large-scale repatriation of Javanese. Additional information on these pseudo-colonies of Javanese on plantations in the Outer Islands may be found in *Verslag . . . betreffende de kolonisatie van Javaansche werklieden* . . . 1920, and in Brokx: Rapport . . . over de randkolonisatie (ms. dated 1933).

harvest hands.[15] A routine report coming from Gedongtataän and Wonosobo in November 1931 contained the statement that some colonists had been anxious to have relatives join them for the rice harvest, but that the migration had not materialized because of the lack of money for railroad and steamer fares.

Therefore in December 1931 the government sent an investigator to

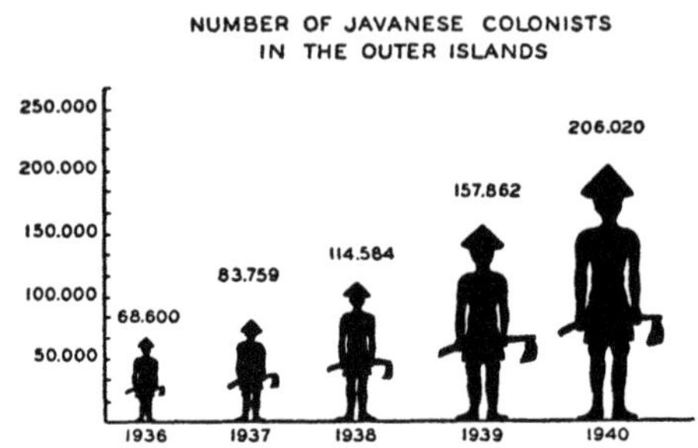

FIG. 165—The numbers of Javanese colonists in the Outer Islands at the ends of the years, 1936 through 1940. (Based on data supplied by the Central Commission for the Migration and Colonization of Natives.)

Sumatra to determine the number of colonists willing to take relatives into their homes and to employ them as harvesters. More than 1,000 households responded, but most of the relatives whose names had been given to the investigator could not be found in Java or were unwilling to migrate. The colonists then agreed to take unrelated families instead, and some 7,000 such persons were sent in 1932. In accordance with Javanese custom, the new arrivals were paid in a share of the crop, a form of payment called *bawon*. In overpopulated Java the share may be only one-eighth or less of the amount that the harvester reaps; but in Sumatra, where labor is scarce, as much as one-fifth or even one-fourth must be promised to the reapers. Thus the newcomers were able to earn enough rice to carry them through the period of clearing and planting their own land and until the maturing of their first crop.

15 To one unfamiliar with Javanese agricultural methods it may come as a surprise that a family having as little as a single hectare (2.47 acres) of land would require help at harvest time. The explanation lies in the fact that in accordance with a custom which the Javanese share with many other groups in Southeastern Asia rice may be cut only head by head. The cutting instrument, called *ani-ani*, consists of a narrow blade, 2 inches long, which is fastened crosswise to a short stick. The use of this instrument is considered the most careful method of harvesting rice, but at the same time one of the least economical from the standpoint of man-hours. The practice is deep-rooted in Javanese tradition, however, and has become a measure of social security on the mother island. Some other provision would have to be made for the poor families of the community and for the widows and orphans, if the more fortunate members should begin to use such labor-saving devices as sickles and scythes.

Maassen credits the officials of 1931 with discovering the possibility of reducing the government's expenditure per settled family by utilizing the settlers' need for harvest hands.[16] It may be true that it was 1931 or 1932 before the officials in Batavia saw the possibility of building the need for harvesters into the colonization scheme, but for about 25 years the reports from the men in charge of Gedongtataän had been calling attention to it.[17] It is therefore clear that the recommendations of the colonization officials working in the Outer Islands had been largely disregarded in Batavia before 1932 and that the new technique of 1932–1941 consisted fundamentally in putting these suggestions into execution.

Colonization on a large scale was finally made practicable by the incorporation into the government's scheme of colonization of the principle of mutual aid and assistance (*tulung menulung*), requiring established settlers to employ newly arrived colonists as harvesters under the bawon form of payment. The voluntary assistance that some of the colonists had given to spontaneous migrants, in return for their help at harvest time, and which had attracted many uncounted newcomers to the colonies each year, was thus made compulsory. Under the new system the burden of supporting newly arrived colonists was essentially shifted from the shoulders of the government to the shoulders of established Javanese settlers. Whenever possible, settlers were sent just before harvest time to the colonies where they were to work as harvesters. When this was not possible, amounts of rice equivalent to what they would have earned were advanced to them in the form of a loan.

The new system was indeed in striking contrast to that adhered to between 1905 and 1911, when each family was given a cash premium for migration and was paid as much as 150 guilders in cash for clearing its land, as well as being allowed free transportation; or to the system followed between 1911 and 1927 when each family was paid a cash premium for migration and could borrow 200 guilders or more at an interest rate of 9 per cent, again with free transportation. The new system represents an even further reduction from the 1928 practice of limiting aid to transportation cost only. In 1932 and thereafter the colonist was given practically nothing free except his land. He was charged between 22 and 25 guilders

16 He writes, "In 1932 the authorities noted that the aid needed by the fresh colonists during the first period of their life in the new country *need not all come from the government; that in fact the older colonists could be counted on to supply a large part* of it, so large a part even, *that in certain cases government aid could be dispensed with altogether*, when once the newcomer had safely reached his destination." (Italics mine). See Maassen: "Javanese Agricultural Migration," 1939, p. 187.

17 As early as 1905 Heyting had pointed out that only the first colonists in each project would require full government support, and that subsequent migrants could be colonized at less cost because of the help they might receive from those already settled. Again, in 1918, Schalkwijk, the manager of the Gedongtataän project, stated that it would be advantageous to bring colonists to Sumatra shortly before the rice harvest because it would give them an opportunity to earn enough food for the first few months of pioneering until their own first crop should be ready for harvesting. See Schalkwijk: "De kolonisatieproeven," 1918, p. 433.

for transportation and household utensils and tools, a sum which had to be paid back within a two- or three-year period. The charge for transportation alone was 12.50 guilders per family, regardless of the size of the family or the distance involved.

The Soekadana Colony

Although most of the settlers in 1932 were absorbed by the desas of Gedongtataän, that colony was filling up so rapidly that it could not have continued much longer to supply new colonists with homesteads. A large new district suitable for Javanese colonies had to be found, preferably in the neighborhood of Gedongtataän, so that this well-established colony might continue to function as a nucleus or, rather, as a bridge. The choice fell upon the Soekadana district, northeast of Gedongtataän, and the first colonists were sent there as early as 1932 (Map, Fig. 166). However, the government decided not to proceed with large-scale migration until more detailed surveys and plans should be completed.

Although the colonists of Gedongtataän applied for 2,300 families, or approximately 9,000 persons, whom they wished to employ during the harvest season in the first quarter of 1933, no new colonists were brought to the Lampoeng Districts in that year.[18] In 1934 the migration of prospective settlers was again supported, but their number was kept to the rather small total of 1,375 persons, who were settled on *marga* land (see p. 207) in the neighborhood of Gedongtataän. Four desas were established in the marga of Wai Lima and 10 in the marga of Wai Semah. At the end of 1934 these had a total colonist population of 4,000 persons and at the end of 1935, of 8,700 persons.[19] These desas may be considered an expansion of Gedongtataän, for they were settled principally by a number of young people from Gedongtataän who took land in the new desas because their own villages no longer had any surplus.

In 1935 detailed plans for Soekadana had been perfected and the sending of settlers on a large scale could begin. During that year 5,500 persons were added to the small group that had been established in the Soekadana district three years earlier. Another 8,784 were sent to that district in 1936, 6,715 in 1937, 13,882 in 1938 and approximately 18,000 in 1939. The plans for 1940 and 1941 provided for an annual gain through migration of 6,650 families, or about 20,000 persons. At the beginning of 1940 the Soekadana colony had a population of 47,000, whereas the Gedongtataän colony had only 38,000 persons. Within 7 years Soekadana (see pp. 221, 222) had acquired through the effectiveness of the government's new program a much larger population than had Gedongtataän in 35 years.

18 Maassen: De Javaansche landbouwkolonisatie, 1937, p. 41. No obstacles were placed in the way of the migration of Javanese who came to the Lampoeng Districts at their own expense, and 827 persons attached themselves to the colonies in 1933.

19 Van der Zwaal: "Nieuwe Javanenkolonies," 1936, p. 331.

The Work of the Central Commission for Migration and Colonization

Increased interest in agricultural colonization was manifested by the government shortly after the appointment of Governor General van Starkenborgh Stachouwer in 1936. One of his first acts was to establish the Central Commission for Migration and Colonization of Natives,[20]

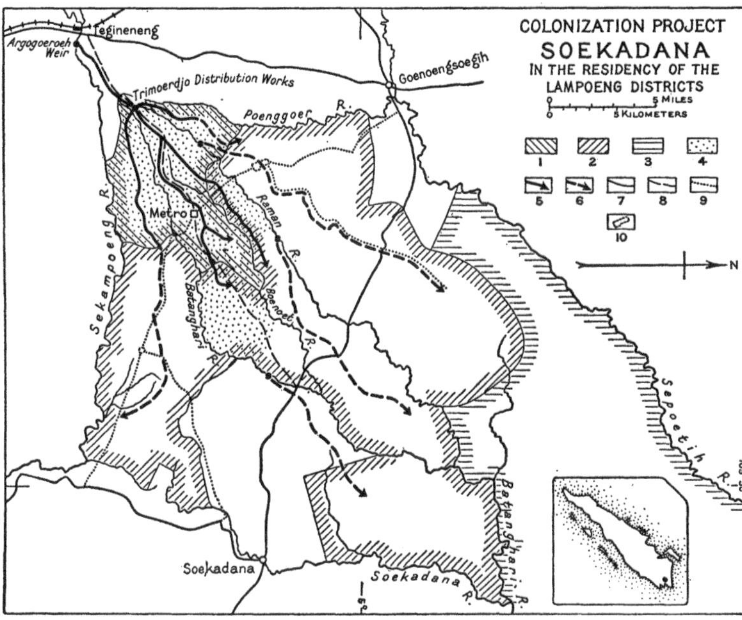

Fig. 166—Soekadana Colony, Residency of the Lampoeng Districts, Sumatra. The numbered symbols indicate: 1—Boundaries of the first Soekadana project (7,000 hectares in area); 2—Boundaries of the area added to the Soekadana project (43,000 hectares); 3—Boundaries of the area considered for further expansion (about 21,000 hectares); 4—Area already settled by colonists at the end of 1939; 5—Irrigation canal in operation; 6—Proposed irrigation canal; 7—Roads existing before the Soekadana project was undertaken; 8—Road constructed to serve the project; 9—Proposed roads; 10—Proposed towns. (Based on data supplied by the Central Commission for the Migration and Colonization of Natives.)

comprised of a member of the Council of the Indies, who served as chairman, and the heads of the Departments of the Interior and of Economic Affairs. The Adviser on Agrarian Affairs in the Department of the Interior was to act as secretary of the commission and to supervise its administration. The commission was to direct all the agricultural coloniza-

20 Government Order No. 30 of January 29, 1937, created the "Centrale Commissie voor Emigratie en Kolonisatie van Inheemschen." Government Order No. 18 of May 31, 1939, changed the name of the commission to "Centrale Commissie voor Migratie en Kolonisatie van Inheemschen."

tion involving Indonesians[21] and to advise and inform the government concerning it. Maassen, who as Adviser on Agrarian Affairs in the Department of the Interior had been in charge of agricultural colonization, became the first secretary of the commission and remained in office until the end of 1939. The success of the large-scale program was due in great measure to his energy, organizing ability, and firm belief in the importance of colonization in the solution of Java's population problem.

In March 1937 the General Secretariat of the Indies[22] circularized all the leading officials in the Outer Islands, notifying them of the governor general's desire to have a systematic search conducted for areas suitable for agricultural colonization and requesting their recommendations.[23] At the same time the governors of Java were requested to impress upon the members of the European and the Indonesian civil services and upon the lurahs the need of encouraging the migration of rural Javanese. This seems to have been the first occasion on which a governor general urged all the branches of the administration to do their utmost to increase Javanese migration.

Before the establishment of the commission, the 1937 budget had provided only 220,882 guilders for colonization. To speed up migration it was therefore decided to use 3,000,000 guilders out of the 25,000,000 guilder Welfare Fund, a gift of the Netherlands Government to the Indies Government. Additional sums were taken out of the Special Native Rubber Export Tax Fund and were allocated to the construction of such irrigation works in Sumatra as would benefit both the local Sumatran population and the Javanese immigrants.[24] The commission also had access to the Madura Welfare Fund for the migration of Madurese.

The increased activities required a large staff of full-time employees. Before 1937 nearly all the work relating to colonization had been done by personnel whose chief duties were connected with other phases of governmental activity. After 1937 the general management was put in the hands of the secretariat of the commission, which coordinated the work of specialists borrowed from various governmental departments. At the end of 1940 the staff consisted of 369 persons, of whom 88 were drawn from the Department of the Interior, 124 from the Department of Economic Affairs, 129 from the Department of Public Works, and 28 from the Department of Public Health.

---

21 The problems involved in the migration and agricultural colonization of unemployed Indo-Europeans and Europeans as well as of Indonesian intellectuals do not come within the frame of this study. These settlers, who numbered at the most a few hundred persons, were under the jurisdiction of the "Kolonisatie Raad," a body created by Government Order No. 23 of January 12, 1937.

22 The General Secretariat conducts the governor general's correspondence and aids him in carrying out his decisions and orders.

23 Circular No. 645/B, March 25, 1937, of the general secretary.

24 For a discussion of the origin of this fund see Vandenbosch: The Dutch East Indies, 1941, pp. 302–303.

Comprehensive Surveys Initiated

One of the fundamental problems involved in launching settlements on the scale that the government considered absolutely necessary was to select suitable areas. In response to the circular requesting recommendations, the commission received data on a great number of areas that seemed to warrant further investigation (Map, Fig. 164). Sixty-two of these, comprising 1,345,000 hectares, lay in Sumatra; 10, with 297,000 hectares, lay in Borneo; 23, with 438,000 hectares, in Celebes; and an additional 21, with 112,000 hectares, were to be found on various other islands. From among these the commission selected for immediate investigation those areas that seemed most promising with respect to geographical location, soil, and irrigation. Each area was subjected to careful surveys by legal experts, soil scientists, and agricultural and irrigation engineers. The various phases of the investigation were undertaken in the order of the costs involved, starting with the least expensive so that unsatisfactory districts could be eliminated with a minimum of expenditure.

First, the questions of who controls the land and whether the owners would be willing to accept Javanese colonists were studied by local administrators. The securing of suitable land offers a complex problem, as much of the Outer Islands is occupied, permanently or temporarily, by the indigenous peoples unless, as is the case especially in Sumatra, it is under lease to plantations or is reserved by the Forestry Service.

The indigenous communities of Southeastern Asia, whether single villages, groups of villages, clans, or tribes, claim territorial sovereignty over extensive areas. Large parts of these areas are not actually used for agriculture but are needed as gathering and hunting grounds, as well as for such forest products as rattan, resin, and lumber. No persons, unless belonging to the community that claims territorial sovereignty over such land, may use it in any way without the consent of the community, and for this permission they usually have to pay a nominal fee. However, in the second half of the nineteenth century and at the beginning of the present one, when the Dutch extended their control over the Outer Islands, they pushed aside these traditional rights in order to open the way for plantation enterprises. This was done, as in Java, through the Agrarian Decree of 1870, which declared that these hunting and gathering grounds were wild land and therefore state property. According to Van Vollenhoven and his school, this declaration violated the sacred land rights of the indigenous communities.[25]

In southern Sumatra, so far the most important area of Javanese agricultural colonization, the highest indigenous political unit in each district is the marga,[26] which comprises a number of villages, or *dusuns*. The

---

25 Van Vollenhoven: De Indonesiër en zijn grond, 1932, pp. 65–82.

26 For a discussion of the marga in southern Sumatra, see Van der Zwaal: Inlandsch gemeentewezen in Zuid-Sumatra, 1936.

margas claim territorial sovereignty, and as the marga lands are contiguous there is no unclaimed land in southern Sumatra. However, for decades the Dutch did not recognize the margas—except in the residency of Benkoelen—but only the dusuns. In this way they released a great deal of the area as no-man's land, which they declared to be state property, and which then could be turned over to plantations under long leases or made available to Javanese colonists. But in recent years the margas have been recognized as indigenous communities with rightful claims to all the marga land, excluding that which had in the meantime been leased to plantations. In 1931 southern Sumatra was divided among the margas, except for the areas occupied by plantations, the city of Palembang, the Javanese agricultural colonies of Gedongtataän and Wonosobo, and several good-sized towns, such as Teloekbetoeng, Kotaägoeng, Lahat, Benkoelen, and Djambi.[27] Thereafter no Javanese agricultural colony could be founded in southern Sumatra without the collaboration and permission of the marga on the territory of which the colony was to be.[28] If there should be reasons why the land should be reserved for the marga members, no colonization of Javanese could take place. Marga officials have shown themselves anxious to support Javanese settlement, however, probably because of the financial advantages connected with it.

The land rights of the indigenous peoples often came to light in unexpected ways. For example, in the colony of Loeboeklinggau, in the residency of Palembang, the land had been given to plantations on long leases but had reverted to the government because the leaseholders had not developed it. When the colonization project was being prepared it was discovered that a part of the land was used by the local people. At first the government expected that they could be turned out because of clandestine occupancy, but a closer examination revealed that the people had been on the land before it was leased and that the leaseholders had been expected to buy them out but had neglected to do so. In the colony of Belitang, it was similarly discovered that the authorities in their first survey had not taken the rights of the local population into account.

Under such circumstances it became necessary for the colony to buy off the rights. This had been done in 1930 by the colony of Gedongtataän when it required for further expansion some land that belonged to the marga of Poegoeng. The colony finally paid 600 guilders in recognition of the rights of the marga and as compensation to those marga members who had fruit trees and pepper gardens on the land that was to be occupied by the colony.

The next phase of investigation, after the colonization officials had ob-

---

27 *Ibid.*, p. 31.

28 Southern Sumatra, which includes the residencies of the Lampoengs, Palembang, Djambi, and Benkoelen, has a territory of 186,310 square kilometers. There are 382 margas; of these 58 lie in the Lampoengs, 174 in Palembang, 58 in Djambi, and 92 in Benkoelen. The average area of a marga is 500 square kilometers.

tained the consent of the existing communities to the establishment of Javanese colonies on their land, was a reconnaissance survey by agricultural and soil specialists, which was followed by a survey made by the Irrigation Service. These scientists collected data on topography, geology, soils, climate, drainage, vegetation, transportation facilities, local agricultural practices, planting seasons, yields, the distribution and density of population—in short, on all the questions relating to the suitability of the area for Javanese settlers. The possibility of irrigation is especially important, as the social prestige of a Javanese is based on the ownership of sawah and it is less difficult to persuade him to migrate to a wet-rice-growing region than to one where he could never look forward to having sawahs. If the area under survey was considered to be irrigable from the technical point of view, there followed a cartographic soil survey, an analysis of soil samples, and an analysis of the water and silt carried by the rivers that were to be sources of the irrigation supply. The final phase involved the detailed planning of the irrigation system, the distribution of the future desas, and the network of roads. These various phases required at least 2 years of intensive work before an area could be thrown open for Javanese colonization (Appendix B, pp. 249,250).

In order to speed up the survey the Department of Public Works made extensive use of aerial photography, employing the services of the Royal Netherlands Indies Airways. The aerial mapping was supplemented with terrestrial measurements of differences in elevation. Only after these, too, had been completed was it possible to decide on the suitability of an area. Table XXXVII summarizes the irrigation engineers' activities in 1939.

Migration Propaganda

One of the most difficult tasks of the commission was to persuade an ever increasing number of Javanese to migrate to the Outer Islands. Each year the number of families who could be obliged to help new colonists to get a start increased. But as long as the readiness of the Javanese to migrate did not keep pace with this absorptive capacity, the commission had to work relentlessly for new recruits. Prospective recruits were systematically approached by both "direct" and "indirect" propaganda. Direct propaganda aimed at immediate recruitment, whereas indirect propaganda strove to call general attention to the colonization enterprise and to put it in the most favorable light (Appendix B, pp. 250,251).

The Ten Commandments for Selecting Colonists

Although the commission was extremely anxious to settle an increasing number of Javanese every year, it nevertheless attempted to have a rigid selection made. Maassen worked out a set of rules, which he called the "ten commandments of colonization," to guide officials entrusted with the selection. These are based on the recommendations made by Heyting, the planner of Gedongtataän, and by his successors (see pp. 200,201). The "commandments" are as follows:

I. Select real tanis; non-farmers are a burden for a colony and endanger its success.
II. Select physically strong people; only they can stand the hardships of pioneering.
III. Select young people; by taking them one reduces future population increase in Java.
IV. Select families; families are the foundation of peace and order in the colonies.
V. Don't select families with many young children; the working members of the family cannot carry that burden at the start.
VI. Don't select former plantation laborers; in 90 per cent of all cases they are the cause of discontent in the colonies.
VII. Don't allow so-called "colonization marriages"; they are a source of unrest in colonies.[29]
VIII. Don't accept expectant mothers; the pioneer settler needs the full help of the wife during the first year.
IX. Don't accept bachelors; sooner or later they will become involved with somebody else's wife.
X. Allow desas or kampongs to migrate as a whole; in such a case the first nine commandments may be ignored.

In spite of the fact that the commission tried to impress upon the officials in charge of selection in Java the absolute necessity of adhering as strictly as possible to these rules, the officials in charge of the projects in the Outer Islands found inferior and unsuitable settlers every year among the migrants (Appendix B, p. 251).

Growth of the Colonies

After 1936 migration increased rapidly (Table XVIII). During 1937, the first year that the commission was in charge, it arranged for the migration of approximately 20,000 Javanese. The goal for 1941, set at 60,000 migrants, was probably achieved, since the migration of 1941 should have been completed by the end of September, several months before the Netherlands Indies were drawn into the war against Japan. Table XVIII records the annual totals of Javanese who migrated after 1932 under government auspices. Some of these figures differ slightly from those in the annual *Statistical Abstracts*.

An annual migration of 60,000 persons, or some 16,000 families, would have been indeed a remarkable achievement in view of the difficulties with which the commission had to struggle; yet even a migration of that size was insufficient to solve the population problem in Java, where the population had been increasing annually at the rate of 600,000 persons. Java and Madura would not require an annual migration equal in size to the

[29] Frequently bachelors marry hastily before emigrating, in order to be acceptable. Their wives are known as voyage wives (*bini djalan*) or tow wives (*bini gandingan*).

annual increase, however, if the migrants were to consist of childless young couples. The Central Bureau of Statistics in Batavia calculated that if Java's rate of natural increase should remain constant at 1½ per cent and if no emigration should take place, the population in the year 2000 would total 116 million and the average density of population would be 879 persons per square kilometer. If, with the same assumption as to

TABLE XVIII—THE OUTER ISLANDS: NUMBER OF JAVANESE SETTLERS TRANSFERRED ANNUALLY WITH THE SUPPORT OF THE NETHERLANDS INDIES GOVERNMENT, 1932–1941

YEAR	NUMBER OF JAVANESE SETTLERS	YEAR	NUMBER OF JAVANESE SETTLERS
1932	7,000*	1937	19,639
1933	751†	1938	32,259
1934	2,756†	1939	45,339
1935	14,710	1940	52,855
1936	13,152	1941	60,000*

* Approximate.
† These figures differ from those given in Maassen: De Javaansche landbouwkolonisatie, 1937, p. 41, to which reference is made on p. 204.
Source: The figures for the years from 1932 through 1938 were given by H. J. Kuneman, chairman of the commission, in a lecture held in Batavia on January 31, 1939, published in *Kolonisatie Bulletin*, No. 4, February 1939. The figure for 1939 is given in *Jaarverslag van de Centrale Commissie . . . 1939*, 1940, p. 13; the figure for 1940 is given in *Nederlandsch-Indische Geografische Mededeelingen*, Vol. 1, 1941, p. 120; the figure for 1941 is an estimate reported in the Netherlands Indies press in January 1941 (for example, *Bataviaasch Nieuwsblad*, January 13, 1941).

natural increase, 80,000 families of 3 persons each (consisting of a couple and one child, the parents' ages being between 15 and 24 and the child's age less than 5 years) should migrate annually, Java's population would reach 74 million in the year 2000. If 120,000 childless couples should migrate annually the population would increase to only 57 million.[30] Although these calculations show how important a role migration can play in population growth, they also show that in 1941 the commission, with its sponsored migration of 16,000 families, was still far below the mark that might represent a solution of Java's population problem.

Fig. 165 depicts the rapid increase in the number of Javanese agricultural colonists in the Outer Islands from 68,600 at the end of 1936 to 206,020 at the end of 1940, while their geographical distribution is brought out in Fig. 167.

During the years 1938, 1939, and 1940 the recruiting was largely limited to central and eastern Java and to Madura; the Sundanese territory in West Java is less densely populated. The leading areas of emigration were the residencies of Pekalongan, Banjoemas, and Kedoe in Middle Java and Madioen and Kediri in East Java, and the two principalities, Jogjakarta and Soerakarta (Table XXXIX). During the same years the commission

30 Maassen: "Javanese Agricultural Migration," 1939, pp. 184–185.

sent most of the Javanese settlers to Sumatra where the majority were directed to the Lampoeng Districts, particularly to Soekadana, and others to the residencies of Palembang, Benkoelen, and Djambi. It proved more difficult to start a stream of migration to Borneo and Celebes, for, although Javanese villagers had often heard of Javanese desas in Sumatra, Celebes and Borneo were still unfamiliar names to them. Nevertheless, the data as

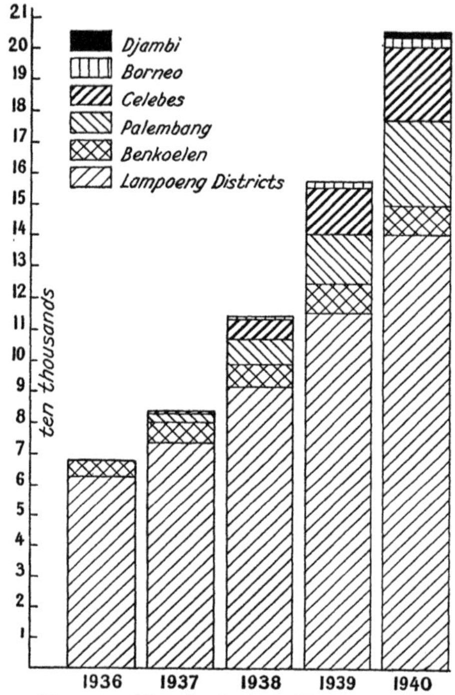

FIG. 167—The distribution of Javanese colonists in the Outer Islands, by residencies, at the ends of the years, 1936 through 1940. (Based on data supplied by the Central Commission for the Migration and Colonization of Natives.)

to destination for the years 1938 to 1940, summarized in Table XL, show a steady increase in the territory in Celebes and Borneo, as well as in Sumatra, that was being drawn into the operational sphere of the commission.

In order to give a detailed description of the establishment and early growth of a typical agricultural colony, Belitang has been singled out for a case study (see pp. 215–221). The remaining areas will be only briefly surveyed.

The Consumption of Food in the Colonies

During the years from 1931 to 1933 Ochse and Terra made detailed studies of diet in the district of Koetowinangoen in the regency of Keboemen, Kedoe Residency, Middle Java.[31] They concluded that the caloric value of the diet of the persons observed was sufficient, although in a few families the margin between the basal metabolic level and the average caloric value was rather small. Many indications suggest that in the most crowded sections of Java, from which a large part of the colonists come, the average diet is not so good as was that of the inhabitants of Koetowinangoen and that many families do not get the number of calories considered necessary for their well-being.

In 1938 and 1939 the commission had surveys made of the settlers' diet in several colonies in southern Sumatra, both old and new, and, for comparison, in two villages of the indigenous population (Table XLI). The survey showed that, with the exception of a single village, the average diet in Javanese colonies was considerably better than that in Koetowinangoen in Java. Colonists in well-established villages ate considerably more meat and fish than did the people in Koetowinangoen and also more rice and other starches such as cassava and sweet potatoes; more coconut products; more vegetables. The only important foods that they consumed in smaller quantities were legumes, including such dishes as *témpé* and *tahu*, made from soy beans. The diet in even such a recent settlement as Kalingrenjeng was higher in calories than that in Koetowinangoen, although less rice and more maize and cassava were used. A comparison of the diet in Kalingrenjeng with that in the other colonies indicates that the consumption of rice increased, and that of cassava decreased, as settlers passed beyond the pioneering stage, but in Bagelen the density of population had reached the point where the consumption of cassava was again large. Health officials were especially pleased by the consumption of animal and vegetable proteins in the colonies and the high intake of vitamin A obtained from leafy vegetables. As vitamin A heightens general resistance to infection, it is particularly important that pioneers living under unhygienic conditions should have an ample supply of it. In an examination of the children in the colonies, only a few were found who showed signs of vitamin-A deficiency.

A comparison of the diet of the Javanese in Sumatra with that of the local people, however, shows that the latter had a considerably higher level of living—indicated by the consumption of meat, fish, and rice—than had the Javanese colonists, largely because they had greater resources of land and cultivated such cash crops as pepper and coffee. Yet, in view of the fact, generally accepted by officials concerned with colonization, that the Javanese settlers had not been able to afford a diet even of the moderately good Koetowinangoen standard before their migration to the Outer Islands, it is apparent that colonization improved their nutrition. People having the "Koetowinangoen standard" are not likely to migrate.

31 Departement van Economische Zaken: Geld- en producten-huishouding . . . in Koetowinangoen, 1934.

The Cost of Colonization, 1932–1940

In 1932 a total of 7,000 Javanese were transported to southern Sumatra at a cost of approximately 48,000 guilders, or an average cost of 7 guilders a person or 28 guilders for a family of four; because of the depression, prices were half what they had been in 1928. In 1935 and 1936 the average cost of moving a family had risen to 30 guilders.

After the establishment of the commission in 1937, the costs rose further as the scale of colonization activities was enlarged; for the extensive surveys greatly increased expenditures. In the years 1938, 1939, and 1940 the commission spent a total of 5,813,006 guilders on colonization (Table XLII), and in the same period it transferred 35,572 families from Java to the Outer Islands. The apparent average cost of 163 guilders a family is misleading, however, for these costs cannot well be allocated on an annual basis. The government's expenditure of nearly 6,000,000 guilders in those 3 years was to prepare the way for the migration of many thousands of families from Java 2 or 3 years later. Moreover, such items in the 1940 budget as the 120,000 guilders for soil surveys and the 905,000 guilders for irrigation surveys and the construction of irrigation works were to benefit not only future colonists but those who had migrated in previous years as well. On the other hand, some of the money spent between 1905 and 1927 on Gedongtataän and Wonosobo also greatly aided the activities of the commission, for by employing new migrants as harvesters these colonies helped to lower governmental costs from 1932 onwards.

The money spent on surveys and on the construction of irrigation systems was to have been recovered in the form of taxes levied on land that had not been subject to taxation before it was brought under permanent cultivation.

The calculation of the governmental costs incurred during the period of large-scale colonization is further complicated by the fact that each settler was expected to repay his transportation and initial supplies—valued at from 22 to 25 guilders a family—to the government. As Table XLIII shows, the proportion of these debts that had to be written off because of death or, more often, because of desertion was decreasing noticeably after 1937. It also shows that many of the colonists paid up within a remarkably short period. The improvements in the selection of settlers, in the supervision of health, and in the management of the colonies seem to have been responsible for the reduction in the losses due to death and desertion.

It is regrettable that so large a sum had to be spent on propaganda. In 1940, for example, expenditures for propaganda accounted for 4.2 per cent of the total budget, thus averaging 7.40 guilders a family. As long as a large percentage of the people of Java remain illiterate and therefore cannot be reached through the printed word, a comparatively inexpensive form of propaganda, it is difficult to see how to reduce this item in the budget. It was hoped, however, that the school booklets and visual aids depicting Tanah Sabrang would so familiarize the village youth with the idea of migration that by the time they became ready to establish their own households, they would be more eager to migrate than were their elders.

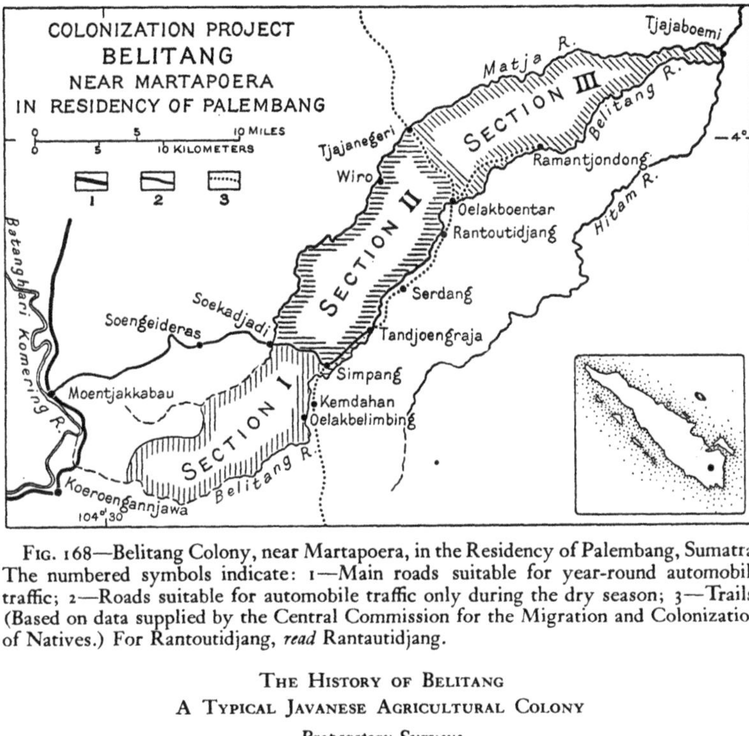

Fig. 168—Belitang Colony, near Martapoera, in the Residency of Palembang, Sumatra. The numbered symbols indicate: 1—Main roads suitable for year-round automobile traffic; 2—Roads suitable for automobile traffic only during the dry season; 3—Trails. (Based on data supplied by the Central Commission for the Migration and Colonization of Natives.) For Rantoutidjang, *read* Rantautidjang.

The History of Belitang

A Typical Javanese Agricultural Colony

Preparatory Surveys

The colonization project of Belitang lies in the division of Komeringhoeloe in the residency of Palembang between the Matja and Belitang rivers (Maps, Figs. 164, 168). Its land belongs to the territories of the margas of Boeaij Pemoeka Bangsa Radja, Madang Soekoe I, Madang Soekoe II, Semendawai Soekoe III, and Belitang. This last marga, the most extensive, has given the project its name. As the members of these margas had no objection to the establishment of a Javanese colony on their land, the marga councils granted their permission in formal resolutions, and at the same time each council made it unlawful for its members to start any new ladangs on land included in the project. The marga of Belitang asked for the right to share in the benefits of the irrigation project by receiving irrigation water for 980 bouws of land to be distributed among the five settlements of Soekadjadi (120 bouws), Tandjoengraja (400 bouws), Rantautidjang (60 bouws), Oelakboentar (180 bouws), and Ramantjondong (220 bouws) (Map, Fig. 168). It was requested that each marga member in these villages be given two bouws of sawah. The government agreed to this in order to enable the local population to intensify its agricultural methods, which until the coming of the Javanese had consisted mostly of shifting cultivation.

Soil scientists found that the extensive clay soils were physically well suited for sawah culture, although poor in the minerals important for plant growth. An analysis of the silt carried by the Batanghari-Komering River showed that irrigation water taken from that river would add rich fertilizing material to the irrigated fields. The clay soils were reserved exclusively for sawahs, while lighter sandy soils were set aside for desas. The soil experts warned against a leveling of the garden land surrounding the houses because the humus-containing topsoil covered a less valuable layer of soil.

The irrigation engineers planned to construct a main canal 40 kilometers in length

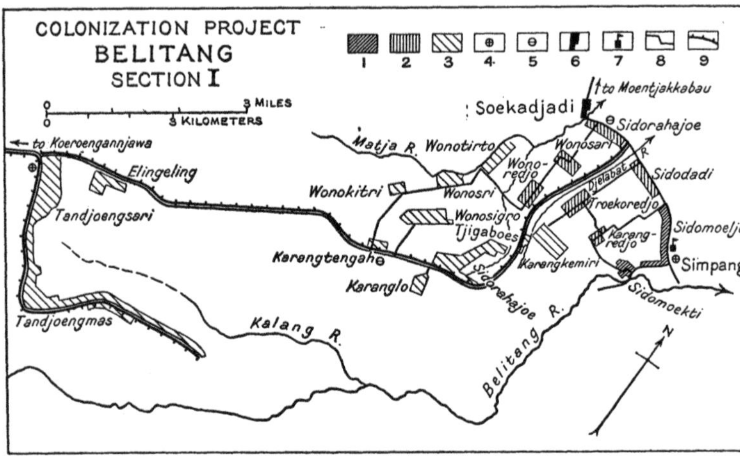

FIG. 169—Section I of the Belitang Colony, near Martapoera, in the Residency of Palembang, Sumatra. The numbered symbols indicate: 1—*Desas* (villages) founded in 1937; 2—Desas founded in 1938; 3—Desas founded in 1939; 4—Clinic; 5—Branch clinic; 6—Malay *dusun* (village); 7—School; 8—Road; 9—Irrigation canal. (Based on data supplied by the Central Commission for the Migration and Colonization of Natives.) For Sidorahajoe near Soekadjadi, *read* Sidohardjo.

which was ultimately to supply 19,000 hectares of sawahs with irrigation water. Its inlet is located at Koeroengannjawa. The primary canal was dug by three different contractors: the Hollandsche Beton Maatschappij, which had a contract for the first 12 kilometers; Volker's Aannemings Maatschappij, which had a contract for 20 kilometers; and Chinese contractors working with Chinese laborers, who were to dig the remaining 8 kilometers. The Hollandsche Beton Maatschappij used a floating dredge placed on a locally built concrete ship (Fig. 148). The first part of the irrigation project was to be finished by October 1, 1941, that is, 4 years after the arrival of the first settlers, and the remainder was to be completed by October 1, 1943. The total costs were estimated at 2,150,000 guilders, of which 1,150,000 were to come from the Special Rubber Export Tax Fund.[32]

Peopling the Colony

Colonists for Belitang were selected in Java on the basis of the so-called "ten commandments of colonization." Like the Soekadana colonists, they traveled by rail from their villages to Tandjoengpriok, the harbor of Batavia, then by steamer to Oosthaven, from there inland by the South Sumatra Railway to Martapoera in the Lampoeng Districts, and finally by bus to the colonization project (Figs. 123–127). The first group of colonists arrived on August 17, 1937. During 1937, a total of 250 families was brought to Belitang; in 1938, 550; in 1939, 1,506; and in 1940, 1,550. It was expected that 1,860 families would come in 1941.

[32] *Jaarverslag van het Departement van Verkeer en Waterstaat over 1939*, 1940, pp. 244–245.

The Javanese colonists sent to Belitang between 1937 and 1941 received more government help than those who went in the same years to Soekadana, because the Belitang project was near no established colony from which it could receive help through the mutual assistance system. Instead, the Belitang settlers of 1937 to 1941 were intended to become the nucleus for further expansion: they were to become for this part of Palembang what the settlers of Gedongtataän were for the Lampoeng Districts.

The total capacity of the Belitang project was estimated at 20,000 families—a quota that was to have been reached by 1951 (Table XLIV). Of these, 5,500, or 27 per cent, were brought during the initial stage between 1937 and 1941 and formed the nucleus around which the future 14,500 families, or 73 per cent of the estimated capacity, were to group themselves. A family that had lived for 3 years in the colony was expected to be ready to help others under the mutual assistance system, and every five old families were to employ one new family during harvest time (Fig. 129) and to take care of it until it should become established. In 1940, therefore, a first group of 50 families was quartered with the 250 colonist families of 1937 and worked for them as harvesters. By 1942 the number of families to be settled under the mutual assistance system was to have been 460 (had the war not interfered), because by that time 2,300 families would have been on the project for 3 years or more. In 1942, also, Belitang would have outgrown the stage of nuclear formation and thereafter was to have received only as many new colonists as it could absorb under the mutual assistance plan. From then on until 1950 the schedule of migration provided for a steady increase in the number of families that could migrate to Belitang (Table XLIV). In 1951 the project was to receive the number of settlers necessary to bring the total to 20,000 families.

The nuclear migrants of 1937 to 1941, who could not be placed as harvesters, were supplied with rice, which they were to pay back in kind, as they were also to pay for their transportation and other incidentals. Each settler received a rice card on which was indicated the amount of rice he had received at regular ten-day intervals. A total of 5.655 *piculs* of husked rice was considered sufficient for a family of four until their first harvest. By the time the settlers had been on the project for 5 months the new gardens were yielding maize and vegetables, and the daily rice allowance was gradually reduced until the first rice crop had been harvested, when the settler was thrown entirely upon his own resources (Table XLV). In addition to rice, each family received salt, oil, and protein-containing foodstuffs in 1937, upon the insistence of the local medical service. In 1938 and 1939 the distribution of these items was replaced by the payment in small installments of 5 guilders, which was added to the debt account.

The families of 1937 to 1941 that were to form the nucleus of Belitang were housed, upon arrival, in sheds, or *bedengs* (Fig. 134), each accommodating several families, and were given about a month to clear their compounds and build simple houses (Fig. 135). In 1940 and thereafter the families migrating under the mutual assistance system were to be placed in homes of older settlers until they had time to build their own shelters. Good building material in the form of standing timber was plentiful in the forest of the colony. The roofs were covered with alang-alang or other hard grasses. In 1940 I noticed three types of houses in the Javanese colonies, indicating stages in the economic improvement of the settlers. The pioneer first cleared a part of his home lot and built a little shack, sufficient to afford his family a shelter against rain and sun. This had to serve until the field had been cleared and the first crop harvested. Then a second, somewhat larger, house was built, consisting of one or two rooms and a kitchen. The construction of this house might have taken from 2 weeks to a month, depending upon the size of the structure and the skill and care of the builder. This building usually served for a few years, until the settler had saved enough money to put up a better house, of several rooms, with a roof made of wooden shingles (a material not commonly used in Java where wood is scarce and expensive), or of tiles. This third stage had, as a rule, been reached in Gedongtataän (Figs. 135, 136, and 139).

Upon arriving in Belitang each family received free of charge old gasoline cans and empty tin cans and beer bottles and on credit such tools as hoes, axes, *parangs* or *bolos*, cooking pots, and small lamps (Figs. 128 and 130). Groups of several families were given whetstones free of charge and were lent two-man saws and handsaws.[33]

The Belitang project was divided into three sections (Map, Fig. 168). The first 250

[33] Vissers: "De landbouwkolonisatie," 1940, p. 10.

families, which arrived in 1937, established the desas of Sidomoeljo and Sidomoekti, near the village of Simpang in the northeastern corner of the southern section (Map, Fig. 169). In 1938 six new desas were formed west of the first settlements, also in the southern section, and received the names Sidodadi, Troekoredjo, Sidohardjo, Wonosari, Wonoredjo, and Karangredjo. All these desas were located along, or near, the road from Tandjoengraja to Soekadjadi and Moentjakkabau. The settlers of 1939, pushing farther into the jungle covering the southern section of the Belitang project, split into two groups of desas; one group was placed immediately south of the desas of 1938 and the other at the southern end of the project. In the first group 1,000 families formed the new desas of Wonotirto, Wonosri, Wonokitri, Wonosigro, Karangtengah, Sidorahajoe, Karanglo, Tjigaboes, and Karangkemiri; in the second group the remaining 500 families made up the desas of Tandjoengsari, Tandjoengmas, and Elingeling. Most of the new desas were located along important roads, then under construction.

In 1940, 500 out of the total of 1,550 families were placed along the highway between Elingeling and Karangtengah to complete the settlement of nuclear families in the southern section. The remaining 1,050 families were to tackle the middle section from two directions: one group, of 500 families, was placed in the northwestern corner of the middle section, near the Malay settlement of Tjajanegeri; and the other group, of 550 families, was settled just north of the road from Simpang to Soekadjadi, that is, at the southern end of the middle section.

The planners of Belitang believed it better to open a pioneer region from several directions rather than from one, in order to secure a larger number of nuclei. This method had the advantage that future settlers could be placed right among, or very close to, their hosts and employers, for whom they could continue to work when the latter needed help, whereas at other times they could work on their own land. This made the construction of bedengs unnecessary. Most of the nuclear desas were filled to not much more than half their capacity, as the rest of the villagers were to come later under the mutual assistance plan.

This system, which may be called the "Palembang system," has one great weakness. Among the various nuclear desas and among the parcels of a single desa patches of forest are left that become the refuge of elephants, wild boars, deer, monkeys, rats, mice, snakes, and other animals that may be destructive or even dangerous. It was in order to prevent this as much as possible that the planners of Soekadana decided against opening their project from various directions; instead of scattering either the desas or the settlers they completely filled a stretch of land, seeing to it that every bit of cultivable ground was cleared during the single season. This "Lampoengs system," however, had the disadvantages that new settlers were eventually located many miles from their hosts and could no longer do occasional work for them, and that a new group of bedengs was required every year.

Shortly after their arrival the settlers of Belitang, like those in most of the other colonization projects of the Outer Islands, received a quarter bouw of garden, or home lot, in a desa and one bouw of irrigable land. This latter they were to cultivate for 3 years in the way that the local population handled their ladangs, planting dry rice, maize, and other crops. In the fourth year the settlers' land was to be irrigated and thus turned from ladang into sawah. Those arriving in 1937 found both home lots and field lots cleared and prepared for planting; those arriving between 1938 and 1941 found that the forest on their home lots had been cut, but not yet burned, whereas their field lots were untouched. This initial aid was given in order to speed up the formation of strong nuclei. Especially during the first years of an entirely new project, time is precious and a great deal may be gained by giving the pioneer a little extra help—so ran the argument of the officials in Palembang. On this point, too, the Lampoeng Districts differed. There the clearing was left entirely to the settlers.

Each desa in Belitang had a foreman, one of the local people, who would advise the Javanese about cutting the forest, as jungle clearing is a completely new experience to most Javanese of today. The foreman would designate the trees to be cut that in falling would bring down several others. He would warn the settlers against the *rengas* tree[34] which, when handled without proper precautions, causes severe itching, swelling, and fever.

[34] *Gluta renghas* (*benghas*) L. belonging to the *Anacardiaceae* family.

The settlers followed more or less the time table of the local people, who practice shifting cultivation and clear forest land fairly regularly. In June they cut the underbrush and in July and August felled the large trees; in early September they began to burn the underbrush and fallen trees and to prepare the land for rice planting, which takes place in late September and early October. Many of the Belitang settlers completed the cutting and burning of their whole field lots during the first year. When the lot was exceptionally heavily forested, the clearing usually took two seasons (Figs. 132, 137, and 138).

After the land had been cleared for cultivation, the staff of the Agricultural Advisory Service (Fig. 144) stepped in to supply the colonists with plant material and advise them on proper methods of cultivation, the timing of farming activities, and all the other problems that might be new to the Javanese, who were accustomed to a different climate and a different soil. Each family was given 5 ears of maize, a can of peanuts, 3 ordinary coconuts, 2 dwarf coconuts, 40 stems of cassava, and 10 young pineapple plants, as well as sweet potatoes, various vegetables, and such fruit-tree seedlings as orange, rambutan, durian, salak, papaya, banana, and others, for the garden. For the field the colonists received an additional five ears of maize and a five-gallon can of seed rice. (In Soekadana the seed rice was still unthreshed—Figs. 131 and 133).

The Agricultural Advisory Service maintained a nursery for fruit trees in Sidomoeljo and seed gardens for maize, cassava, peanuts, soy and other beans, eggplant, red peppers, and other vegetables in Karangredjo. Additional seed material—such as kapok and the variety of peanut known as Schwarz 21, as well as salak from Buitenzorg and tobacco from Magelang—was brought in from Java or other parts of Sumatra.

In 1939 the personnel of the Agricultural Advisory Service carried out experiments with fertilizer and collected data on yields by making a number of test harvests on colonists' fields. Fifty-four tests were made on land planted with rice for the first time and twenty-four on land producing a second rice crop. The yields of the first group were slightly higher than those of the second, as had been expected (Table XLVI). Test harvests made in other projects during the third and fourth year of occupancy showed a considerable decrease in yield and the necessity for irrigation was evident. If for some reason irrigation should not then be available, settlers ought either to be given new fields or to be supplied with fertilizer. In the fall of 1940 the government distributed half a picul of double-superphosphate among the Belitang settlers of 1937 because the irrigation water could not be supplied on time, that is, during the fourth planting season.

Marketing

Considerable trading went on between Javanese colonists and the local people. From the beginning the Javanese women found a good market for the delicacies they are in the habit of preparing. Simpang and the desas of Wonoredjo and Karangtengah developed into active marketing places, the last because of the presence of a number of Javanese from Jogjakarta who traditionally have an excellent business talent. I was frequently amazed at the distances the Javanese colonists carry their surplus produce. I met settlers from Tandjoengsari who told me that they were on their way to Moentjakkabau, a walk of 12 kilometers each way. Judging from observations in Gedongtataän and Soekadana, I feel certain that a lively traffic of trucks and buses will develop in the Belitang colony as soon as the road network is completed.

Health Service

The commission was especially on guard against malaria, dysentery, and intestinal influenza. Moreover, settlers were given a combined cholera-typhus-dysentery injection before leaving Java. Medical field workers (*mantriverpleger*) on the project supervised health conditions under the guidance of an Indonesian government doctor. They were in charge of the clinics, of which there were three in the Belitang project by the end of 1940 (Fig. 145). These received the simpler cases for hospitalization, whereas serious cases were taken to the regular hospital at Batoeradja.

Dutch officials in Sumatra claimed that it was useless to distribute mosquito nets because the Javanese were not familiar with their use, did not handle them properly, and sometimes even cut them up to make garments.[35] The fight against malaria was there-

---

[35] In Celebes the officials in charge of the projects did distribute mosquito nets and instructed the medical field workers to enforce proper use of them.

fore limited to the proper drainage of the land, the spraying of breeding grounds, and the distribution of quinine. Each desa had its quinine distributor chosen from the ranks of the settlers. It was his duty to visit every house regularly, distribute quinine tablets to all persons suffering from malaria, and report them to the doctor. Sick persons had to be given a prescribed dose at regular intervals. From experience the commission had learned that it was not enough just to distribute quinine tablets, since the Javanese would frequently sell them rather than consume them. For a time the tablets distributed at the cost of the commission were colored black so that their presence could easily be detected in the market places. Even that method did not assure the actual consumption of the quinine by malaria sufferers, and the final solution was found only when the quinine distributors were required to push the tablets down the throats of the sick persons with their index fingers and see that the patients did not cough them up again. The desas established in 1940 suffered severely from malaria, and a second doctor had to be sent to the project.

The commission found that health conditions in newly established desas were always worse than in older settlements, because of the unusually heavy physical exertion involved in the task of clearing the land, together with the general undernourishment prevalent to a greater or lesser degree among settlers coming from the depressed regions of Java. The colonists had also to adjust themselves to a climate frequently different in many respects from that of their native islands. In addition, on all the new projects some settlers had to be hospitalized as a result of attacks by wild boars.

Education

At the time of my visit to Belitang, the project had only one school, located at Sidomoeljo, but there were plans for two additional schools. The school had three grades and was attended by 30 girls and 91 boys. The commission engaged the teachers in Java. Each received a house and a garden of a quarter of a bouw plus a monthly salary of 11 guilders if he was a *guru bantu*, or assistant teacher, and of 30 guilders if a *guru kepala*, or head teacher.

On August 1, 1940, a two-year teachers' training course that was expected to graduate 15 teachers annually was started in Gedongtataän.

Recreation

Javanese are extremely fond of their *gamelan* orchestras and *wajang kulit*, or shadow puppet plays, and are not content without them (Figs. 156 and 157). Each desa received from the commission a set of gamelan instruments and a set of puppets for the wajang kulit, both usually of a cheap type. The colonists of Gedongtataän, however, were proud of their beautiful gamelan instruments, the gift of Paku Buwono X, Susuhuman of Soerakarta. Metro, the administrative center of Soekadana, had received a set of wajang kulit figures from the Sultan of Jogjakarta and a gamelan from the Prime Minister of the same principality.

Village Administration

The Javanese desas in Belitang, as in other colonies, are modeled after those in Java. At the head of each desa is a lurah under whom is a *kebajan*, or messenger and policeman, and a *tjarik*, or scribe. A *modin* looks after the religious and spiritual needs of the settlers and gathers them for prayers (Figs. 153, 154).

The whole colony was under an assistant wedana who was brought over from Java. Under him were two *mandurs* and one assistant mandur. The immediate superior of the assistant wedana was the *controleur* in Martapoera.

Debt Payments and Taxes

The settlers were obliged to make repayment in the following amounts: 12.50 guilders for transportation, 1 guilder for seed rice, 5 guilders for salt, cooking oil, and protein-containing food, 3.50 guilders for tools and household utensils—a total of 22 guilders.[36] They could pay either in cash or in rice. By October 1939 the colonists of 1937 had paid 1,254 guilders and those of 1938, 1,453 guilders, totaling 2,707 guilders.

All settlers were exempt from government taxes and compulsory labor, or *heerendienst*,

[36] According to the commission's report of 1940, the amount that it collected from each colonist varied slightly but averaged 25 guilders.

for the first 3 years.[37] In 1940 the first 250 families were obliged to meet this tax obligation. Marga taxes had to be paid after 2 years. All Javanese colonists in southern Sumatra, with the exception of those in the colonies of Wonosobo and Gedongtataän, were members of the margas on the territories of which they had been settled, and they had to pay the same amount of marga taxes as the local population (Fig. 152). Even the village administration was to be fitted into the marga structure. It was planned, for example, to withdraw the assistant wedana when the Belitang project had been completed. The lurahs were then to come under the jurisdiction of the *pasirah*,[38] or head of the marga, and were to have the same rank as the *krio*, or head of the local dusun, or village.

Other Javanese Agricultural Colonies in the Outer Islands

Other Colonies in the Lampoeng Districts

The largest Javanese agricultural colony is the Soekadana colony located in the Lampoeng Districts, between the Wai Sekampoeng and the highway connecting Teloekbetoeng with Goenoengsoegih and Soekadana (Maps, Figs. 164, 166). Formerly some of this land had been used by the local population for the making of ladangs, but most of it had been forest reserve, which the government released for colonization. The first quick survey in 1932 seemed to indicate that this project could be brought under irrigation through the construction of simple desa irrigation works of the kind the Javanese are capable of building without the assistance of engineers; but a later, more careful, survey showed that the district required large-scale irrigation works that would need to be constructed by the Irrigation Service of the Department of Public Works. In 1934 plans for the construction of irrigation works to supply 7,000 hectares with water were approved. Further surveys showed that 43,000 hectares could be brought under irrigation, and at the time of my visit in September 1940 the engineer in charge of Soekadana was investigating the possibility of additional expansion that would give the colony a total of from 58,000 to 64,000 hectares of irrigated land, thus accommodating some 83,000 families; for each family would receive one bouw of irrigable land.

The map in Fig. 166 shows the location of the Argogoeroeh weir, which raises the water in the Wai Sekampoeng and leads it into the primary canal, which in turn carries it to the main distribution works at Trimoerdjo. There it is divided among three secondary canals, which finally deliver it to the tertiary canals. The map shows the canals that were in operation at the end of 1939, as well as those that remained to be constructed (Figs. 146, 147, 149, and 151).

Although it had been planned that the irrigation system at Soekadana should supply 7,000 bouws with water during the 1940–41 season, actually only 5,000 to 6,000 bouws received water, according to a statement by the irrigation engineer.

In 1940 the commission settled Javanese colonists in the subdistricts[39] of Teloekbetoeng, Kotaägoeng, Kotaboemi, and Menggala, as well as the subdistrict of Soekadana. According to plans which I was shown at the time of my visit, Soekadana was to receive 6,650 families in that year, whereas the other subdistricts were to receive a total of only 850 families (Table XLVII). Of the 7,500 families to be settled in 1940, 2,292 families were to receive rice on credit and 5,208 families were to come under the mutual assistance system. Of the latter, the small colonies were to employ 841 families, Gedongtataän, 1,750 families, and Soekadana, 2,617 families. As the Soekadana colony was growing so rapidly, it was expected to be able to take care of all the newcomers under the mutual assistance system after 1942. Thereafter Gedongtataän was no longer to be forced to accept harvesters.

In Gedongtataän the ratio of harvesting families to old settlers was as 1:3 on sawahs and 1:4 on tegalans. From the old settlers' point of view, this compared unfavorably with the Belitang ratio of 1:5 on tegalans. In the Lampoeng Districts a colonist had to

[37] The land taxes and the heerendienst are discussed in Vandenbosch: The Dutch East Indies, 1941, Chapter 17.

[38] The word pasirah is of Sanskrit origin. It has the same root as the English "sire" and the French *seigneur*.

[39] *Onderafdeelingen*.

shoulder his share in the support of newcomers in his second year of residence and regularly thereafter but in Belitang not until his fourth year. In these matters the "Palembang system" (see p. 218) burdens the old colonist less than does the "Lampoengs system." In the Lampoengs, the absorptive capacity, expressed by the number of families that could be settled under the mutual assistance system, was increasing at so rapid a rate that, had not war interfered, it would have been possible to receive all incoming families under that system by 1943. The rate of increase would have been even greater than it was had it not been for the bottleneck of irrigation. The absorptive capacity of a colony depends not only upon the number of established families but also upon the time required to construct irrigation systems.

Writing in 1939, Maassen urged that "the settler's need for outside help should be stimulated. . . . This means that each colonist should be granted as much land as possible—so much in fact that he is unable to attend to his crops single-handed, particularly at harvest time."[40] In actual practice, however, each colonist was given only one bouw of land but was nevertheless forced to accept harvester-guests whether or not he needed assistance.

Palembang

In addition to Belitang, the residency of Palembang had another colony 20 kilometers from the town of Loeboeklinggau in the subdivision of Moesihoeloe, known as the Toegoemoeljo colony (Map, Fig. 164). Like Belitang, it was established in 1937. Its area comprised 35,000 hectares, of which only 6,575 hectares were irrigable, a small proportion as compared with that of Belitang or Soekadana. Its land, which belonged to the marga of Proatin Lima, was to be irrigated with water from the Klingi River (Fig. 150). The marga population was to receive 2,735 hectares of irrigated land; hence only 3,840 hectares would be available for the Javanese. The main canal, which was to be 3.15 kilometers in length, required a special kind of construction to prevent heavy losses of water in the sandy soils. The whole irrigation system of this colony was to have been completed by the end of 1942 at an estimated cost of 765,000 guilders.

In October 1940 this project had 15 desas, according to an oral statement by the controleur at Loeboeklinggau. Six hundred and fourteen families had arrived in 1937, 859 in 1938, 423 in 1939, and 590 in 1940. An additional 2,500 families would fill the colony completely.

As in Belitang, the Javanese colonists came under the marga laws, although in 1940 the Javanese assistant wedana had not yet been replaced by the local pasirah. Marga taxes, which the settlers had to pay by the beginning of the third year, amounted to two guilders per family, plus one picul of rice, which went to the lurah as his remuneration (Figs. 152 and 153). In other Javanese villages the lurah received a piece of land, the so-called *ambtsveld*, as remuneration, but this system was not carried over to the marga of Proatin Lima.

The colony of Toegoemoeljo was considered a failure by the commission because of the small size of its irrigable area as compared with the total amount of land. The preliminary surveys had led the government to believe that the colony included far greater irrigation possibilities. Moreover, the fact that the local population made rather large demands was disappointing to the commission, which always tried to reserve as much land as possible for the colonization of Javanese. This disappointment might have been prevented if the surveys preceding the initiation of the project had been carried out with greater care.

Djambi

In 1940 the residency of Djambi received its first colony. It was established in the subdistrict of Bangko and was named after the Tabir River, which is to become the source of irrigation water for the colony's 17,500 hectares of potential sawah (Map, Fig. 164). The first group of settlers, comprising 510 families, or 1,955 persons, narrowly escaped a major catastrophe during the first months, for plague was brought into the subdistrict by pilgrims returning from Mecca; fortunately, a rigid quarantine saved the colony from infection. No sooner had this danger passed than the project suffered heavily from a serious drought that destroyed the young upland rice. This meant that the settlers had to borrow rice from the government for a second season.

Benkoelen

Colonization in Benkoelen was started in 1908 but differed radically from that in the

---

[40] Maassen: "Javanese Agricultural Migration," 1939, p. 188.

Lampoeng Districts. In the latter, at least until 1932, the government was anxious to prevent close contact between the colonists and either the indigenous people or the plantations. In both Gedongtataän and Wonosobo the Javanese colonies were kept as separate entities and were administered by Javanese officials. In Benkoelen, however, the Javanese colonists at once became marga members. As a matter of fact, the first colony was financed out of marga funds. Only after this experiment had been successful did the government assume financial responsibility for colonization in Benkoelen. From the outset the settlers depended greatly upon employment on local plantations, and as little was done to supply them with sawahs, they had to practice ladang culture in order to raise sufficient food. All of the Benkoelen colonies are considerably smaller than those in the neighboring residencies. Because of its topography, Benkoelen lacks areas that would allow large-scale irrigation systems capable of watering thousands of farms. In 1910 there were 596 persons in its colonies. Fifteen years later, in 1925, the number had increased only to 1,326 persons.[41]

Sumatra's West Coast

The first colonization project of the residency of Sumatra's West Coast was opened early in 1941 by 500 families, which were brought there in two groups. The colony was named Batahan after the Batahan River that is to supply the irrigation water, and is located near Air Bangis in the district of Ophir in the northern part of the residency.

South and East Borneo

Large parts of southern Borneo are composed of extensive marshes the water level of which is affected by tides or monsoons, or both. The local people of Bandjermasin and Hoeloesoengai in the drainage basin of the Barito River have developed special methods of cultivating rice in these swamps, which they have converted into tidal sawah, west-monsoon sawah, or east-monsoon sawah, according to the source of the water supply. The cultivation methods differ for each type of sawah.[42]

In inland swamps, where the water level is very high during the rainy season and very low during the dry season, the cultivation of lowland rice is a risky undertaking. As the fields are covered by too much water to allow rice cultivation during the rainy season, rice must be planted when the flood water has dropped to a height of from 20 to 25 centimeters. In order to have seedlings ready by the time the water level has fallen sufficiently, the cultivators construct small floating seedbeds. The floats, made of bamboo, thick banana stems, or the leaf ribs of the sago palm, are covered with aquatic plants and mud, and then the seed rice is sown on them. After 3 weeks or so the seedlings must be transplanted. As the water does not usually drop to the proper planting level in all of the fields at the same time, there should then be at least one field where the level has dropped sufficiently to permit the seedlings to be transplanted to it temporarily in thick bunches. The bunches may stay there, if necessary, for another month before they are divided and the seedlings finally transplanted to the field where they are to mature. This planting method enables a family to cultivate two or three bouws of rice without outside help, as the transplanting may be spread over a long period. In Java, on the other hand, one family cannot cultivate much more than one bouw without outside help as the rice there is transplanted only once, and this job must be done within a short time.

The water level of the swamps near the sea is more uniform and is influenced more strongly by the daily tidal movements and prevailing winds in the estuaries than by the monsoons. However, the preparation of seedbeds and the transplanting methods are similar to those inland. Late-maturing varieties that require 7 to 8, or even as much as 10 months, are cultivated near the sea, whereas only early-maturing varieties are used in the inland areas.

Although in recent years more and more swampland has been planted with rice in the region around Bandjermasin and Marabahan east of the Barito River and along the Serapat Canal (connecting the Barito with the Moeroeng River), southern Borneo still has extensive swamps that have not yet been utilized. When the commission's attention was called to these possibilities, it decided to experiment with a small group of settlers in order to find out how rapidly the Javanese, who is not familiar with the cultiva-

---

[41] Wellan: Zuid-Sumatra, 1932, p. 109.

[42] Beusechem: "De sawahbouw," 1939. See also Schuitemaker: "Korte schets van den landbouw," 1938, and Schophuys: Het stroomgebied van de Barito, 1936.

tion of rice in tidal swamps, would be able to familiarize himself with this new type of cultivation. Recognizing a similarity between southern Borneo and Cochinchina, the commission sent a rice specialist to Cochinchina in 1939 to study cultivation methods in the swamp regions of the Mekong Delta.

In July 1938, a first group of 95 Javanese families was placed near the kampong of Kertakhanjar to cultivate 100 bouws of swamp sawah that had been leased from the local population. The experiment was successful, and the commission felt sufficiently

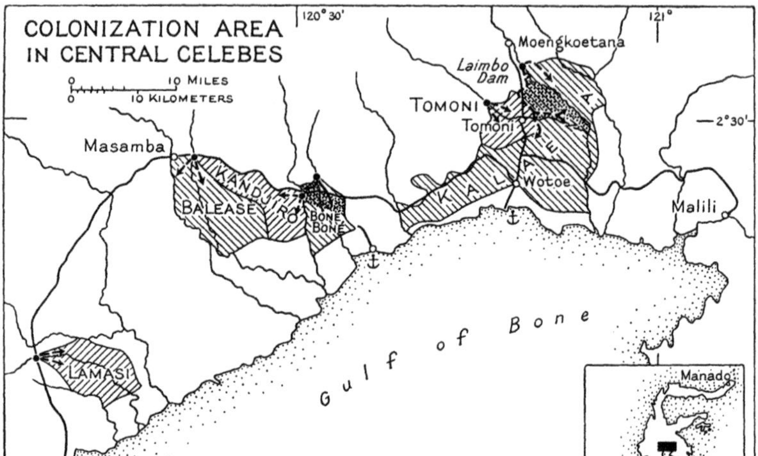

FIG. 170—Colonization area in central Celebes. The numbered symbols indicate: 1—Areas selected for agricultural colonies; 2—Irrigation dam; 3—Existing roads; 4—Areas already occupied by Javanese colonists in January, 1940; 5—Disembarkation points for colonists; 6—Irrigation canals in operation in January, 1940; 7—Proposed irrigation canals. (Based on data supplied by the Central Commission for the Migration and Colonization of Natives.)

encouraged to send a second group of 106 families in 1939. This group came early enough to help with the harvest, and the settlers of 1938 were then transferred across the Barito River and settled on their own land south of the Serapat Canal and southwest of the town of Bandjermasin. The colony, given the name of Poerwosari, comprises an area of 34,000 hectares (Map, Fig. 164). The second group was brought to a kampong of local people on the banks of the Lauk River to learn the new technique before they joined the settlers of Poerwosari. In May 1940, the third year, a third group of 198 families was added to the Poerwosari project. The planners of this colony estimated that 15 per cent of the land would be occupied by villages, canals, and roads, leaving 28,900 hectares of swamps for the cultivation of rice. This would be sufficient for 19,-200 families, as each family was allowed 1½ hectares, or about twice as much as in Sumatra. The colony already had some drainage canals but more were required, and plans had been made to dig them with dredges. The Poerwosari experiment is especially important in view of the fact that southern Borneo has 1,400,000 hectares of swamps that could be made available for settlement by the construction of drainage canals.

Like Poerwosari, the Madurese colony of Madoeredjo differs fundamentally from the colonies in Sumatra. In Madoeredjo dry land is cultivated exclusively. The colony, located near Pengaron in the Meratoes Mountains, includes 16,000 hectares of cultivable land, none of which can be irrigated. But this is no drawback in the eyes of the Madurese settlers, since the peasants of Madura surpass all other Indonesians in the tilling of tegalans. As in their native island, the Madoeredjo settlers terrace their tegalans to protect them against accelerated erosion, and include green manures, above all *crotolaria*, in their crop rotation. Each Madoeredjo family was given two hectares of contiguous tegalan land, whereas the Javanese family in Sumatra was given two separate parcels aggregating 0.9 hectare. The Madurese are excellent stock breeders, also. The goal in Madoeredjo is therefore the establishment of mixed farming. The colony was started in 1938 with 100 families, and these were joined in 1939 by 194 additional families.

Celebes

In Celebes the commission initiated three colonization projects for Javanese, all following the pattern that had been developed in Sumatra. Two of the projects were small ones located on the Gulf of Mandar; the third and most extensive one, the Masamba-Malili project, lies at the head of the Gulf of Bone.

When the commission began its work its attention was drawn to the Sadang plain in southwestern Celebes, north of the town of Parepare, where a large irrigation system, designed to irrigate 60,000 hectares of sawah, was under construction. Investigation showed, however, that all of this land was needed for the inhabitants of Celebes itself because parts of that island are already so densely settled that outlets must be created in other parts of the island that are not yet utilized agriculturally. But northeast of the Sadang plain there were two smaller areas available for the settlement of Javanese, the Boengi and the Mapili plains. The fact that both of these were covered, not by forests, but by alang-alang (*Imperata*), necessitated the immediate construction of irrigation systems, as without irrigation water a single bouw of grassland does not produce enough to sustain a peasant family. In the Philippines—where grass-covered areas were used extensively for pioneering—more land was allowed per family, and the National Land Settlement Administration broke the soil with tractors and disk-plows and furnished carabaos and steel plows to the colonists. The Dutch, however, preferred to settle Javanese colonists at the digging-stick and hoe level.

The first settlers were brought to Boengi and Mapili in the summer of 1937. The colony on the Boengi plain—immediately north of the Sadang plain– named Paria (Map, Fig. 164), received 147 families (471 persons) in the first year, another 40 families (112 persons) during the second, and 7 families (28 persons) during the third year. The slow growth of the Paria project may be explained by the fact that drainage and irrigation proved to be more difficult than had been anticipated. As trained surveying personnel was not available the project could not be surveyed and divided into parcels until 1940, and the construction of the required drainage and irrigation works could not be started until 1941. This delay was no excessive handicap for the settlers already located in the colony, as each of them had a good garden plot and some sawah. Moreover, it was possible for them to earn additional rice by working as harvesters for the local Buginese. They also earned cash by helping to construct the Sadang irrigation system.[43] The delay did, however, slow down the growth of this colony to a rate much lower than the usual record of the commission.

The Mapili project, located on the highway between Mapili and Polewali, grew considerably faster than did Paria. It was started with 115 families (380 persons) in 1937. These were joined by 710 families (2,335 persons) in 1938 and an additional 538 families (1,838 persons) in 1939.[44] The Mapili colony received its irrigation water from the Maloso River through a provisional canal for three years. The final irrigation canal was scheduled to begin operation at the end of 1940.

The most extensive settlement areas in Celebes were found north of the Gulf of Bone, where a plain stretches from Malili in the east to Palopo in the west over a distance of nearly 125 miles. It is from 6 to 20 miles wide and comprises 2,000 square kilometers (Map, Fig. 170). Abendanon, a geologist who visited this region in 1909, recognized its potentialities at that early date. He wrote:

> With the exception of a few patches of wet-rice fields it [the Masamba-Malili plain] is not cultivated, although it can compete with the best irrigated plains of

[43] *Jaarverslag van de Centrale Commissie . . . 1939*, 1940, p. 34.

[44] These figures are taken from the annual reports of the Commission and differ slightly from those given in the *Indisch verslag 1939*, 1939, Vol. 2, pp. 50–53.

Java in respect to the abundance of water courses that come out of the mountains and cross it. It could be turned into the granary of Central Celebes. . . . As the exportation of rice also offers no difficulties, it would be desirable to take earnest and vigorous steps toward this goal, possibly by means of an immigration of Javanese. . . . Immigration seems to me to be eminently desirable for Celebes. Emigration should help Java and immigration should help Celebes. This problem deserves serious governmental consideration and calls for a solution without delay.[45]

Little attention was paid to Abendanon's observations until the governor general's request in 1937 for suggestions of new settlement areas. At that time the Resident of Celebes reported possibilities in two sections of the Masamba-Malili plain, but a pre-

TABLE XIX—CELEBES: GROWTH OF THE JAVANESE COLONIES ON THE MASAMBA-MALILI PLAIN, 1938–1940

AREA	ARRIVING COLONISTS IN— 1938 Families	1938 Persons	1939 Families	1939 Persons	1940 Families	1940 Persons	RESIDENT FAMILIES DECEMBER 31, 1940
Kalaena	477	1,553	1,237	?	1,987	6,769	2,625
Masamba	487	1,569	936	?			2,135
Lamasi	—	—	—	—	249	832	249
Total	964	3,122	2,173	7,359	2,236	7,601	5,009

Source: Compiled from: *Kolonisatie Bulletin*, No. 8, 1940, pp. 21, 22; *Jaarverslag van de Centrale Commissie . . . 1938*, 1939, p. 42; and *Jaarverslag van de Centrale Commissie . . . 1940*, 1941, Table.

liminary survey showed that practically the whole plain might be utilized. It was but thinly settled by Buginese and Toradjas: Malili had a population density of 3.88, and Masamba one of 7.58, persons per square kilometer. A total of 43,000 hectares of irrigable land were finally selected in three different parts of the plain: an eastern part, north and west of Malili, called the Kalaena Area; a central part, south and east of Masamba, known as the Masamba Area; and a western part, north of Palopo, known as the Lamasi Area (Map, Fig. 170). These three parts could accommodate approximately 56,000 families.

As there were no old Javanese colonies in this region, the commission formed two strong nuclei around which later colonists were grouped. In 1938 the two colonies were started, Bonéboné[46] in the Masamba Area, with 487 families (1,569 persons) and Kalaena in the Kalaena Area, with 477 families (1,553 persons), as shown in Table XIX. The colonists made the six-day trip by steamers of the K. P. M.[47] from Soerabaja to the Gulf of Bone, where they were brought ashore in small motor boats. Each settler was given a quarter of a bouw of land for his house and garden and one bouw of irrigable land, which he was to use as tegalan for two years before it was to be transformed into sawah.

Unfortunately the first settlers lost a great deal of valuable time because, although they arrived in July and were immediately given their garden land, they did not receive their fields until late in September in Bonébоné and the middle of October in the Kalaena project. The clearing of the tegalans was therefore from 1 to 2 months late and the burning was not successful, as the rainy season had already started. Consequently only a small part of the tegalan area was planted with dry rice and maize. The settlers of Bonébоné were able to cut but 83 per cent of their tegalan area, burn 73 per cent, and

[45] Abendanon: Geologische . . . doorkruisingen van Midden-Celebes, quoted in unsigned article entitled "Nieuwe kolonisatiemogelijkheden," in *Kolonisatie Bulletin*, No. 2, 1938, pp. 17–18.

[46] Also known as Tamoekoe colony.

[47] Royal Packet Navigation Company.

finally plant only 12½ per cent. The yield from even this small cultivated area suffered damage by mice and wild animals. The Kalaena colony did somewhat better. Its settlers brought 56 per cent of their tegalan area under cultivation and had a satisfactory harvest, although the crops here, too, were harmed by animals.

The progress reports seem to indicate that the personnel in charge in Celebes were still experimenting and had not yet worked out as smooth a working program as that in southern Sumatra where years of experience stood the administration in good stead. Despite the weaknesses in the operations on the Masamba-Malili plain, the commission continued to dispatch large groups of settlers to this project, as may be seen by a comparison of Tables XIX and XL.

The Kalaena subdivision (Map, Fig. 170) was to receive its irrigation water from the Kalaena River, which was to be dammed at Laimbo; the Tomoni subdivision was to be supplied with water by the Tomoni River, for which also a dam was planned. The construction of the Laimbo dam was begun in 1940, and the Department of Public Works employed a large number of the colonists as laborers. In Sumatra the employment of settlers for such tasks had been discontinued because tanis who were not accustomed to heavy earth work were less efficient than regular laborers. The local labor shortage in Kaleana and the costs of importing laborers accustomed to earth work from Java must have been the reason for the disregard of the Sumatran experience. The total cost of irrigation works for the Kalaena area was estimated at 1,500,000 guilders or about 75 guilders per hectare of irrigated land, an amount that cannot but be considered a reasonable investment for a hectare of well irrigated riceland.

The Masamba Area (Map, Fig. 170) is divided into the three subdivisions of Bonébone, Kandjiro, and Balease, each one named after the river that is to become the source of irrigation water. The dam on the Bonébone was finished in 1940. The map shows the areas settled from 1938 to 1940.

The Lamasi Area (Map, Fig. 170) received its first colonists in 1940. It was estimated that this part of the Masamba-Malili plain included some 13,500 bouws of irrigable land (of which 3,500 bouws were set aside for the indigenous population) and that the irrigation works would involve a cost of about 500,000 guilders.

Summary and Appraisal of the Colonization Work of the Netherlands Indies Government

We have seen that, although the need for measures to relieve Java's population pressure had been clearly recognized as early as the first decade of this century, the government did not give serious attention to agricultural colonization until the 1930's. It had taken about a quarter of a century of halting experimentation to develop a colonization scheme that was workable and promising, and that reduced government expenditure per settled family to the barest minimum. Many mistakes had been made in Gedongtataän, but each one, though costly, proved valuable since it contributed toward the development of a more efficient procedure.

The Gedongtataän project had been started without previous systematic surveys or a careful mapping of the land and without a plan as to the location of the future desas, sawahs, and irrigation systems. As a result, a number of desas were laid out on low land that should have been designated for sawahs. It had been assumed that the Javanese would be able by themselves to develop irrigation systems, but this proved to be a more difficult undertaking than had been anticipated. In 1911, six years after the founding of Gedongtataän, the government authorized a survey of the land; however, it was not until 1916 that the survey was actually made. It was also in 1916 that the government decided to finance the construction of the necessary irrigation works. At that time a number of badly placed desas had to be relocated, with the result that the villagers lost their gardens and fruit trees, then several years old. All of this was, of course, a time-wasting and expensive procedure. The growth of the Gedongtataän colony suffered, furthermore, from the poor selection of settlers and from the lack of adequate medical supervision. The costs of the project were increased by a credit policy that was far too liberal—credits were used by the settlers for non-constructive purposes—and by additional expenditures necessary to correct initial mistakes.

The history of Gedongtataän between 1905 and 1932 reflects the attitude of the government during that period. Batavia was not convinced of the feasibility of large-scale agricultural colonization. Instead, it was concentrating its attention on a multitude of measures designed to improve the condition of the masses in Java itself. Among these were attempts to intensify Javanese agriculture by the construction of irrigation systems, the educational work of agricultural advisers, the distribution of selected seeds, and the demonstration of improved methods of cultivation.

Another important factor affected the attitude of the government. During the period from 1910 to 1929 the Indies, on the whole, enjoyed prosperity. The expanding Western plantations gave employment to many tens of thousands of workers both in Java and in Sumatra. The estates in Sumatra depended almost exclusively upon workers coming from Java. Therefore, the government was more interested in encouraging and supervising the migration of plantation laborers than in the establishment of Javanese agricultural colonies. The latter were costly, whereas the employment of Javanese by Western estates in the Outer Islands relieved the population pressure in Java without causing the Treasury any expense. This solution, however, was not permanent; in times of declining prices the planters would dismiss many laborers and send them back to Java, to be reabsorbed in the overcrowded desas.

When the depression of the early 1930's forced great curtailment in the plantation industries and when from all the signs it seemed obvious that some of these would never again employ as many people as they had in the past, the government turned again to agricultural colonization, for the pressure of population in Java had become really dangerous. Gedongtataän, for years considered proof that the government-sponsored agricultural settlement of Javanese was far too costly and slow to contribute substantially to the solution of Java's problem of overpopulation, became once more the subject of examination. The study revealed the mistakes of the past and showed how the work might be organized to keep down the costs per settled family to a minimum and yet achieve satisfactory results. Thus, in the end, the experiment of Gedongtataän proved to have been a valuable testing ground for the government.

The key to the success of the new colonization scheme was the introduction of preparatory surveys, supplemented by a discriminating selection of colonists, and the abandonment of the attempt to force Western concepts upon the Javanese peasants. Long-term bank loans, for example, were discontinued; instead, the new administrators made a full use of practices, such as mutual assistance and the bawon form of payment, that were in harmony with Indonesian ways of thinking. The government threw a part of the financial burden of settlement upon the colonists themselves. This required of them the utmost in strength, ingenuity, and perseverance, but when they had overcome the worst of the difficulties they became more firmly attached to their new home than if they had not had to work for everything themselves.

Colonization expenses were low between 1932 and 1936, the first years of the scheme's operation, since new settlers could be grouped around existing colonies. By 1937, however, it had become necessary to go farther afield, so that large sums had to be invested in extensive surveys and in the founding of new nuclei. The higher expenditures from 1937 to 1941 were to benefit many more families than were actually settled during that period; therefore it would be misleading to include the entire outlay when calculating the costs per settled family in those years.

The efficiency with which agricultural colonization has been handled since 1937 deserves a full measure of admiration. The procedure was worked out with great care. Each step was planned to avoid unnecessary expenditures of time and money. The officials, both Indonesians and Netherlanders, who were in charge of the work, threw their whole strength behind the task.

The number of Javanese settled in the Outer Islands in the single year of 1941 was three times as large as the entire population of Gedongtataän had been 20 years after its founding. This was indeed a remarkable advance. Nevertheless, in spite of the rapid increase in migration, the movement was still less than one-fourth as large as it would have had to be in order to absorb the increase in Java's population. It is not certain that colonial migration can ever be brought up to the estimated required level of 120,000 childless young couples a year. It is certain, however, that migration on such a scale could not be carried on indefinitely. It would eventually become difficult to find land suitable for the cultivation of rice as well as to find an indigenous population willing to share its land with Javanese settlers. The population of the Outer Islands is also increasing at a rapid rate, and in a number of districts—notably the Toba Batak country, the Padang Highlands, Bali, and parts of Celebes—the density is already so high that migration to neighboring areas has become necessary.

Realizing that agricultural migration has its definite limitations and confronted with the problem of arresting a steady decline in the welfare of the rural population of Java,[48] the government encouraged the growth of

[48] See, for example, the statement made by Resident J. A. Ruys, reported in *Het B. B. Congress 1938*, 1938, p. 51.

manufacturing industries producing consumers' goods for the Indonesian masses, which were started in Java between 1934 and 1941. The development and peopling of the Outer Islands, however, is of the utmost importance for these young industries, for the greater the market in the Outer Islands, the larger the number of Javanese who can be employed in the new industries and who therefore are not forced to join the stream of migrants overseas. The value of the flow to Sumatra and other parts of the Indies of goods manufactured in Java increased rapidly after 1935, as can be seen in Table XX. This flow is a profitable one for Java's industries, for the average buying power of the population in the Outer Islands is considerably higher than that of the Javanese villager, who has a cent-and-half-cent economy.

TABLE XX—JAVA AND MADURA: EXPORTS OF MANUFACTURED GOODS TO OUTER ISLANDS

Year	1935	1936	1937	1938	1939	1940	1941
Value of goods in millions of guilders	34	40.1	64.1	54.6	58.6	73.4	96.0

Source: Sitsen: The Industrial Development of the Netherlands Indies, 1942, p. 4.

The level of living of those parts of the Outer Islands, especially in Sumatra, Borneo, and Celebes, that have been drawn into world trade has been considerably higher than that of Java because the farming population on a large scale turned to the production of export crops without, except in some minor instances, neglecting the raising of food crops. Yet, through all the discussion of agricultural colonization in the Indies runs the thought that the ultimate aim should be to create villages and communities in the Outer Islands that would be exact duplicates of those in Java, where the people would behave exactly as they did in the home island, and where they would have the same type of economy. Each colonist is given one hectare of land, on which he grows only food crops. This carries over into the Outer Islands a pattern of living that is far from ideal. The mode of life in Java is, to a large extent, an outgrowth of the great density that has existed there for generations and the result of several centuries of colonial rule. Although in apportioning one hectare of land to a colonist the government improved his economic status over that which he, as an individual, had had in Java, it did not lift him above the level of subsistence farming. He is able to make ends meet and pay the land tax, but his buying power cannot compare with that of the indigenous population.

We may therefore raise the question: would it not be preferable to give the Javanese more land so that, in addition to rice and other food crops, he might grow a number of cash crops such as tobacco, coffee, pepper, and rubber, even if this would mean that fewer Javanese could be settled in the Outer Islands? This should forestall the danger of transplanting to the

Outer Islands those very conditions that have made the situation in Java nearly unbearable.

It should further be pointed out that by giving each family only one hectare of land the government immediately creates high densities of population in the colonization areas. If we assume that in a settlement area a solid block of 1,000 square kilometers is divided into farms of one hectare each, distributed among colonists whose families will ultimately average 5 persons, the density will amount to 500 persons per square kilometer of cultivated land within the boundaries of the colony. In as short a period as a generation such a colony will already be faced with a scarcity of land, and some members of the new generation will have to look for free land outside.

It is argued that any deviation from the Javanese way will endanger the success of agricultural colonization. This assumes, however, that the Javanese is static and has not the capacity to profit from the stimulus of his new neighbors' example; also, this is not a complete deviation, since even in Java itself peasants produce export crops whenever they have sufficient land. It is the increasing population pressure that has forced more and more concentration on the production of food crops for local consumption.

The effect of extensive agricultural colonization on the communities in the Outer Islands is much more immediate, noticeable, and permanent than on the communities in Java. In the latter the gaps caused by emigration are quickly filled up, whereas in the Outer Islands colonization brings radical changes. Land that in the past was either entirely unutilized or was used only occasionally by a hunter, gatherer, or shifting cultivator, is fundamentally transformed. The forest disappears, and in its place, as far as the eye can see, stretch irrigated rice fields, interspersed here and there with villages set among gardens and groves of fruit trees (Figs. 140–143). The colonies are strange isles, microcosms of Java, standing in the midst of a land whose peoples still practice methods of cultivation unknown on the plains of Java for generations.

Colonization has another important effect on the Outer Islands, or rather, on their population. Under the stimulus of the Javanese example, indigenous peoples who heretofore have practised shifting cultivation change over to permanent cultivation. In an increasing number of instances the local communities have been willing to allow the founding of Javanese settlements on their land only on the condition that they themselves receive a share in the use of the irrigation systems to be constructed, so that they may cultivate sawahs instead of ladangs. Such an intensification of land use in the Outer Islands is imperative if immigration of Javanese is to continue for many decades.

Chapter VIII

CONSIDERATIONS FOR THE FUTURE

Most of the outstanding differences between agricultural colonization in the Philippines and that in the Netherlands Indies derive from the fact that the Philippines is a young nation working out its own destiny whereas the Netherlands Indies have been a colony whose administrators were anxious to maintain law and order and at the same time improve the condition of the peasantry.[1]

In the Philippines colonization was a national adventure. It was one of the first large-scale enterprises undertaken by the Philippine Commonwealth. American advisers had assisted in drawing up the initial plans, and in Koronadal there were two American settlers who were on an equal footing with Filipinos; the execution of the project, however, was entirely in the hands of an all-Filipino staff. All of the participants had a national, as well as a personal, stake in the success of the National Land Settlement Administration. Everyone shared the glory as well as the responsibility. The leading Manila newspapers and magazines carried pictures of the settlers and of the prize-winning houses and gardens. Colonization problems and their relation to the future of the Commonwealth were discussed over the radio. All the participants shared also the discomforts and risks.

Javanese colonization in *Tanah Sabrang*, on the other hand, was an experiment within the framework of colonial administration; it was not a people's movement. It was not in any sense a step in self-government for the Indonesian people. The Dutch realized that the *Pax Nederlandia* had removed what Malthus called the "positive checks" and this had resulted in the tremendous population increase in Java; they therefore felt an obligation to remedy the consequent drop in the standard of living. Agricultural resettlement was a project of the Dutch for the Javanese. Although Indonesians participated in the work of the colonization commission the ultimate responsibility was entirely in the hands of Netherlanders.

In the Philippines defense considerations played an important role in the choice of the settlement sites and of General Santos as manager of the National Land Settlement Administration. Young army reservists were given preference over other applicants. The settlement of large numbers of young men with military training in such strategic locations as Cotabato and northern Davao was planned in order to neutralize the Japanese colony in Davao, which consisted largely of young men who were reservists of the Japanese Army or Navy. In the Netherlands Indies defense considerations did not influence the planning of agricultural settlements, for, although

1 The colonial character of the Netherlands Indies was not altered by the constitutional change of 1922, which made the Indies an integral part of the Kingdom of the Netherlands. Despite the new position of the Indies there was, for example, until early in 1945 a Minister of the Colonies.

the colonists were subjects, they were not citizens of the Kingdom of the Netherlands and therefore were not expected to defend the Indies in case of attack.

For many years the school youth of the Philippines, especially those in agricultural schools, learned about Mindanao and its opportunities and were urged to contribute to the national effort to make this island an integral part of the country. This fact may explain the large number of settlers with more than average schooling. Also, the Filipino press kept Mindanao before the eyes of its readers as a treasure awaiting development but endangered by sinister plans of American imperialists—in the days before the establishment of the Commonwealth—and later by the threat of Japanese expansion. The Outer Islands, on the other hand, traditionally had a bad reputation among the self-respecting villagers in Java. For decades migration to the Outer Islands was tantamount to an admission of being a failure, a misfit, a person without standing. Although the reputation of the Outer Islands was improving as a result of the work of the colonization commission, the stigma had not completely disappeared. Also, no patriotic consideration figured in the decision of the Javanese peasant to move to Tanah Sabrang. Among Javanese villagers the term for government is "Company." The memory of the injustices suffered by their ancestors under the Dutch East India Company and the memory of the hardship of the "forced culture system" of the past century are still alive and account for much of the distrust of the Javanese whenever he knows that the government is sponsoring something. He identifies the government and government officials with fads of cultivation, sanitation, or other aspects of his life, with enterprises which in his eyes are too often either foolish, unreasonable, or just a plain nuisance. One day a high Dutch official, out hiking in mufti through the Javanese countryside, fell into conversation with a *tani* working his ricefield. When the Netherlander asked the Javanese how he liked the local Dutch administrators, the peasant good-humoredly voiced his irritation at their constant interference: "One week they come and tell us to hoe with our backsides toward the sun and the next week they tell us we should be hoeing with our backsides away from the sun." The Javanese tani's wariness of all governmental suggestions was a great obstacle to migration.

In order to overcome the distrust of the villagers, the colonization commission of the Netherlands Indies had to send colonists back to their home villages as propagandists. In the Philippines this kind of propaganda was unnecessary because of the different relationship between the government and the potential settlers. The Filipino knew that it was his government that was undertaking colonization work, and he was anxious to apply for a share in this work and its benefits. I was in the office of the chief of the N.L.S.A. in Manila when two spokesmen of a tenant organization in the distant province of Tarlac inquired in person about the possibilities of settlement in Mindanao. The chief told them about a settlement project to be opened in northern Luzon. But the two expressed their preference for Mindanao because they had read in the press that Koronadal was free from typhoons. Such a scene would have been impossible in the Indies, where a humble Javanese villager would not have access to a policy-making European official in Batavia. Nor would the Javanese tani be accustomed to getting information through the press. Moreover, he would be sent to a place designated by the government and would certainly not be allowed to make his own choice except under very unusual circumstances.

Therefore it is understandable that colonization found a heartier response among Filipino *taos* than among Javanese tanis. The colonization commission was continually faced with the problem of filling the annual quota of settlers, whereas the N.L.S.A. was more concerned with opening up new territory fast enough to take care of those who wished to migrate. Among these latter were a number of graduates of agricultural schools, as well as men who had once worked in the sugar industry of Hawaii or in the fruit orchards and vegetable fields of California but who had found it impossible, upon returning to their home provinces (especially along the Ilocos coast), to acquire land at a reasonable price. Such men readily made use of the offer of the N.L.S.A. and have shown energy and enterprise.

Whereas only migrants from seriously overpopulated areas of Java were recruited for agricultural colonies in the Outer Islands, every province of the Philippines had the right to be represented in the settlement projects of the N.L.S.A. The Dutch method was of course more efficient in relieving pressure in localities where this relief was most needed. The democratic way in the Philippines sacrificed some efficiency for the broader objective of giving citizens from all parts of the Commonwealth the right to share in a national undertaking; at the same time it achieved the goal of breaking down provincialism and promoting national solidarity by bringing together as many divergent groups as possible.

The average sizes of the individual farm holdings in the settlement projects of the two countries were strikingly different. In order to accomodate the largest possible number of families, the colonization commission allowed each settler only a subsistence farm of one hectare, whereas it was the aim of the N.L.S.A. to create medium-sized farms of 6, 8, or 12 hectares on which the owners would grow not only food crops needed by their households but also commercial crops in demand on the domestic market and heretofore imported. The larger size of the farms in the Philippines justified the granting of credits up to 1000 pesos per settler. The N.L.S.A. felt confident that the settlers would be able to meet the payments of both interest and amortization out of the sale of their cash crops. Since the settlers were obliged to sell their crops through the N.L.S.A., the latter had firm control over them and could therefore enforce regular payments. Javanese subsistence farming on a small farm plot, on the other hand, did not justify the granting of large credits, as the peasant would not be able to pay off the loan. It was, indeed, hard enough for him to repay the 25 guilders for his transportation and equipment.

A study of the work of the N.L.S.A. and of the Netherlands Indies Government brings out certain prerequisites for successful agricultural colonization. All settlement projects should be carefully studied in advance by specialists. It is most important to give an adequate amount of time and effort to surveys covering land rights and land requirements of the indigenous population, topography, soil, climate, vegetation, drainage, irrigation possibilities and needs, suitable crops, proper cultivation methods, and necessary soil-conservation measures. In the Netherlands Indies the preparatory studies and surveys for each project have in recent times required approximately two years.

The settlement area should be accessible: if necessary it should be opened by roads. The land should be subdivided and the sites of the future villages and fields should be decided upon *before* the arrival of the settlers; otherwise too much valuable time is lost and the period during which the settler must receive support is prolonged.

Great care must be taken to protect the health of the settler at the outset. Malaria, dysentery, and hookworm are especially frequent in tropical pioneer settlements. The last two are most effectively combated by constructing wells and installing pumps for the supply of safe drinking water, by constructing latrines, and encouraging the wearing of shoes. The fight against malaria, one of the greatest scourges of the tropics, is more difficult. Clearing the forest often creates the very conditions under which certain dangerous species of anopheles thrive and causes the outbreak of malarial epidemics. Nothing demoralizes a colonization project more effectively than a high incidence of malaria. Government help is absolutely necessary and the settlers should be supplied with medical care, medicines, and hospital facilities. Wherever conditions permit, the fight by all known means should be directed against the carrier of the disease. In addition, individuals who have contracted malaria must be properly treated. Experience in the Netherlands Indies has shown that the frequency of malarial infection will decline when a colony has passed the pioneer stage and that then the public health personnel may be reduced.

Particular attention should be directed to the selection of settlers. Although the standard of living is low throughout the congested rural regions of Southeastern Asia and although nearly everything in a Philippine *barrio* or a Javanese *desa* can be duplicated in a short time in a pioneer community, yet not every Filipino or Javanese villager has the stamina needed by a settler. Young couples with farming experience make the best pioneers. They must be healthy and energetic.

The amount of financial support given to individual families will depend upon such factors as the size of farm holdings, the amount of cash crops to be grown in addition to subsistence food crops, marketing possibilites, and the education and experience of the settler. A graduate of an agricultural school in Luzon who comes from a farming family is of course better prepared to work with credit than is an illiterate Javanese tani.

Wherever possible the organizers of colonization projects should encourage the development of cooperatives. These would handle credit requirements, the marketing of surplus crops, and the purchase of seeds, fertilizers, and livestock for breeding purposes. Individual settlers usually have not the funds to buy good breeding animals. Moreover, if these are purchased by a cooperative all members can share in the opportunity of improving their stock. Credit and marketing cooperatives of the type planned in the Philippines would be effective means of protecting the peasants against exploitation by usurers and middlemen.

After the present war both the Indies and the Philippines will have to resume the task of agricultural colonization unless the war and epidemics should so reduce the population that pressure in the congested regions will have been eliminated. When and to what extent the work will be carried on will depend upon financial conditions and upon the priority given to the various reconstruction and rehabilitation tasks. The transfer of large numbers of landless families that have barely been able to make a miserable living, together with the resultant development of untapped material resources, should become an important part of the rehabilitation program in both countries. A long war will create new competing producers of the agricultural export crops of Southeastern Asia and may affect untold numbers of families which in prewar years obtained their livelihood by performing wage labor in agricultural export industries. Many of these laborers will probably have to become subsistence farmers and will therefore have to be settled on suitable land. Fortunately both these countries—as well as Indochina, Thailand, Malaya, and Burma—have considerable possibilities for further agricultural colonization. The settlement regions of the Indies are spread over the Outer Islands from Sumatra in the west to New Guinea in the east. Sumatra, Borneo, and Celebes, because of their proximity to Java, are far more important than New Guinea. The potential settlement area would be greatly increased if non-irrigable, as well as irrigable lands, could be drawn into the colonization scheme. In Java and Madura the *tegalan* area is greater than the *sawah* area. If Javanese and Madurese settlers were to be supplied with work animals and plows they could cultivate tegalans year after year in the Outer Islands, provided that they practiced crop rotation and utilized animal manure. The experimental settlement of Madurese at Madoeredjo in Borneo should supply the necessary experience on which to base large-scale settlement of Madurese and Javanese accustomed to tegalan culture.

In the Philippines the potential pioneer land is located for the most part in the provinces of Cotabato, Davao, Agusan, and Bukidnon in Mindanao and to a lesser degree in southern Negros, Mindoro, Palawan, and the eastern sections of Luzon. In my opinion, it would be desirable to concentrate chiefly on Mindanao after the war, since this island has the greatest possibilities of settlement expansion, and also because it lies outside the typhoon belt. The other potential settlement areas have long been receiving a small flow of settlers, migrating of their own accord from nearby communities. A concentration on Mindanao would, moreover, allow a more efficient use of scientific and managerial personnel, which could then divide its time among projects located not too distantly from each other. It would also make the shipment of supplies and the marketing of products considerably easier and would justify the operation of chartered ships between Manila and Mindanao, or even of ships owned by the N.L.S.A., if commercial companies should charge excessive freight and passenger rates. A scattering of projects over the whole Philippine archipelago has the disadvantage that heavy equipment must be transported over great distances when transferred from one colony to another.

At the end of the war in the Pacific there will be in the Philippines and other Pacific war zones large quantities of heavy equipment, such as tractors, jeeps, bulldozers, and ships and boats of various types, that the American forces are not likely to transport to the United States again and that could be turned to peaceful and constructive purposes, thus speeding the rehabilitation of the islands.

In the peace settlement after the present war the Philippine Commonwealth will insist upon a considerable reduction, if not the elimination, of the Japanese colony at Davao. The Philippine Government would be more than justified in taking over all those Japanese holdings in Davao that were developed in violation of the Philippine land laws. Such a step would make from 30,000 to 60,000 hectares of productive land available for immediate settlement.

Although the founding of new agricultural colonies on a large scale is an important part of the reconstruction and rehabilitation program both in the Indies and in the Philippines, agricultural colonization alone will not solve the economic and population problems of these two countries. Additional measures will have to be taken.

In the Philippines these will probably be agrarian reform and an industrialization program. Agrarian troubles, signalized before the war by outbreaks among the tenants of large *haciendas* in the crowded provinces of central Luzon, will not end until the large landed estates have been broken up and the land has been divided among the peasants who till it. In many instances the tenants claim that their forefathers were deprived of the land through ignorance of Spanish land laws, through trickery, and through the disregard of the Spaniards for customary land rights. After the American occupation, the Philippine Government bought a total of 410,000 acres of land from the friars in order to eliminate one annoying source of agrarian unrest. It has been reported that the Japanese have appropriated large estates and have turned this land over to the peasants in a move to gain the good will of the Filipino tao. Should this prove to be true the Philippine Government will be practically obliged to purchase the haciendas belonging to the Church, to corporations, and to private individuals. Thus to acquire the estates and then sell them to the tenants would be a most important step forward in the reconstruction of the Philippines and would remove a constant and serious threat to internal peace. It would be in line with the policy initiated by the Commonwealth in 1937 and with Article XII, Section 4 of the Constitution of the Philippines, which authorizes the National Assembly to expropriate land upon payment of just compensation in order to subdivide it into small lots and sell it to farmers at cost. This would ultimately give the peasants the total yield of their fields instead of the half or even less that they receive under the tenancy system. If at the same time the population in those crowded provinces where most of the landed estates are located were to be reduced by colonization measures, the peasants remaining there could be given more land to cultivate.

A third way of attacking the population problem of the Philippines is through the development of secondary industries. A start was made in this direction between 1935 and 1941. The leaders of the Philippine Commonwealth, foreseeing economic repercussions when political independence should become effective, set out to prepare the nation for the time when, because of the loss of tariff exemption, it could no longer sell its products to the United States.[2] Their goal was to make the country less dependent upon foreign sources of consumers' goods, especially food and clothing. It was realized that many commodities heretofore imported could be manufactured locally from raw materials which were already at hand or which could be produced in the islands. Examples of the latter are rubber and cotton, two products which the N.L.S.A. tried to develop on a large scale. On the one hand, industrialization tends to make the country self-sufficient, at least in those goods which can be produced locally without great effort. On the other hand, each new industry creates new employment opportunities, draws laborers from agriculture, and contributes thereby to the solution of the problem of rural overpopulation and unemployment. Each tenant farmer or sharecropper who becomes a full-time laborer in an industrial enterprise reduces the competition for land. Even if his employment is only seasonal his income and level of living are raised. At the same time those tenants who remain on the land have a chance of getting better terms from their landlords.

The industrialization movement in the Philippines between 1935 and 1941 was characterized by the government's direct participation in the establishment and operation of new industries. Local private capital was mainly invested in land and the production of the traditional export crops of the Philippines. On the whole these producers have exhibited a procrastinating attitude, hoping against hope that somehow an eleventh-hour miracle would preserve the old economic structure. A purchase of landed

2 For a detailed discussion of the industrialization movement in the Philippines see Shepherd: Industry in Southeast Asia, 1941, pp. 86–122.

estates by the government would force their former owners to search for new investment possibilities and a carefully planned industrialization program should attract this capital. Thus the same funds would contribute in two ways to a solution of the social, agrarian, and economic problems of the islands. The Philippine Commonwealth will not have the necessary capital for such an undertaking after the war; instead it will have to turn to the United States for the financing of its reconstruction and rehabilitation program. It is this country's moral obligation to give the coming political independence of the Philippines a firm economic basis.

In the Indies the customary systems of land tenure, together with Dutch land policy, fortunately prevented a concentration of cultivated land in the hands of a few. There is therefore no need for an agrarian reform of the type that may come in the Philippines. In some regions, however, the Indies Government granted concessions of public land for 75-year periods to individual entrepreneurs as well as to companies, without making sufficient allowance for population growth. In such instances the government may refuse renewal of the concessions after their expiration—some of those in question are expiring between 1940 and 1950—in order to make the land available for an expansion of the peasant crop area and for agricultural settlement.

Industrialization has been given a foremost place among the means of alleviating Java's population pressure. In the desire to create as many employment opportunities as possible, the government from 1934 to 1941 encouraged the growth of small-scale industries in rural districts rather than the development of large mechanized plants producing great quantities of goods with a relatively small number of laborers. Whether agricultural colonization or industrialization is the more important measure has been a subject of controversy. My own opinion is that agricultural colonization, although it cannot go on indefinitely, will nevertheless be able to take care of the population increase of Java and Madura for many decades, whereas industrialization possibilities, although important, may more quickly reach the saturation point. During the last decade prior to the war, industrialization surpassed colonization as a means of providing livelihood for needy Javanese. The development of industries in established communities that abound in unemployed labor proceeds much faster than the creation of pioneer agricultural settlements in remote forest areas. During the first years after the war industrialization will probably again be in the lead but, I believe, for only a limited period. When the development of secondary industries has reached the point where it meets the local demand, colonization may come to the forefront. For in all probability the Indies, like the Philippines, will remain primarily an agricultural country with its wealth chiefly in the soil and climate. Many of the natural resources and other requirements for the development of primary industries are lacking, and secondary industries will probably work for the domestic market only and will not be able to compete successfully with other Asiatic manufacturing countries on foreign markets. In the past the purchasing power of the bulk of the domestic population was low. This set the limits for secondary industries.[3] However, extensive agricultural colonization in the Outer Islands should have a stimulating effect upon Java's industries. The more rapidly the landless Javanese are settled in the Outer Islands under conditions that raise their level of living and make it possible for them to buy the goods they had to do without in the home island, the better for the industrialization program of Java. The future of secondary industries will depend also, however, upon the economic policy of the government after the war. These developed under protective measures, and, although they expanded especially during 1940 and 1941 because of the difficulty of obtaining goods from the customary sources, they will continue to require protection against outside competition. The price level of raw materials exported from the Indies, especially of those produced by peasants, will also be of the greatest importance. In recent years the Indies, like all countries producing raw materials primarily, have received low prices for these exports while at the same time they have had to pay comparatively high prices for imported finished goods. Sitsen points out that in 1913, for example, a peasant could obtain a can of imported salmon in exchange for two pounds of copra whereas in 1939 he had to give six pounds. Nobody can now foretell what the future holds in store for the peasant producers of the Indies and other parts of Southeastern Asia. Will there be a market for their rubber, tapioca, kapok, and copra? All planning, moreover, will have been in vain and there will be no improvement in the standard of living among the masses unless the industrial countries are willing to pay a fair price for tropical raw materials. For only if the Indies again export their agricultural and mining products will they be able to continue their struggle to relieve the difficulties of overpopulation in Java at a rate that will keep the balance between population and the means of existence.

In the discussion of agricultural colonization I limited myself to the Indies and the Philippines for two reasons. First, because these were the only countries where agricultural colonization had proceeded beyond the initial stage of experimentation and, secondly, because the war restricted my own field work to these archipelagos. But Indochina also suffers seriously from congestion and maldistribution of population. As its problems bear many similarities to those found in the Philippines and in the Indies, Indochina should be able to profit from Dutch, and in many instances also from Filipino, experience. The delta of the Red River in Tongking, as well as the coastal plains of Annam, are comparable to Java, while the

3 Hart reports that "already in 1934, however, the rapid expansion in several branches of industry was becoming alarming and there was real danger that within a few years productive capacity would exceed demand with disastrous competition and price cutting as unavoidable consequences" (Hart: Towards Economic Democracy in the Netherlands Indies, 1941, p. 96).

thinly settled areas of Laos, Cambodia, western Cochinchina, and the mountainous regions of Annam and Tongking are comparable to the Outer Islands. Moreover, the Annamese is as reluctant as the Javanese to leave his native village and settle among people who do not speak his language nor follow his customs. The few colonization experiments[4] that have been made in Indochina in recent years have brought out many of the same lessons as did the early trial period in the Indies.

> The necessity was felt for organizing this movement carefully, preferably by groups of families from the same village headed by their chiefs, who alone are capable of maintaining the cohesion necessary if the Tonkinese are to settle down to their ordinary way of living immediately after their establishment in the settlement areas.[5]

Here, too, the need for population adjustment has become so pressing that resettlement measures will have to be started on a large scale after the war.

Industrialization also has been advocated for Indochina. In contrast to the Philippines and the Indies, Indochina has most of the basic raw materials for heavy as well as for secondary industry. Mercantilistic considerations, however, deterred France from industrializing the country. It was not until 1938 that growing unemployment and population pressure finally led the government to initiate a compromise program of promoting secondary industries.

Migration across international boundaries in Southeastern Asia was, and probably will be, negligible; almost all such migrants who had before the war entered the countries encircling the South China Sea came from China or India, the principal exception being the Javanese in Malaya. Most of the latter entered the country as plantation laborers but ultimately settled on the land and engaged in rice cultivation. This development was beneficial to both countries; on the one hand the laborers and their families did not return to their overcrowded villages in Java and on the other hand they increased the rice production of Malaya. The Malays accept the Javanese readily as both are Moslem and of the same racial stock. There would be no reason to advocate a settlement of Javanese in the Philippines or Thailand or of Annamese in Malaya or Borneo, or of Filipinos in Cambodia or New Guinea so long as there was sufficient suitable agricultural land available within the boundaries of the native countries of the prospective pioneers.

In the past, Southeastern Asia did not to any great extent enjoy a free exchange of ideas and information. Identical problems were often tackled without the benefit of knowledge gained from the experience of a neighbor-

---

4 International Labour Office: Labour Conditions in Indo-China, 1938, pp. 216–241. See also Brévié: "Conseil supérieur de la colonisation," 1938, pp. 715–756; and Services du Protectorat: "Activité colonisatrice du Tonkin," 1938, pp. 757–779.

5 International Labour Office: *op. cit.*, p. 236.

ing country. However, in the post-war period close collaboration will be desirable to facilitate reconstruction and rehabilitation throughout this area. One way of bringing this about would be by a frequent exchange of officials and technical specialists who could study the problems related to those in their own country and who could also be called upon to give advice.

After the war there will be an urgent need for effective measures swiftly taken to ameliorate suffering. At the same time there will probably be, at least temporarily, a scarcity of specialists familiar with problems of tropical agriculture, land settlement, and related fields. A pooling of scientific manpower, as well as of experience, will therefore be beneficial, if not essential, to the whole of Southeastern Asia. Such a pooling could probably be best effected by an international agency, which might take the form of a broadly representative regional council.

APPENDIX A

SUPPLEMENTARY NOTES ON THE WORK OF THE NATIONAL LAND SETTLEMENT ADMINISTRATION

Colonization of Northeastern Luzon

Agricultural colonization in northeastern Luzon and in the island of Mindoro was planned to supplement the Mindanao project and to attract settlers from the Ilocos coast and the central plain of Luzon. Although colonization had not been started in Mindoro by the time of the Japanese invasion, President Quezon, upon the recommendation of the directors of the N.L.S.A., had issued proclamation No. 610, on August 29, 1940, reserving the Mallig plain in Isabela province for a settlement project. The area covered by the proclamation embraced 64,597 hectares and included the municipalities of Cabagan, Gamu, Ilagan, and Tumauini.

The Mallig plain is a long, narrow, upland plain in the western part of Isabela province. In the west it borders the foothills of the central cordillera of Luzon and in the east is separated from the Cagayan Valley proper by low ridges. It is about 12 kilometers wide by 50 kilometers long. The plain is crossed by a number of rivers such as the Siffu and the Mallig that come out of the central cordillera and flow into the Cagayan River. The N.L.S.A. planned to utilize these rivers for irrigation systems.

Although specific climatic data are lacking, it is known that the Mallig plain has a short, dry season from February to April, a season of occasional rains from May to August, and a rainy season from September into the middle of January. The plain is covered with grass, except for thin belts of forest along rivers and creeks. Silibon is the most important of the grasses, and its presence indicates the existence of a hardpan, which is very extensive and lies at a depth of from 15 to 30 centimeters. It developed after the shifting cultivators of Kalinga had deforested the land, thus opening the way for the grasses. It would be necessary to break this hardpan with subsoilers before cultivation on an extensive scale might be possible. The topsoil of Mallig is especially deficient in nitrogen, but this could be rectified with the help of legumes.

In contrast to the procedure at Koronadal, the Mallig region had already been surveyed by the Bureau of Lands and divided into homesteads of 4, 6, and 8 hectares before the settlers arrived. A large number of these homesteads had been applied for by the time that the N.L.S.A. obtained control over the region. The first settlers had come in 1931, and there had been a second wave after 1938. A check-up in 1941 showed, however, that many of the applicants for homesteads had not yet begun to develop their land, much less to live on it. A number of reasons seem to have been responsible for this situation: the lack of a road, which was finally constructed during 1940 and 1941; the scarcity of easily accessible drinking water, which made deep well-boring necessary; a lack of capital on the part of the applicants; and, last but not least, the prevalence of malaria-spreading anopheles in the streams. Public-health officials found a splenic index of more than 75 per cent in the Mallig region. The entire local population of Kalinga was infected with malaria, and many homesteaders who took up residence there died from the disease.

The N.L.S.A. had two objectives in Mallig: first, to help homesteaders who had already made a start, or who had filed applications, to establish themselves successfully, and second, to bring in settlers for those farm lots that were still free.

In January 1941 the N.L.S.A. set a small group of employees and laborers to work in this region. As a result of its experience in Koronadal the administration did not wish to admit large numbers of settlers until the necessary preparations had been made. For this reason the Mallig project had not been officially opened to settlers at the beginning of April 1941, when I visited the district. Only a small number had been accepted. These pioneers were working for the administration and used their free hours to build their future homes. An administration building, a warehouse, a machine shed, and living quarters for the employees and laborers had been erected; an administration seed farm had been started. Surveyors were laying out the townsite; others were searching for the wooden markers, indicating the boundaries of the individual farm lots to be replaced with cement markers. At the same time surveyors were looking for suitable locations for *barrios*. The drinking-water problem had not yet been solved. A truck had to haul water from a small spring about 6 miles from the administration center. The drilling of deep wells would be necessary before large groups of settlers could be admitted.

To supply Mallig with equipment and provisions was considerably more difficult than to supply Koronadal. Goods could be shipped by water to the latter place, whereas everything had to be hauled to Mallig by truck, either from Manila over a distance of 250 miles, or at least from San José, in the province of Nueva Ecija, at the head of the railroad, a distance of 140 miles. The road leading from Nueva Ecija through Nueva Vizcaya into the Cagayan Valley is mountainous and difficult.

The overseer of the Mallig project, Dr. Samonte, an Ilocano, anticipated a rapid influx of Ilocanos as soon as the district should be opened. Settlers were expected to come especially from Pangasinan, Tarlac, and Nueva Ecija, where they were working as laborers or as tenants; but migrants were also expected directly from the crowded villages on the Ilocos coast.

The Text of the Act Creating the National Land Settlement Administration (Commonwealth Act No. 441)

Be it enacted by the National Assembly of the Philippines:

SECTION 1.

There is created a corporation composed of five persons to be known as "National Land Settlement Administration." They shall act as members of the Board of Directors in charge of the management of the Corporation and shall be appointed every three years by the President of the Philippines, with the consent of the Commission on Appointments of the National Assembly. The Corporation shall enjoy the general powers mentioned in the Corporation Law, shall be subject to the provisions thereof insofar as the same are compatible with the provisions of this Act, and, for attaining the purposes of the same, may perform any act which a corporation, copartnership, or natural person is authorized to perform under the laws existing or which may be hereafter enacted. This Corporation shall have its main office in the City of Manila.

SECTION 2.

The Corporation shall serve as an agency of the Commonwealth Government for the attainment of the following objectives:

(a) To facilitate the acquisition, settlement and cultivation of lands whether acquired from the Government or from private parties;

(b) To afford opportunity to own farms to tenant farmers and small farmers from congested areas, and to trainees who have completed the prescribed military training;

(c) To encourage migration to sparsely populated regions, and facilitate the amalgamation of the people in different sections of the Philippines;

(d) To develop new money crops to take the place of the present export crops which may suffer from the loss of preferences which they enjoy in the American market.

SECTION 3.

To carry out these purposes, the Corporation shall have the following powers:

(a) To hold without limitation as to area public agricultural lands for a period not exceeding twenty-five years, renewable by the President of the Philippines for another period of not exceeding twenty-five years;

(b) To recommend to the President of the Philippines the reservation of public lands, preferably those situated along national highways, and to provide for the clearing, breaking, and cultivation of the lands so reserved for subsistence farming, or for money crops, or for both, on a cooperative basis or under such arrangements as may prove beneficial to the settlers, and for their survey into convenient lots, and to set aside such area or areas as may be deemed desirable for townsites, roads, government building sites, parks, and other public improvements. The expenses incurred in connection with the survey and subdivision of lots for allotment to settlers or for their initial cultivation shall be chargeable to the fund herein appropriated and shall be prorated among the lots of the subdivisions and charged against the corresponding settlers.

(c) To dispose of the lands so reserved, held, surveyed, or subdivided to persons qualified under the Constitution and the Public Land Act and who possess such other qualifications as may be prescribed by the Board of Directors. The disposition and final grant of the land shall be made subject to the limitations prescribed in the Constitution and the Public Land Act and to the compliance of all conditions as may be imposed by the Board of Directors. Within a period of ten years after such final grant, the land shall not, except by inheritance, be encumbered, alienated or transferred, nor shall it become liable to the satisfaction of any debt contracted prior to the expiration of said period; but the improvements or crops on the land may be mortgaged or pledged to the credit agencies created under the provisions of this Act. No settler shall be entitled to hold a lot more than twenty-four hectares, and no officer or employee of the Corporation shall be permitted to acquire, directly or indirectly, any land within the reservation, unless with the specific approval of the Board of Directors in each case. The applicants shall be recruited from all provinces in proportion to their respective population, and in case a province shall not be able to fill the quota assigned, the unfilled portion of the quota may be covered from other provinces having a greater number of applicants;

(d) To acquire from private parties those lands that are necessary to enable it to carry out the purposes for which it is created, for roads, highways, streets, and avenues, or those private lands surrounded by or adjacent to the public land acquired by the Corporation;

(e) To make contracts and enter into such arrangements or contracts as it may consider convenient and advantageous to the common interest of the settlers and the Government for the development, exploitation, and operation of any of its properties;

(f) To establish and operate credit agencies to extend credit to the settlers upon the security of rights acquired by them as settlers on the crops raised or improvements made by them. The rate of interest on such loan shall not exceed six *per centum per annum*.

(g) To establish and operate electric light and water plants, water supplies, irrigation systems, trading stores or cooperatives to engage in the buying and selling of commodities, and other services or improvements which in the judgment of the Board will promote the well-being of the settlers;

(h) To act as agent, broker, commission merchant, or representative of the settlers in the marketing of the products raised or made by such settlers;

(i) To borrow, issue bonds, or otherwise raise funds for carrying out the objects of this Corporation, whenever it is deemed necessary for the interest of the settlers, giving its property as security therefor;

(j) To engage in manufacturing, milling, lumbering, retailing, and in such business enterprises or industries as may be necessary and desirable to insure the success of the land settlement projects;

(k) To adopt such uniform rules and regulations as may be necessary to carry out the purposes of this Act;

SECTION 4.

The Board of Directors shall elect its chairman from among its members. Three members of the board shall constitute a *quorum* for the transaction of business. The members, if not Government officials or employees, shall each receive such per diems as may be determined by the Board, not to exceed thirty pesos for each day of meeting actually attended by them.

SECTION 5.

Subject to the approval of the President of the Philippines, the board shall appoint a manager and fix his compensation, which shall not exceed eighteen thousand pesos per annum. The manager shall, subject to the approval of the board, appoint such technical, clerical, and other employees as may be necessary: *Provided* that all appointments for positions with compensation of two thousand four hundred pesos per annum or more shall be subject to the approval of the President of the Philippines.

SECTION 6.

The capital stock of the Corporation shall be twenty million pesos, to be subscribed by the Government of the Commonwealth of the Philippines.

SECTION 7.

There are appropriated out of the proceeds of the coconut oil excise tax or out of any other available funds in the Philippine Treasury such sums as may be necessary to pay for such part or parts of the capital as may be required from time to time by the Board of Directors with the approval of the President of the Philippines, but the aggregate value of which shall not exceed four million pesos in any one year.

SECTION 8.

The Corporation shall submit its annual report and balance sheet to the President of the Philippines and the National Assembly as provided for in sections five hundred and seventy-four to five hundred and seventy-seven inclusive, of the Administrative Code.

SECTION 9.

This Act shall take effect upon its approval.

Approved, June 3, 1939.

THE TEXT OF THE STANDARD CONTRACT BETWEEN THE N.L.S.A. AND THE SETTLER

KNOW ALL MEN BY THESE PRESENTS:

That the NATIONAL LAND SETTLEMENT ADMINISTRATION, a Corporation created by Commonwealth Act No. 441, hereinafter called the ADMINISTRATION, through its ...... and .... of the Municipality of .... province of .... of legal age, married /single /widower, a citizen of the Philippines, hereinafter called the SETTLER, for and in consideration of the mutual covenants and benefits hereinafter specified, hereby enter into a contract of agricultural settlement subject to the following terms and conditions:

SECTION I, RIGHTS AND OBLIGATIONS OF THE SETTLER

That the Settler hereby agrees and binds himself:

1. To move to .... Settlement District in the province of ...., and become a member of the agricultural settlement established therein;
2. To plant and cultivate the land allotted to him to such plants and crops, and to raise and breed such livestock as the Administration shall prescribe;
3. To devote himself exclusively to the cultivation of the parcel of land allotted to him and not to leave the settlement or engage in any other trade or occupation without the approval of the Administration;
4. To comply with all the rules and regulations which the Administration may from time to time promulgate for the settlement, until such time as the Administration shall have withdrawn its control over the settlement;
5. Not to sell, assign, lease, encumber or otherwise transfer, except by inheritance, the land allotted to him, until after the expiration of a period of ten (10) years immediately following the final grant of title to such land. This obligation shall extend to and bind likewise the successors and assigns of the Settler;
6. To pay to the Administration in the manner prescribed in par. 3, Section III hereof, with interest thereon at a rate not to exceed four per cent (4%) per annum, all the advances made by the Administration covering:
   a. The cost of transportation and other expenses incidental thereto for the Settler, his family or dependents, his belongings, work animals and agricultural implements, from his place of residence to .... Settlement District in the province of ....;
   b. The cost of materials and skilled assistance furnished the Settler for the construction of his dwelling-house and farm structures;
   c. The cost of all assistance rendered the Settler in accordance with par. 4, Section II of this contract; and
   d. The cost of subsistence and clothing furnished the Settler and his household, unless such Settler is employed upon a daily wage basis, in which case the cost of his subsistence and clothing and that of his household shall be periodically deducted from his earnings.
7. To deposit for marketing purposes in a settlement warehouse to be established by the Administration all his agricultural and other products of a non-perishable kind, which are in excess of his needs for subsistence and planting purposes.

SECTION II, RIGHTS AND OBLIGATIONS OF THE ADMINISTRATION

The Administration agrees and binds itself:

1. To assign to the Settler a parcel of twelve (12) hectares of agricultural land within .... Settlement District in the province of ...., for his occupation and

cultivation, subject to the condition that full title to such land shall not be granted until after the Settler has fully discharged all his obligations to the Administration and placed the entire land under cultivation;

2. To furnish transportation and advance such expenses incidental to the journey as the Administration may consider necessary, for the Settler and his family or dependents, his belongings, work animals and agricultural implements, from his place of residence to .... Settlement District in the province of ....;

3. To supply the Settler with the necessary materials and furnish him with skilled assistance for the construction of his dwelling-house and such farm structures as may be necessary, on condition that all such buildings shall conform to the types and dimensions approved by the Administration;

4. To furnish the Settler with:
  a. The use of tractors, plows or other mechanical means for the preparation of the land;
  b. Manures, seedlings and other planting materials;
  c. Livestock, work animals and agricultural equipment, subject to the condition that the Administration shall retain a lien upon such livestock and work animals until the Settler shall have fully discharged his indebtedness to the Administration.

5. To furnish subsistence for the Settler and his household, until such time as he shall have produced and harvested sufficient food crops for his requirements;

6. To sell in behalf and for the benefit of the Settler all the products deposited by the latter in the settlement warehouse established by the Administration, in the most favorable markets, and credit the full proceeds of such sales to the Settler's account after deducting the cost of storage, insurance, freight and handling.

7. To grant upon the Settler full title to the land assigned to him, upon payment of all his obligations to the Administration and after completely cultivating his land.

SECTION III, GENERAL PROVISIONS

1. Within ten (10) years after the final grant of the title under par. 1, Sec. II hereof, the land shall not be liable for the satisfaction of any debt contracted by the Settler prior to the expiration of the said period of ten (10) years.

2. All agricultural tools and equipment, work animals, seeds, fertilizers, and other plant materials, building materials and supplies furnished the Settler by virtue of this contract shall, upon delivery to the Settler, be considered to be completely at his risk, and shall not be transferred, sold or otherwise disposed of by the Settler without the authorization of the Administration.

3. The payment of all the Settler's obligations to the Administration under this contract shall be made in annual amortizations beginning at the end of the third year after his arrival at the settlement and shall be completed before the end of the twentieth year, on condition that:
  a. In the fourth year and thereafter, until the Settler's account shall have been fully liquidated and closed, the Administration shall retain and credit to the Settler's account, such proportion of the proceeds of the sales of the products marketed under par. 6, Sec. II hereof, as the Administration may from time to time determine; in no case, however, to exceed 30 per cent of the total proceeds of such sales; and
  b. The balance of the proceeds of such sales or any portion thereof may be paid to the Settler in cash, unless the Settler should elect to have the same credited to his account.

4. Failure of the Settler to comply with any of the conditions of this contract shall be a sufficient cause for the forfeiture of all his rights as a Settler in the .... Settlement District.

5. If after due inquiry by the Administration the Settler should be declared incompetent to perform the work of the Settlement, or if there should exist any other good and sufficient reason, this contract shall be deemed cancelled and all the rights, improvements and interests of the Settler in and to the land allotted to him under this contract shall revert to the Administration, which shall cause such rights, improvements and interests of the Settler to be assessed by a Board of Assessors appointed by the Administration, and shall apply their value thus assessed to the payment of all his obligations to the Administration, any balance in excess of said obligations to be paid to the Settler.

6. Except as provided in the two next preceding paragraphs, this contract shall be effective between the parties for a period of twenty (20) years unless sooner terminated by the early compliance of the Settler with the conditions and stipulations covenanted herein.

IN WITNESS WHEREOF, the parties have hereunto set their hands this .... day of ...., 194...., at ...., ...

NATIONAL LAND SETTLEMENT ADMINISTRATION
By:

....
Manager Settler

IN THE PRESENCE OF:

....

ACKNOWLEDGEMENT

UNITED STATES OF AMERICA
COMMONWEALTH OF THE PHILIPPINES } ss.
PROVINCE OF .... }

Personally appeared before me this .... day of ...., 194...., ...
....
.... known to me and to me known to be the same persons who executed the foregoing instrument, and who declared that the same is their voluntary act and deed.

This certifies that the foregoing instrument is a contract of agricultural settlement of a parcel of land and consists of three (3) pages including this one, the first of which has been signed on the left margin by the parties and the witnesses, and the last page at the foot of the deed proper.

Doc. No. ....
Page No. ....
Book No. .... Notary Public
Series of 194............ Until December 31, 194....

APPENDIX B

SUPPLEMENTARY NOTES ON THE WORK OF THE COMMISSION FOR MIGRATION AND COLONIZATION

Surveys

The surveys (pp. 207,209) were primarily the task of soil scientists and irrigation engineers. Each group followed definite time tables that had been worked out in advance. For instance, the soil surveyors' time table for 1940 provided for investigations in 13 districts: 6 in Sumatra, 4 in Borneo, 2 in Celebes, and 1 in New Guinea (Table XXXIV).

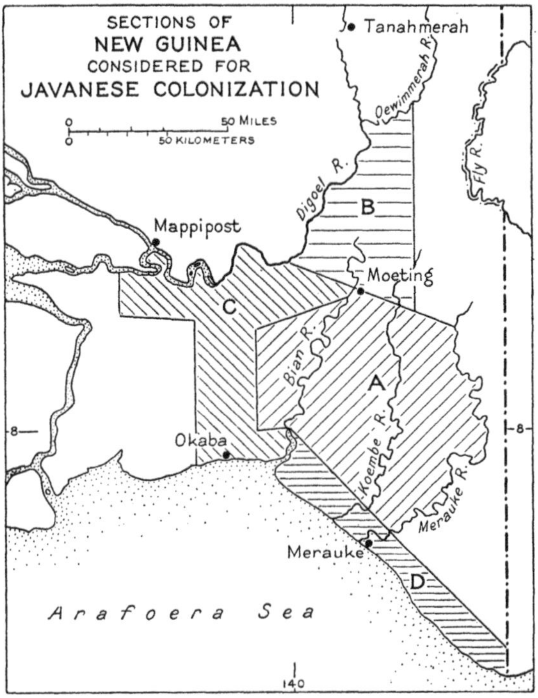

Fig. 171—Sections in the southwestern part of Netherlands New Guinea considered for Javanese colonization. The shaded areas on the map include the following: Area A (approximately 1,000,000 hectares); Area B (approximately 360,000 hectares); Area C (approximately 550,000 hectares); Area D (approximately 225,000 hectares). Based on data given in *Nederlandsch-Indische Geografische Mededeelingen,* 1941, p. 100.

In June 1941 the commission sent its soil specialists to southwestern New Guinea to carry out a reconnaissance survey in three areas and map the soils of another that had already been the subject of a reconnaissance survey in October 1940. This work was to be finished by the end of November 1941. The areas to be reconnoitered were (Map, Fig. 171): I. the land between the Bian and the Merauke rivers, comprising 1,000,000 hectares; II. the land between Moeting and the Digoel River, about 360,000 hectares; III. the land between the Digoel River and Okaba on the coast, 550,000 hectares. The area to be mapped was: IV. the coastal region from the mouth of the Bian River to the Dutch-Australian border, comprising about 225,000 hectares.[1] Tables XXXV and XXXVI show the amount of work done by the soil surveyors between 1934 and 1940. During these 7 years they approved 970,000 hectares for colonization; because of lack of data it is, however, impossible to say how much of this was also surveyed by irrigation engineers.

If we compare the amount of land approved by the Soil Institute with the expenditure we find that the soil survey alone of one hectare of approved land cost an average of 0.70 guilders in 1938 and of not quite half a guilder in 1939.[2] The irrigation engineers of the Department of Public Works could not keep up with the soil specialists during 1938 because of lack of personnel, but in 1939 they made considerable progress. At the end of 1939 they had investigated 44 out of a total of 140 proposed projects and had recommended 25 areas for further investigation; but of the latter, 5 had to be turned down by soil specialists upon their second survey, while the local population raised objections to the settling of Javanese in 3 others. That left 17 projects in which the surveys could be pursued further. Of the remaining 96 projects, 3 had already been rejected by soil specialists and the survey of 38 more had been indefinitely postponed for various reasons; so that the irrigation engineers had 55 projects on their calendar. In December 1939, the Department of Public Works estimated that all of these could be handled by the end of 1943.

Propaganda

In order to persuade a conservative and wary Javanese *tani* to leave his native island one must approach him at the psychological moment and in a convincing manner. For this purpose a few successful, well-dressed, and satisfied colonists were brought back to their home districts with samples of soils, agricultural produce, and numerous photographs of life in the colony. With such concrete evidence they often dispelled the fear and scepticism of their former fellow villagers. Peasants who would be too shy to ask questions in a large assembly were given an opportunity to talk confidentially with the visitor over a leisurely smoke or cup of coffee. The visitor was usually accompanied by his wife, who would discuss her experiences with the women of the village. After having selected a satisfied colonist as their propagandist, the officials left him entirely to his own devices. Recruiting by this means met with the greatest success when undertaken shortly before the scheduled date of departure; if it was attempted too far in advance the volunteers might change their minds. It was also important to time the recruitment during *patcheclick*, the annual period of food scarcity in Java, that is, during the last weeks before harvest. Another direct propaganda method was to send Javanese villagers on excursions to the colonies in the Outer Islands at the government's expense.

Methods of indirect propaganda included lantern-slide talks by a professional Javanese lecturer and theatrical performances. There was scarcely a railroad station, pawnshop, or public building in Java that did not display colorful posters inviting the Javanese to "Tanah Sabrang." The work of the commission was kept constantly before all Dutch as well as Indonesian officials by means of beautifully illustrated wall calendars showing photographs of typical scenes in the colonies and memorandum books full of slogans and facts about colonization. Two editions, in Malay and in Dutch, of a quarterly bulletin on colonization were also issued.

Special attention was directed towards familiarizing Javanese village youth with the idea of migration, having in mind that "he who has the young people has the future." The commission presented the pupils of village continuation schools (grades four and five) with textbooks[3] on Tanah Sabrang (Fig. 161). Each village school in central and

---

[1] *Nederlandsch-Indische Geografische Mededeelingen*, 1941, pp. 99–100.

[2] In the year 1938, 100,000 guilders were expended and 142,000 hectares were approved; 151,921 guilders were expended and 318,000 hectares were approved in 1939.

[3] Hardjawisastra: Bojong Menjang Tanah Sabrang, 1938; and also Hardjawiraga and Koesrin: Ajo Menjang Kolonisasi, 1940.

eastern Java received reproductions of three paintings depicting the embarkation of colonists, the clearing of the forest, and a market scene in Gedongtataän. Settlers were constantly urged to keep in touch with their relatives and friends by mail and were provided with photographs, attractive illustrated postcards (Fig. 158), stamps, and, if they were illiterate, with scribes, without charge; during harvest time they were urged to send envelopes with samples of rice. Every month thousands of these messages reached Javanese villages. In 1939 colonists in the Lampoeng Districts sent 28,019 postcards, 1,842 letters, 428 envelopes containing rice, and 12,246 photographs to Java, according to the records of the Commission. The many money orders sent by settlers to their relatives in Java also had propaganda value, as proof that better economic conditions prevailed in the Outer Islands than in Java. The records showed that Javanese colonists in the Lampoeng Districts sent 165,000 guilders to Java during 1935 and 115,000 guilders during the first eight months of 1936.[4]

The most powerful weapon of modern propaganda, the motion picture, played a prominent part in the program. A. Jonkers, an official of the Department of the Interior familiar with Javanese customs and ideas, supplied the script for a film, "Tanah Sabrang," which was directed by Mannus Franken. As most Javanese villagers had never seen a motion picture, the action is slow and very detailed. The picture shows all the steps in migration, from recruiting to the final well established *desa* where every settler has his own house, garden, and fertile sawah. The picture cleverly portrays a *wajang* play, the traditional Javanese entertainment. Three of the characters are the beloved old wajang figures of Semar and his two sons Petruk and Gareng, to whom the Javanese has been accustomed since childhood. A special point was made in the film of the fact that the familiar characters of the wajang go along to Sumatra, so that the Javanese might feel more at home in the new land. The other characters are modern Javanese villagers and officials who make the journey step by step from central Java to the Lampoengs.

Three trucks with projectors and sound machinery made the rounds of central and eastern Java, presenting the picture each night in a new desa. The performances were held in the open air and were attended by thousands of Javanese of all ages who came in from the surrounding desas. Many saw it several times, delighted at being able to attend a free movie. In this way the film's persuasive powers had a chance to take hold of their imagination.

Recruitment

I was told by a leading official on the staff of the commission that about 30 per cent of the 1938 and 1939 migrants were of "poor quality" and should not have been selected, because they were either too old or too weak to make good pioneers, or they were bachelors accompanied by "voyage wives," or they were not real tanis but former plantation laborers or people who had made their living by doing any kind of work they could find. Half of these turned out well, however, whereas the other half lost courage during the months of clearing and building and deserted the colonies in order to seek work as wage laborers. The main reason for the shortcomings of the selection was that the *lurahs* were asked to supply a certain number of migrants but found great difficulty in filling their quota if they adhered strictly to the rules of selection; they therefore ignored the latter in order to carry out the order. Cases of desertion not only meant a loss for the government because the deserters did not meet their financial obligations, but also slowed down the growth of the projects, for each settled family was expected to help additional families under the mutual assistance system. However, the commission did not persecute the deserters. If they were discovered on a plantation, an arrangement was made with the management for the payment of their debts, but otherwise the deserters were left alone. As long as they did not return to Java, the commission felt that one of its main functions had been fulfilled.

Migration Timetables

The season for dispatching colonists depends upon the climate of the colonization area, which in turn determines the time of the rice harvest and that for clearing and burning the forest. Colonists destined for the Lampoeng Districts and Palembang must

[4] Rookmaaker: "De Javanenkolonisatie," 1937, p. 427.

be sent during February and March; for Sumatra's West Coast in April and May; for Djambi in May; for Borneo in June and July; for Celebes in June and July or in November; those for Benkoelen in November. The commission drew up detailed time tables well in advance. For instance, Table XXXVIII, giving a part of the time table for migration to the Lampoeng Districts during February 1940, was prepared during the summer of 1939, long before recruiting had been finished. In each case departure and arrival were so timed that a whole day was available to bring the settlers from their home villages to the steamer and another from the steamer to their destination.

APPENDIX C

STATISTICAL TABLES

TABLE XXI—THE KORONADAL VALLEY PROJECT: HOME PROVINCES OF SETTLERS, AS OF JANUARY 27, 1941

HOME PROVINCE	NUMBER OF SETTLERS THEREFROM IN— Lagao	Tupi	Marbel	Polomolok	Total
Abra	2	2	—	—	4
Agusan	2	—	—	—	2
Albay	3	4	3	3	13
Antique	27	—	9	3	39
Bataan	—	2	—	—	2
Batangas	1	19	—	—	20
Bohol	57	37	11	68	173
Bulacan	11	—	2	1	14
Cagayan	1	—	1	—	2
Camarines Norte	1	—	2	—	3
Camarines Sur	—	1	3	—	4
Capiz	51	59	70	17	197
Cavite	4	4	1	1	10
Cebu	96	48	94	15	253
Cotabato	14	1	5	13	33
Davao	1	—	—	—	1
Ilocos Norte	4	—	3	—	7
Ilocos Sur	—	1	2	1	4
Iloilo	77	179	438	128	822
Laguna	9	11	3	—	23
Lanao	9	3	—	—	12
La Union	7	3	1	3	14
Leyte	61	85	72	42	260
Manila	17	21	6	3	47
Marinduque	1	1	—	—	2
Masbate	1	—	—	1	2
Misamis Occidental	5	—	2	—	7
Misamis Oriental	10	9	—	6	25
Mountain Province	2	—	1	1	4
Negros Occidental	14	15	40	—	69
Negros Oriental	14	4	1	2	21
Nueva Ecija	4	5	26	—	35
Pampanga	27	31	31	11	100
Pangasinan	19	10	68	3	100
Rizal	8	—	1	—	9
Samar	1	1	1	4	7
Sorsogon	3	—	1	4	8
Tarlac	8	23	52	1	84
Tayabas	6	3	—	15	24
Zambales	3	—	3	1	7
Zamboanga	1	—	3	—	4
TOTAL SETTLERS	582	582	956	347	2,467

Source: Unpublished records of the N.L.S.A.

TABLE XXII—JAVA AND MADURA: GROWTH OF THE INDONESIAN POPULATION, 1815–1930

CENSUS YEAR	TOTAL INDONESIAN POPULATION	AVERAGE RATE OF YEARLY INCREASE	CENSUS YEAR	TOTAL INDONESIAN POPULATION	AVERAGE RATE OF YEARLY INCREASE
1815	4,499,250	——	1890	23,609,312	2.19
1845	9,374,477	2.48	1895	25,370,545	1.45
1860	12,514,262	1.94	1900	28,386,121	2.27
1870	16,233,100	2.64	1905	29,978,558	1.10
1880	19,540,813	1.87	1920	34,428,711	0.93
1885	21,190,626	1.63	1930	40,891,093	1.79

Source: *Volkstelling 1930*, Vol. 8, 1936, Table 3. The data given by J. van Gelderen in his paper "The Numerical Evolution of Population," 1931, vary slightly from the above.

TABLE XXIII—JAVA: LAND USE AND DENSITIES OF INDONESIAN POPULATION IN SELECTED DISTRICTS, 1930

DISTRICT	AREAS (SQ. KM.) Total	Cultivated by Indonesians	Irrigated Riceland (*sawahs*)	PROPORTION CULTIVATED LAND TO SAWAH*	DENSITY PER SQ. KM. CULTIVATED LAND
West Java					
Ploembon, Cheribon Res.	103.8	92.4	57.9	1.00 : 0.627	1,192
*Percentages**	*100.0*	*89.1*	*55.8*		
Anjer, Bantam Res.	297.6	236.1	42.7	1.00 : 0.181	200
*Percentages**	*100.0*	*79.3*	*14.4*		
Middle Java					
Adiwerno, Pekalongan Res.	92.8	85.0	54.3	1.00 : 0.639	1,765
*Percentages**	*100.0*	*91.6*	*58.5*		
Wiradesa, Pekalongan Res.	94.6	80.9	49.7	1.00 : 0.614	1,135
*Percentages**	*100.0*	*85.5*	*52.5*		
East Java					
Tegalombo, Madioen Res.	415.9	319.2	51.1	1.00 : 0.160	185
*Percentages**	*100.0*	*76.8*	*12.3*		
Poenoeng, Madioen Res.	357.7	332.4	17.9	1.00 : 0.054	193
*Percentages**	*100.0*	*92.9*	*5.0*		

* Percentages and proportions computed by the author.

Source: Data for West Java from *Volkstelling 1930*, Vol. 1, Table 8; data for Middle Java from *Ibid.*, Vol. 2, Table 9; data for East Java from *Ibid.*, Vol. 3, Table 8.

TABLE XXIV—JAVA AND MADURA: PERCENTAGE DISTRIBUTION OF NUMBER OF FARMS, ACCORDING TO SIZE, BY RESIDENCIES, 1903

SIZE OF FARMS	Percentage distribution of number of farms in—								
	Bantam	Batavia	Preanger Regencies	Cheribon	Banjoemas	Pekalongan	Semarang	Kedoe	Rembang
	100.0	100.0	100.0	100.0	100.0	100.0	100.0	100.0	100.0
Less than 0.18 hectare	33.2	15.8	18.1	12.9	6.2	8.3	19.6	12.9	18.7
0.18 to 0.349 "	23.3	11.9	13.1	18.3	9.2	15.2	17.8	21.4	18.9
0.35 to 0.529 "	15.0	8.1	11.1	17.5	15.3	15.4	16.3	20.2	10.3
0.53 to 0.709 "	13.2	24.5	15.5	27.4	36.6	27.4	26.5	23.1	20.1
0.71 to 1.419 hectares	11.2	24.8	18.1	17.6	20.7	22.2	14.6	15.0	18.3
1.42 to 2.839 "	3.2	8.1	12.8	4.0	8.4	7.3	3.7	5.3	9.7
2.84 hectares or more	0.9	6.8	11.3	2.3	2.6	4.2	1.5	2.1	4.0
Less than 0.71 hectare	84.7	60.3	57.8	76.1	67.3	66.3	80.2	77.6	68.0

Source: Hasselman: Algemeen overzicht van de uitkomsten van het welvaart-onderzoek, 1914, Appendix R. This gives the actual number in each group, from which the percentages above have been computed by the author.

TABLE XXIV—JAVA AND MADURA: PERCENTAGE DISTRIBUTION OF NUMBER OF FARMS, ACCORDING TO SIZE, BY RESIDENCIES, 1903, *Continued*

SIZE OF FARMS	Percentage distribution of number of farms in—						
	Madioen	Soerabaja	Madura	Kediri	Pasoeroean	Besoeki	JAVA (Exclusive of Madura and principalities)
	100.0	100.0	100.0	100.0	100.0	100.0	100.0
Less than 0.18 hectare	16.0	21.6	19.4	9.8	10.0	9.6	15.8
0.18 to 0.349 "	18.5	14.1	17.9	22.8	13.7	14.5	17.0
0.35 to 0.529 "	11.4	8.5	13.9	13.2	13.0	12.7	14.4
0.53 to 0.709 "	22.9	20.2	22.0	24.4	31.4	27.4	23.7
0.71 to 1.419 hectares	20.6	24.1	17.7	18.3	22.1	22.8	18.2
1.42 to 2.839 "	6.9	7.3	6.5	7.7	7.5	9.6	7.0
2.84 hectares or more	3.7	4.2	2.6	3.8	2.3	3.4	3.9
Less than 0.71 hectare	68.8	64.4	73.3	70.2	68.1	64.2	70.9

Source: As given above.

TABLE XXV—JAVA AND MADURA: NUMBER OF LANDOWNERS AND OF LANDLESS MEN, IN CERTAIN RESIDENCIES, 1903

RESIDENCY	LANDOWNERS (*a*)	LANDLESS MEN (*b*)	RATIO OF *a* TO *b*
*West Java**			
Bantam	157,518	25,851	6.1 : 1.0
Batavia	48,121	23,810	2.0 : 1.0
Preanger Regencies	367,479	303,850	1.2 : 1.0
Cheribon	252,466	109,154	2.3 : 1.0
*Middle Java**			
Banjoemas	178,881	262,186	0.7 : 1.0
Pekalongan	272,925	219,021	1.2 : 1.0
Semarang	421,174	72,762	5.8 : 1.0
Kedoe	356,039	228,065	1.6 : 1.0
Rembang	227,922	76,949	3.0 : 1.0
*East Java**			
Madioen	189,157	109,575	1.7 : 1.0
Besoeki	142,409	35,695	4.0 : 1.0
Madura	317,026	81,961	3.0 : 1.0

* The residencies have been arranged approximately according to their present provincial divisions, although these were not in effect in 1903.
Source: Hasselman: *op. cit.*, Appendix S.

TABLE XXVI—JAVA AND MADURA: PAYMENTS MADE TO INDONESIANS BY THE SUGAR INDUSTRY, 1920–1938

YEAR	AREA IN SUGAR CANE (HECTARES)	PAYMENTS (GUILDERS) FOR— Wages: Total (000's omitted)	Wages: Per hectare	Deliveries*: Total (000's omitted)	Deliveries*: Per hectare	Ground rent (000's omitted)	Indemnities‡ (000's omitted)
1920	156,059	79,270	507.95	7,581	48.58	14,383†	None
1922	160,909	96,970	602.63	7,628	47.41	16,765	"
1924	172,311	92,850	538.85	6,743	39.13	19,927†	"
1926	179,703	94,560	526.20	5,376	29.92	21,064	"
1928	194,937	106,268	545.14	5,008	25.69	22,868†	"
1930	198,642	97,866	492.68	2,986	15.03	25,044	"
1931	201,159	83,978	417.47	2,207	10.97	25,026	59
1932	171,603	53,077	309.30	911	5.31	21,703	755
1933	88,635	21,866	246.70	384	4.33	11,854†	2,691
1934	38,811	9,712	250.25	189	4.86	6,526†	2,654
1935	29,453	7,270	246.84	233	7.90	3,811†	2,763
1936	35,880	7,517	209.50	244	6.81	3,183†	1,950
1937	85,710	15,298	178.49	476	5.55	5,531†	721
1938	85,309	16,499	193.41	387	4.54	5,613†	153

* Payments were made by the plantations to the Indonesians for deliveries of all kinds of building material, packing material, animal manure, lime, firewood, bamboo, etc. Such deliveries represented a source of income for the local population in the neighborhood of the plantations.

† In these years the data on ground rent are incomplete. In 1920 and 1924 a single plantation failed to report; in 1928, five plantations failed to report; in 1933, one; in 1934, two; in 1935, two; in 1936, two; in 1937, eight; and in 1938, six.

‡ Because of the depression some sugar plantations could not carry out their contracts and so had to pay indemnities.

Source: *Berichten . . . van de suikerindustrie in Nederlandsch-Indië*, No. 33, 1939, p. 110.

TABLE XXVII—JAVA AND MADURA: NUMBER OF INDONESIANS OWNING MORE THAN 17.7 HECTARES OF LAND, BY RESIDENCIES, 1905 and 1925

RESIDENCY	YEAR 1905	1925	NET CHANGE
*West Java**			
Bantam	5	157	+152
Batavia	7	376	+369
Preanger	556	1,226	+670
Cheribon	18	268	+250
*Middle Java**			
Pekalongan	212	106	−106
Semarang	95	250	+155
Banjoemas	28	207	+179
Kedoe	20	80	+ 60
Rembang	23	43	+ 20
*East Java**			
Madioen	45	78	+ 33
Kediri	25	107	+ 82
Soerabaja	104	79	− 25
Pasoeroean	38	137	+ 99
Besoeki	18	223	+205
Madura	15	50	+ 35
JAVA AND MADURA†	1,209	3,387	+2,178

*The residencies have been arranged approximately according to their present provincial divisions, although these were not in effect until 1927.

†Exclusive of the principalities.

Source: Scheltema: Deelbouw in Nederlandsch-Indië, 1931, p. 275.

TABLE XXVIII—JAVA AND MADURA: NUMBER OF INDONESIAN TENANTS COMPARED TO THE TOTAL NUMBER OF PERSONS ENGAGED IN AGRICULTURE, BY RESIDENCIES AND PRINCIPALITIES, 1905

RESIDENCY OR PRINCIPALITY	INDONESIAN TENANTS Total number	As per cent of all engaged in agriculture
*West Java**		
Bantam	27,411	10.6
Batavia	42,285	11.7
Preanger	66,794	10.0
Cheribon	4,915	1.2
*Middle Java**		
Pekalongan	17,062	4.2
Semarang	12,478	2.2
Rembang	3,492	1.2
Banjoemas	14,594	5.4
Kedoe	33,970	5.3
Jogjakarta	36,304	9.8
Soerakarta	22,773	6.9
*East Java**		
Soerabaja	4,511	0.9
Madioen	3,783	1.3
Kediri	13,195	3.8
Pasoeroean	16,665	3.7
Besoeki	8,804	4.2
Madura	12,074	2.9
JAVA & MADURA	341,110	5.3

*The residencies have been arranged approximately according to their present provincial divisions, although these were not in effect until 1927.

Source: Scheltema: *op. cit.*, p. 272.

TABLE XXIX—JAVA: NUMBERS OF PEASANTS HOLDING FIELDS AND OF PEASANT LESSORS, BY SCOPE OF LEASE, IN SELECTED DESAS, IN THE REGENCY OF TOELOENGAGOENG, KEDIRI RESIDENCY, 1939

DESA	PEASANT HOLDERS: TOTAL NUMBER	PEASANT LESSORS: Total		Leasing entire holding		Leasing more than half of holding	
		Number	Per cent of holders	Number	Per cent of holders	Number	Per cent of holders
Baleredjo	100	76	76.0	18	18.0	50	50.0
Kalangbret	108	56	51.8	30	27.8	14	12.9
Kaoeman	145	85	58.6	5	3.4	70	48.2
Batangsaren	162	113	69.8	29	17.9	22	44.4
Modjosari	111	101	91.0	16	14.4	85	76.5
Sepatan	104	102	98.1	35	33.6	11	15.7
Gondosoeli	143	141	98.6	8	5.6	56	39.1
Modjoaroem	131	131	100.0	62	47.3	23	17.5
Notoredjo	193	27	14.0	21	10.9	2	1.0
Djarakan	84	77	91.7	—	—	7	8.3
Sidomoeljo	128	117	91.4	44	34.4	25	19.5
Bendo	129	64	49.6	32	24.8	21	16.2
Poendoengan	131	131	100.0	56	42.7	55	42.0
Redjosari	141	94	66.7	47	33.3	25	17.7
Wonokromo	113	113	100.0	7	6.2	27	23.9
Boengoer	210	135	64.3	87	41.4	19	9.0
Gondanglor	102	97	95.1	94	92.2	—	—
Soekoredjo	75	34	45.3	22	29.3	4	5.3
Soekowijono	162	134	82.7	114	70.4	9	5.5
Babadan	91	68	74.7	68	74.7	—	—
Sembon	177	136	76.8	134	75.7	—	—
TOTAL	2,740*	2,032	74.2	929	33.9	575	20.9

* This total is made up of 260 *bebaus* and 2,480 *gogols*. A bebau is a village official who receives the use of a certain amount of land as remuneration.
Source: Soenario: *op. cit.*, p. 125.

TABLE XXX—JAVA AND MADURA: QUANTITIES OF MAJOR FOOD CROPS AVAILABLE PER HEAD OF THE INDONESIAN POPULATION IN SELECTED YEARS FROM 1913 TO 1940

YEARS	QUANTITIES OF FOOD CROPS AVAILABLE PER HEAD (KILOGRAMS): Rice (husked)	Maize (husked)	Cassava (fresh)	Sweet potatoes (fresh)	Peanuts (shelled)	Soybeans
1913	102	29	71	25	3.7	4.3
1916–1924*	96	38	134	34	3.7	3.7
1925–1928*	88	45	133	27	2.7	4.4
1933	86	46	125	26	2.4	4.9
1934	78	37	123	24	2.4	4.1
1935	85	42	132	30	2.3	2.9
1936–1939*†	82	41	151	27	2.4	5.4
1940	87	37	159	32	2.7	5.6

* Annual average of quantities available for consumption.

† Exceptionally good rice harvests in 1938, 1939, and 1940, rather than a fundamental improvement in the situation, are reflected in the statistics of these years. The quantities of husked rice, husked maize, fresh cassava, and fresh sweet potatoes available per head in the years from 1936 to 1939 were as follows (kilograms):

	Rice	Maize	Cassava	Sweet Potatoes
1936	83	45	146	29
1937	78	40	138	26
1938	85	39	160	26
1939	83	39	159	27

Source: Data for 1913, 1916–1924 (the average), 1933, and 1934 from Scheltema: The Food Consumption of the Native Inhabitants of Java and Madura, 1936, p. 16; the averages for 1925–1928 and 1936–1939 from Departement van Economische Zaken: Voedselproblemen, 1940, p. 21; the annual data for the years 1935 through 1939 from *Economisch Weekblad*, Vol. 9, 1940, p. 721; the annual data for 1940 from Verslag van de Commissie tot de bestudeering van staatsrechtelijke hervormingen, 1941, Part I, p. 15.

TABLE XXXI—JAVA AND MADURA: INDONESIAN INTERDEPARTMENTAL MIGRANTS BY RESIDENCY OR PRINCIPALITY OF BIRTH AND BY PLACE OF HABITATION, 1930

RESIDENCY OR PRINCIPALITY	EMIGRANTS (BORN IN RESIDENCY NAMED; ENUMERATED IN ANOTHER)				IMMIGRANTS (ENUMERATED IN RESIDENCY NAMED; BORN IN ANOTHER)			
	Total		1930 Habitation (per cent of *a*)		Total		Place of birth* (per cent of *e*)	
	Actual number	Per cent of 1930 population in residency named	Java or Madura	Outer Islands	Actual number	Per cent of 1930 population in residency named	Java or Madura	Outer Islands
	(*a*)	(*b*)	(*c*)	(*d*)	(*e*)	(*f*)	(*g*)	(*h*)
West Java								
Bantam	76,777	7.5	39.6	60.4	24,387	2.4	79.4	6.0
Batavia	83,491	3.4	56.5	43.5	273,496	11.2	91.9	5.1
Buitenzorg	131,991	6.1	87.5	12.5	87,982	4.1	94.4	2.6
Priangan	169,053	5.0	77.2	22.8	117,719	3.5	92.0	6.0
Cheribon	83,451	4.1	75.4	24.6	55,921	2.8	96.7	1.8
Middle Java								
Banjoemas	143,994	5.9	49.8	50.2	88,104	3.6	93.5	5.1
Pekalongan	107,169	4.1	69.3	30.7	33,192	1.3	94.5	2.5
Kedoe	333,951	15.9	54.6	45.3	47,827	2.3	80.1	12.1
Semarang	85,705	4.4	68.5	31.5	115,535	5.9	95.0	3.0
Djapara-Rembang	84,407	4.6	71.1	28.9	39,841	2.2	95.8	1.6
Jogjakarta	160,616	10.4	60.2	39.8	38,775	2.5	91.2	4.8
Soerakarta	135,546	5.4	68.1	31.9	66,540	2.6	84.1	1.8
East Java								
Bodjonegoro	122,637	6.2	81.6	18.4	29,648	1.5	93.6	1.3
Madioen	188,948	10.0	77.8	22.2	39,127	2.1	93.1	1.5
Kediri	281,269	11.5	70.5	29.5	150,996	6.2	97.5	0.7
Soerabaja	146,839	8.1	63.9	36.1	155,441	8.6	93.0	3.1
Malang	151,054	5.6	80.1	19.9	216,249	8.0	93.5	2.2
Besoeki	15,203	0.7	68.4	31.6	500,122	24.4	97.5	0.3
Madura	270,826	13.9	96.0	4.0	13,489	0.7	29.2	20.5
JAVA AND MADURA	2,772,927				2,094,391			

* Incomplete, as the census returns in some cases did not give the residency of birth.

Source: The data of columns *a*, *b*, *e*, *f* from *Volkstelling 1930*, Vol. 8, 1936, Table 7. The percentages given above in columns *c*, *d*, *g*, and *h* have been computed from the data given in this source.

TABLE XXXII—THE OUTER ISLANDS: DENSITY OF POPULATION, BY RESIDENCIES AND GOVERNMENTS, 1930

RESIDENCY OR GOVERNMENT	DENSITY PER SQ. KM.	RESIDENCY OR GOVERNMENT	DENSITY PER SQ. KM.
Sumatra	*17.4*	*Borneo*	*4.0*
Atjeh & Dependencies	18.1	West Borneo	5.5
East Coast of Sumatra	17.9	South and East Borneo	3.5
Tapanoeli	26.7	*Groote Oost (Great East)*	
Sumatra's West Coast	38.4	Celebes	22.4
Djambi	5.5	Celebes & Dependencies*	30.8
Palembang	12.7	Manado	12.8
Benkoelen	12.3	Bali and Lombok	175.2
Lampoeng Districts	12.6	Timor & Dependencies	26.2
Riouw & Dependencies	9.4	The Moluccas	1.8
Bangka and Billiton	16.6	OUTER ISLANDS	*10.7*

*Southern Celebes.

Source: *Volkstelling 1930*, Vol. 4, Table 9, and *Ibid.*, Vol. 5, Table 9.

TABLE XXXIII—SUMATRA: JAVA-BORN INDONESIAN INHABITANTS, BY RESIDENCIES, 1930

RESIDENCY	NUMBER OF JAVA-BORN INDONESIANS Male	Female	Total	PER CENT OF JAVA-BORN IN TOTAL INDONESIANS
Atjeh	28,220	19,948	48,168	4.9
East Coast of Sumatra	256,654	205,706	462,360	31.4
Tapanoeli	7,637	5,723	13,360	1.3
Sumatra's West Coast	22,079	12,794	34,873	1.9
Djambi	8,252	4,264	12,516	5.3
Palembang	36,995	23,471	60,466	5.7
Benkoelen	14,634	10,193	24,827	7.8
Lampoeng Districts	46,944	43,927	90,871	25.9
Riouw	7,600	5,338	12,938	5.0
Bangka and Billiton	4,025	3,166	7,191	4.7
SUMATRA	433,040	334,530	767,570	9.9

Source: *Volkstelling 1930*, Vol. 4, 1935, pp. 33, 182, 183.

TABLE XXXIV—THE OUTER ISLANDS: SCHEDULE OF SURVEYS BY THE SOIL INSTITUTE FOR THE CENTRAL COMMISSION FOR MIGRATION AND COLONIZATION OF NATIVES, 1940

AREA TO BE SURVEYED	TYPE OF SURVEY	MONTHS OF SCHEDULED WORK: January	February	March	April	May	June	July	August	September	October	November	December
Sumatra													
Atjeh: Various Districts	Reconnaissance				x	x							
Atjeh & East Coast of Sumatra: Alas Pamah	Cartographic Soil									x	x		
Sumatra's West Coast: Air Bangis	Cartographic Soil		x										
Palembang: Koengkoe	Cartographic Soil						x	x					
Lampoeng Districts: North Batanghari	Cartographic Soil				x	x							
Sragi	Cartographic Soil								x	x			
Borneo													
West Borneo: Sanggau-Ledo	Cartographic Soil					x							
Sambas Koemba	Cartographic Soil						x						
South and East Borneo: Pengaron	Cartographic Soil		x	x									
Rapak Area	Reconnaissance								x	x	x		
Groote Oost													
Manado: Paloe	Reconnaissance								x				
Doemoga	Reconnaissance									x			
The Moluccas: Legari, New Guinea	Reconnaissance		x										

Source: *Jaarverslag van de Centrale Commissie . . . 1939*, 1940, p. 19.

TABLE XXXV—THE OUTER ISLANDS: TOTAL AREAS EXAMINED AND APPROVED ANNUALLY BY THE SOIL INSTITUTE, 1934–1940

YEAR	AREAS (THOUSAND HECTARES)— Examined: Reconnoitered	Examined: Surveyed	Approved
1934		22	22
1935		40	36
1936		58	50
1937		118	79
1938	200	308	141
1939	954	615	331
1940	344	438	311
TOTAL	1,498	1,599	970

Source: *Jaarverslag van de Centrale Commissie . . . 1940*, 1941, p. 20.

TABLE XXXVI—THE OUTER ISLANDS: TOTAL AREAS EXAMINED AND APPROVED BY THE SOIL INSTITUTE, BY RESIDENCIES, DURING THE PERIOD FROM 1934 to 1939

RESIDENCY OR GOVERNMENT	AREAS (THOUSAND HECTARES)— Examined	Approved
Sumatra		
Atjeh .	None	—
East Coast of Sumatra	400	102
Tapanoeli	262	None
Sumatra's West Coast	130	19
Djambi	243	109
Palembang	140	58
Benkoelen	40	40
Lampoeng Districts	112	99
Riouw	None	—
Bangka and Billiton	None	—
Borneo		
West Borneo	72	None
South and East Borneo	230	94
Groote Oost		
Celebes & Dependencies	370	121
Bali and Lombok	None	—
Timor & Dependencies	"	—
The Moluccas	"	—
OUTER ISLANDS	1,999	642

Source: *Jaarvers ag van de Centrale Commissie . . . 1939*, 1940, p. 21.

TABLE XXXVII—THE OUTER ISLANDS: COLONIZATION PROJECTS SURVEYED BY THE IRRIGATION SERVICE, 1939

PROJECT	AREA (THOUSAND HECTARES)*— Surveyed	Tentatively Approved	IMPRESSION
Sumatra			
Atjeh: Alas Pamah	11	11	Favorable
East Coast of Sumatra:			
Loeboe	56	23	Fairly favorable
Sosa	15	7	Unfavorable: practically nonirrigable
Bila	25	25	Favorable as far as irrigation is concerned; rejected because of soil
Natas	22	20	Same as for Bila
Koealoe	25	25	Same as for Bila
Tapanoeli: Natal	25	20	Same as for Bila
Aek Si Oetam	11	?	Not unfavorable
Palembang: Moesi-Koengkoe	100	80	Favorable
Lampoeng Districts: Djapara	17	15	Favorable; reservoir irrigation
Borneo			
South and East Borneo:			
South Serapat	70	30	Favorable; drainage project

* All these areas are rough approximations.
Source: *Jaarverslag van de Centrale Commissie . . . 1939*, 1940, p. 23.

TABLE XXXVIII—TIMETABLE FOR DISPATCHING JAVANESE COLONISTS TO THE LAMPOENG DISTRICTS DURING FEBRUARY 1940

STEAMSHIP				SCHEDULE OF TRIPS			
Name	Maximum capacity		Minimum quota of	Departure, Tandjoengpriok		Arrival, Oosthaven	
	Families	Persons	passengers	Day	Hour	Day	Hour
*Tobelo**	193	543	350	February 11	9 p.m.	February 12	10 a.m.
Rooseboom	90	250		" 12	5 p.m.	" 13	6 a.m.
*Tobelo**	193	543	350	" 13	8 p.m.	" 14	9 a.m.
"	193	543	350	" 15	7 p.m.	" 16	8 a.m.
D. v. Twist	90	250		" 15	5 p.m.	" 16	6 a.m.
*de Weert**	300	838	600	February 16	7 p.m.	February 17	6 a.m.
*Tobelo**	193	543	350	" 17	7 p.m.	" 18	8 a.m.
"	193	543	350	" 19	7 p.m.	" 20	8 a.m.
Rooseboom	90	250		" 19	5 p.m.	" 20	6 a.m.
D. v. Twist	90	250		" 22	5 p.m.	" 23	6 a.m.
*Patras**	445	1,252	850	February 23	12 noon	February 24	6 a.m.
"	445	1,252	850	" 25	12 "	" 26	6 a.m.
Rooseboom	90	250		" 26	5 p.m.	" 27	6 a.m.
*Patras**	445	1,252	850	" 27	12 noon	" 28	6 a.m.
"	445	1,252	850	" 28	7 p.m.	" 29	6 a.m.

*Steamship especially chartered; all others listed were regular ships of Line No. 2; which did not require a minimum quota of colonist passengers.
Source: Unpublished working plan.

TABLE XXXIX—JAVA AND MADURA: INDONESIAN COLONISTS MIGRATING TO THE OUTER ISLANDS, BY RESIDENCY OR PRINCIPALITY OF ORIGIN, 1938, 1939, and 1940

RESIDENCY OF ORIGIN	COLONISTS DEPARTING IN — 1938			1939			1940		
	Families	Persons		Families	Persons		Families	Persons	
	No.	No.	%	No.	No.	%	No.	No.	%
West Java									
Cheribon	195	589	1.5	60	209	0.5	121	377	0.7
Middle Java									
Banjoemas	743	2,620	8.1	864	3,151	6.9	1,308	5,423	10.3
Pekalongan	1,015	3,702	11.5	1,008	3,714	8.2	795	3,087	5.8
Kedoe	940	3,626	11.2	1,653	6,093	13.4	2,156	8,567	16.2
Semarang	14	55	0.1	313	1,143	2.5	552	1,913	3.6
Djapara-Rembang	45	162	0.5	117	438	1.0	65	240	0.5
Jogjakarta	11,810	6,561	20.3	2,230	8,126	17.9	1,955	7,704	14.6
Soerakarta	877	3,158	9.8	933	3,545	7.8	982	3,880	7.3
East Java									
Bodjonegoro	214	731	2.3	269	889	2.0	789	2,364	4.5
Madioen	679	2,346	7.3	1,084	3,705	8.2	839	2,990	5.7
Kediri	2,132	7,281	22.6	3,443	12,799	28.2	3,179	12,209	23.1
Soerabaja	None	None	—	217	768	1.7	437	1,200	2.3
Malang	226	822	2.5	41	134	0.3	82	287	0.5
Madura	207	606	1.9	194	625	1.4	116	381	0.7
Unspecified	None	None	—	None	None	—	673	2,233	4.2
TOTALS	9,097	32,259	100.0	12,426	45,339	100.0	14,049	52,855	100.0

*Percentages computed by the author.
Source: *Jaarverslag van de Centrale Commissie . . . 1938*, 1939; *Jaarverslag van de Centrale Commissie . . . 1939*, 1940; *Jaarverslag van de Centrale Commissie . . . 1940*, 1941

TABLE XL—THE OUTER ISLANDS: INDONESIAN COLONISTS MIGRATING FROM JAVA AND MADURA, BY RESIDENCY OR GOVERNMENT OF DESTINATION, 1938, 1939, and 1940

RESIDENCY OF DESTINATION	COLONISTS ARRIVING IN — 1938 Families No.	1938 Persons No.	1938 Persons %	1939 Families No.	1939 Persons No.	1939 Persons %	1940 Families No.	1940 Persons No.	1940 Persons %
Sumatra	*7,005*	*25,452*	*78.8*	*9,334*	*34,842*	*76.8*	*11,191*	*43,003*	*81.3*
Djambi	None	None	—	None	None	—	510	1,955	3.7
Palembang	1,588	5,726	17.7	2,091	7,761	17.1	2,361	9,296	17.6
Benkoelen	None	None	—	61	240	0.5	145	579	1.1
Lampoeng Districts	5,417	19,726	61.1	7,192	26,841	59.2	8,175	31,173	58.9
Borneo	*220*	*759*	*2.4*	*300*	*1,036*	*2.3*	*339*	*1,310*	*2.5*
Groote Oost (Celebes & Dependencies)	*1,872*	*6,048*	*18.8*	*2,792*	*9,461*	*20.9*	*2,519*	*8,542*	*16.2*
TOTALS	*9,097*	*32,259*	*100.0*	*12,426*	*45,339*	*100.0*	*14,049*	*52,885*	*100.0*

Source: Data for 1938 from *Jaarverslag van de Centrale Commissie . . . 1938*, 1939; data for 1939 from *Jaarverslag van de Centrale Commissie . . . 1939*, 1940; data for 1940 from *Nederlandsch-Indische Geografische Mededeelingen*, 1941, p. 120.

TABLE XLI—SUMATRA AND JAVA: COMPARISON OF AVERAGE DAILY DIETS IN CERTAIN VILLAGES

VILLAGE	AVERAGE CONSUMPTION PER HEAD PER DAY† By weight (grams): Meat & fish	Grains	Tubers	Legumes‡	Coconut products	Other vegetables	Fruits	Calories (total)	Vitamins (international units) A	B
Sumatra (Lampoeng Districts)										
Javanese Colonial Villages	*9.8*	*378.0*	*94.7*	*28.6*	*42.0*	*169.3*	*29.7*	*1780*	*5056*	*236*
Bagelen (1905)*	13.1	345.6	117.2	39.1	70.0	84.4	33.2	1816	2345	230
Pringsewoe (1925)*	10.5	398.7	49.5	27.7	62.5	147.6	7.7	1829	3728	239
Goenoengsari (1931)*	6.9	431.3	81.0	27.7	44.5	129.5	11.0	1970	10160	266
Karangmoeljo (1935)*	13.1	382.4	47.3	33.0	35.5	264.3	66.1	1758	7040	246
Trimoerjo (1935)*	8.9	350.4	77.3	22.7	26.0	95.6	42.4	1567	3303	212
Kalingrenjeng (1939)*	6.4	358.9	196.7	21.6	14.4	125.3	17.7	1740	3672	224
Indigenous Villages	*31.3*	*496.0*	*16.4*	*11.1*	*20.4*	*17.6*	*7.8*	*1974*	*582*	*291*
Menggala	46.5	484.0	15.8	11.6	31.6	43.9	5.0	2062	419	300
Panoemangan	16.2	508.1	17.1	10.7	9.3	28.3	10.7	1886	745	283
Java (Kedoe Residency)										
Indigenous Village Koetowinangoen	4.5	322.0	50.8	32.2	35.6	57.0	53.7	1561	§	§

* The year in which the village was founded. The surveys in Sumatra were made in 1940, and that in Java was made in 1934.

† The consumption of meat and fish, grains, and tubers was further analyzed as follows:

Village	Average consumption per head per day by weight (grams) Meat	Fish	Grains Rice	Maize	Tubers Cassava Fresh	Dried	Sweet potatoes	Other tubers
Sumatra (Lampoeng Districts)								
Javanese Colonial Villages	*3.0*	*6.8*	*366.0*	*12.0*	*77.6*	*9.0*	*4.5*	*3.6*
Bagelen (1905)	5.8	7.3	345.2	0.4	102.8	0.8	6.4	7.2
Pringsewoe (1925)	4.9	5.6	397.5	1.2	40.1	0.3	5.0	4.1
Goenoengsari (1931)	1.4	5.5	428.0	3.3	77.2	0.7	2.5	0.6
Karangmoeljo (1935)	2.7	10.4	380.0	2.4	39.0	None	2.6	5.7
Trimoerjo (1935)	2.3	6.6	349.4	1.0	65.2	None	8.8	3.3
Kalingrenjeng (1939)	0.8	5.6	295.0	63.9	141.5	52.7	1.5	1.0
Indigenous Villages	*4.1*	*27.2*	*496.0*	*0.5*	*14.4*	None	*1.8*	*0.2*
Menggala	7.2	39.3	484.0	None	13.7	None	2.1	None
Panoemangan	1.0	15.2	508.0	0.1	15.2	None	1.5	0.4
Java (Kedoe Residency)								
Indigenous Village								
Koetowinangoen	2.9	1.6	322.0	None	— Not analyzed —			

‡ Including young pulses and soybeans.
§ Data unavailable.
Source: Unpublished data of the Central Commission.

TABLE XLII—NETHERLANDS INDIES: ANNUAL EXPENDITURES OF THE CENTRAL COMMISSION FOR MIGRATION AND COLONIZATION, 1938, 1939, AND 1940

ITEM	EXPENDITURES IN — 1938 Guilders	1938 Percent	1939 Guilders	1939 Percent	1940 Guilders	1940 Percent
General Expenses						
Secretariat	37,080	2.7	39,630	2.1	41,750	1.7
Propaganda	82,553	5.9	69,500	3.6	104,000	4.2
Recruitment in Java	19,715	1.4	22,050	1.2	34,250	1.4
Railroad Transportation	115,087	8.2	164,360	8.6	179,900	7.2
Steamer Transportation	114,170	8.1	174,260	9.1	180,750	7.2
Tools for Settlers			49,160	2.6	81,600	3.2
Soil Surveys	274,660	19.6	132,000	6.9	120,000	4.8
Irrigation	180,508	12.9	474,700	24.9	905,000	36.2
Medical Services	*		*		67,500*	2.7
Local Expenses						
Sumatra						
Sumatra's West Coast	None	—	None	—	8,200	0.3
Djambi	"	—	333	0.0	80,000	3.2
Palembang	177,885	12.7	183,242	9.6	146,500	5.8
Benkoelen	1,252	0.1	8,400	0.4	2,800	0.1
Lampoeng Districts	168,923	12.0	339,560	17.8	262,000	10.5
Borneo						
South and East Borneo	42,500	3.0	68,880	3.6	79,942	3.2
Groote Oost						
Celebes & Dependencies	188,436	13.4	182,970	9.6	207,000	8.3
TOTAL	1,402,769	100.0	1,909,045	100.0	2,501,192	100.0

*Each residency spends a part of its allotted money on medical service. In 1938 and 1939, all the medical expenditures were from these sources.

Source: Data for 1938 from *Jaarverslag van de Centrale Commissie . . . 1938*, 1939, p. 7; data for 1939 from *Jaarverslag van de Centrale Commissie . . . 1939*, 1940, pp. 7, 8; data for 1940 from *Jaarverslag van de Centrale Commissie . . . 1940*, 1941, p. 7.

TABLE XLIII—THE LAMPOENG DISTRICTS: THE TOTAL DEBT INCURRED ANNUALLY BY THE JAVANESE COLONISTS IN EACH OF THE YEARS FROM 1935 THROUGH 1940 AND THE YEAR-END STATUS OF THESE DEBTS IN 1940

YEAR OF LOAN	TOTAL DEBT INCURRED IN YEAR (GUILDERS)	STATUS OF THE ANNUAL DEBTS AT THE END OF 1940: Written off — Value (guilders)	Written off — Per cent of total debt*	Repaid (guilders) — Total paid to date	Repaid (guilders) — Paid 1940	Still owing (guilders)
1935	42,138.15	16,350.31	38.8	25,684.38	551.14	103.46
1936	45,529.15	17,339.16	38.1	27,865.24	2,073.02	324.75
1937	61,064.97	21,600.85	35.3	36,762.37	10,975.12	2,700.75
1938	93,441.18	19,732.46	21.1	52,335.54	33,155.16	21,373.18
1939	172,456.25	29,063.87	16.8	48,616.07	47,714.07	94,776.31
1940	137,211.24	21,690.81	15.8	828.42	828.42	114,692.01

*Percentages computed by the author.
Source: *Jaarverslag van de Centrale Commissie . . . 1940*, 1941, p. 34.

TABLE XLIV—BELITANG COLONY: SCHEDULE OF THE NUMBER OF FAMILIES TO BE ADDED TO THE COLONY ANNUALLY IN EACH YEAR FROM 1937 THROUGH 1951

YEAR	NUMBER OF INCOMING FAMILIES: Nuclear	Under the mutual assistance system*	Annual total
1937	250		250
1938	550		550
1939	1,500		1,500
1940	1,500	50	1,550
1941	1,700	160	1,860
1942		460	460
1943		770	770
1944		1,142	1,142
1945		1,234	1,234
1946		1,388	1,388
1947		1,616	1,616
1948		1,863	1,863
1949		2,140	2,140
1950		2,464	2,464
1951		1,213	1,213
TOTAL	5,500	14,500	20,000

*For a description of the "Mutual Assistance System" see pp. 202–204 of the text. In 1941 Belitang had 800 families of 3 years' residence. Each five of these families were obliged to take in one incoming family at harvest time. The colony could therefore absorb 160 families under the "mutual assistance system" in 1941.
Source: Unpublished working plan.

TABLE XLV—BELITANG COLONY: SCHEDULE OF DAILY ALLOWANCES OF RICE TO NEWLY ARRIVED COLONISTS*

RECIPIENT	ALLOWANCE OF UNHUSKED RICE (KATIS†) March to August	September	October	November	December	January	February
Man	0.90	0.70	0.60	0.50	0.30	0.20	0.20
Woman	0.70	0.50	0.45	0.40	0.25	0.20	0.20
Older Child (3-12 yrs.)	0.50	0.35	0.30	0.25	0.20	0.15	0.15
Young Child	0.30	0.25	0.20	0.15	0.15	0.10	0.10
Average Family of Four	2.40	1.80	1.55	1.30	0.90	0.65	0.65

*Rice rations were distributed every 10 days. The average family received over the year a total of 5.655 *piculs* of unhusked rice. (1 picul = 100 katis).

†One kati is equivalent to 617.613 grams.

Source: Vissers: "De landbouwkolonisatie van Javanen te Belitang," p. 10.

TABLE XLVI—BELITANG COLONY: DISTRIBUTION OF YIELDS OF TEST HARVESTS IN FIRST AND SECOND PLANTINGS ON TEGALANS, 1939

CLASSES OF YIELDS (PICULS PER BOUW)	NUMBER OF HARVESTS IN FIELDS PLANTED* — For first time	For second time	Total
Less than 15	0	0	0
15 or more but less than 20	0	2	2
20 " " " " " 25	14	3	17
25 " " " " " 30	15	8	23
30 " " " " " 35	6	3	9
35 " " " " " 40	13	6	19
40 " " " " " 45	2	1	3
45 " " " " " 50	2	1	3
50 " " " " " 55	1	0	1
55 " " " " " 60	1	0	1
60 or more	0	0	0
TOTAL	54	24	78

*In the fields planted for the first time the mean yield was 31.46 piculs per bouw; the highest yield was 58.8 piculs per bouw; and the lowest yield was 20.4 piculs per bouw. In the fields planted for the second time the mean yield was 30.76 piculs per bouw; the highest yield was 45.1 piculs per bouw; and the lowest yield was 18.5 piculs per bouw.

Source: Vissers: "De landbouwkolonisatie van Javanen te Belitang," p. 22.

TABLE XLVII—LAMPOENG DISTRICTS: COLONIZATION PLAN FOR THE DISTRIBUTION OF FAMILIES BY SUBDISTRICTS, COLONIES, AND SECTIONS, 1940

ADMINISTRATIVE DIVISIONS	NUMBER OF SETTLERS' FAMILIES — Resident in section October 31, 1939	To be added in 1940
Soekadana Subdistrict	*12,955*	*6,650*
Soekadana Colony		
Wai Raman Section	12,863	1,233
Gedongdalem Section		300
Batanghari Section		5,067
Tjandiredgo Colony	92	50
Teloekbetoeng Subdistrict		*150*
Pasisir Barat Colony		
Wates Section	235	20
Hergodadi Section		30
Wai Semah Colony		
Adimoeljo Section	2,077	20
Pedjambon Section		20
Kalianda Colony		
Hargopantjoeran Section	884	15
Pasoeroean Section		20
Sidomoeljo Section		25
Kotaâgoeng Subdistrict		*500*
Wonosobo Colony		
Wonosobo Lama Section	1,850	50
Srikaton Section		50
Sripoernomo Section		100
Marga Poegoeng Colony		
Wai Ngison Section	106	100
Tjandiretno Section		50
Oeloe Beloe Colony	192	50
Pampangan Colony	62	50
Antarberak Colony	61	50
Kotaboemi Subdistrict		*79*
Pakoeanratoe Colony		
Gedongmenang Section	218	24
Djabi-Djabi Section		35
Ketopang Colony	86	20
Menggala Subdistrict		*121*
Toeloengboho Colony	54	30
Talangtembosoe Colony	33	16
Karta Colony	53	75
TOTAL		*7,500*

Source: Unpublished working plan.

LIST OF REFERENCES CITED

ALLEN, J. S. Agrarian Tendencies in the Philippines. *Pacific Affairs*, New York, Vol. 11, 1938, pp. 52–65.

Annual Report of the Director of Lands for the Period from January 1 to November 14, 1935. Manila, 1935.

Annual Report of the Drainage and Irrigation Department of the Federated Malay States and the Straits Settlements 1937. Kuala Lumpur, 1938.

Anti-Usury Board, Department of Justice, Commonwealth of the Philippines. *See* M. VILLALOBOS.

Atlas van Tropisch Nederland. Uitgegeven door het Koninklijk Nederlandsch Aardrijkskundig Genootschap. Batavia and Amsterdam, 1938.

BALLY, W. The Employment of Green Manures in Rice Growing. *Monthly Bull. of Agricultural Science and Practice, in International Rev. of Agriculture*, Rome, Vol. 27, 1936, pp. T10–T31.

BARBER, ALVIN. British Malaya on a War Footing. *Far Eastern Survey*, New York, Vol. 9, 1940, pp. 1–5.

BECKING, J. H. De djaticultuur op Java: Een vergelijkend onderzoek naar de uitkomsten van verschillende verjongingsmethoden van den djati op Java. *Mededeelingen van het Proefstation voor het Boschwezen, No. 22*. Dept. van Landbouw, Nijverheid en Handel in Nederlandsch-Indië, Wageningen and Batavia. 1928.

BENITEZ, CONRADO. History of the Philippines, Economic, Social, Political. Boston, New York, etc., 1940.

Berichten aan de leden der centrale organisaties van de suikerindustrie in Nederlandsch-Indië, No. 33.

BERKHOUT, A. H. Terrassenaanleg in verband met de ontginningsordonnantie (art. 5 van Staatsblad 1874, No. 79). *Tijdschrift voor Nijverheid en Landbouw in Nederlandsch-Indië*, Batavia, Vol. 43, 1892, pp. 109–141.

BEUKERING, J. A. VAN. Het Ladangvraagstuk. Unpublished ms. dated Batavia, 1940.

BEUSECHEM, D. VAN. De sawahbouw in de moerasstreken van Zuider- en Oosterafdeeling van Borneo. *Kolonisatie Bull.*, Batavia, No. 7, October 1939, pp. 2–7.

BEYER, H. O. Philippine Archaeology and Its Relation to the Origin of the Pacific Islands Population. *Proc. Sixth Pacific Sci. Congr. Pacific Sci. Assn. 1939*, Berkeley and Los Angeles, Vol. 4, 1941, pp. 157–164.

BISHOP, C. W. Origin and Early Diffusion of the Traction-Plough. *Antiquity*, Gloucester, Vol. 10, 1936, pp. 261–281.

BLAIR, E. H., and J. A. ROBERTSON, eds. The Philippine Islands 1493–1898: Explorations of Early Navigators, etc., as Related in Contemporaneous Books and Mss. Translated. Cleveland, 1903–1909. In 55 vols.

BLEEKER, P. Nieuwe bijdragen tot de kennis der bevolkingstatistiek van Java. Uitgegeven door het Koninklijk Instituut voor Taal-, Land- en Volkenkunde van Nederlandsch-Indië, The Hague, 1870.

BOEKE, J. H. The Structure of Netherlands Indian Economy. Institute of Pacific Relations, International Research Series. New York, 1942.

BRASS, L. J. Stone Age Agriculture in New Guinea. *Geogr. Rev.*, New York, Vol. 31, 1941, pp. 555–569.

BRÉVIÉ, JULES. Conseil supérieur de la colonisation. *Bull. Economique de l'Indochine*, Hanoi, Vol. 41, 1938, pp. 715–756.

BROEK, J. O. M. Diversity and Unity in Southeast Asia. *Geogr. Rev.*, New York, Vol. 34, 1944, pp. 175–195.

———. The Economic Development of the Netherlands Indies. Institute of Pacific Relations, Inquiry Series. New York, 1942.

———. The Economic Development of the Outer Provinces of the Netherlands Indies. *Geogr. Rev.*, New York, Vol. 30, 1940, pp. 187–200.

Brokx, W. Rapport van den Assistent-Resident voor Kolonisatie ter Oostkust van Sumatra over de Randkolonisatie. Unpublished ms. dated Medan, 1933.

Buck, P. H. (Te Rangi Hiroa). Vikings of the Sunrise. New York, [1938].

Burger, D. H. De desa Ngablak, Regentschap Pati in 1869 en 1929. *Koloniale Studiën*, Weltevreden, Vol. 17, 1933, pp. 226–240.

———. Rapport over de desa Pekalongan in 1868 en 1928. *Economische Beschrijvingen I*, Weltevreden, [1929.]

Burkill, I. H. A Dictionary of the Economic Products of the Malay Peninsula, with Contributions by William Birtwistle and Others. Published on behalf of the governments of the Straits Settlements and Federated Malay States. London, 1935. In 2 vols.

Census of the Philippines 1939. Commonwealth of the Philippines, Commission of the Census, Manila, 1940–1941. In 5 vols.

Commonwealth of the Philippines. *See* Philippines (Commonwealth)

Cook, O. F. Milpa Agriculture, A Primitive Tropical System. *Annual Report of the Smithsonian Institution 1919*, Washington, 1921, pp. 307–326.

Copeland, E. B. Rice. London and New York, 1924.

Credner, Wilhelm. Siam, Das Land der Thai: Eine Landeskunde auf Grund eigener Reisen und Forschungen. Stuttgart, 1935.

Danhof, G. N. Bijdrage tot de kennis van het alang-alangvraagstuk in de Lampongsche Districten. *Tectona*, Buitenzorg, Vol. 33, 1940, pp. 197–223.

Dawson, O. L. Philippine Agriculture, A Problem of Adjustment. *Foreign Agriculture*, Washington, Vol. 4, 1940, pp. 383–456.

Departement van Economische Zaken. Geld- en producten-huishouding, volksvoeding en -gezondheid in Koetowinangoen: Part 1, J. J. Ochse and G. J. A. Terra: Het onderzoek naar den economischen en landbouwkundigen toestand en het voedselverbruik te Koetowinangoen; Part 2, W. F. Donath: De voedingswaarde van het inheemsche menu te Koetowinangoen; Part 3, C. C. de Langen: De gezondheidstoestand van de bevolking te Koetowinangoen. Buitenzorg, 1934.

———. Voedselproblemen en overheidspolitiek op Java en Madoera. Batavia, 1940. This study also appeared in *Koloniaal Tijdschrift*, The Hague, Vol. 29, No. 7, 1940.

Deventer, C. T. van. Een Eereschuld. *De Gids*. Amsterdam, 1899, Part 3, pp. 205 ff.

Dobby, E. H. G. Settlement Patterns in Malaya. *Geogr. Rev.*, New York, Vol. 32, 1942, pp. 211–232.

Enquête betreffende het ladangvraagstuk, ingesteld door de Studie-Commissie voor het Ladangvraagstuk. *Landbouw*, Buitenzorg, Vol. 6, 1930–1931, pp. 1022–1033.

Faulkner, O. T., and J. R. Mackie. The Introduction of Mixed Farming in Northern Nigeria. *Empire Journ. of Experimental Agriculture*, Oxford, Vol. 4, 1936, pp. 89–96.

———. West African Agriculture. Cambridge, England, 1933.

Forbes, W. C. The Philippine Islands. Boston, 1928. In 2 vols.

Forde, C. D. Habitat, Economy and Society: A Geographical Introduction to Ethnology. London, 1934.

FRIEDERICY, H. J. De taak van het bestuur en de inheemsche rechtsgemeenschappen bij de intensiveering van den ladangbouw. *Tectona*, Buitenzorg, Vol. 30, 1937, pp. 488–501.

———. Ponre, bijdrage tot de kennis van adat en adatrecht van Zuid-Celebes. *Bijdragen tot de Taal-, Land- en Volkenkunde van Nederlandsch-Indië*, The Hague, Vol. 89, 1932, pp. 1–34.

GELDEREN, J. VAN. The Numerical Evolution of Population with Particular Reference to the Population of Java. *Proc. International Congr. for Studies on Population Rome 1931*, Rome, Vol. 1, 1933, pp. 275–287.

GONGGRIJP, G. Over de invloed van het westerse grootbedrijf op de inheemse samenleving in Nederlandsch-Indië. Haarlem, 1930.

GONGGRIJP, J. W. Soil Management and Density of Population in the Netherlands Indies. *Comptes Rendus du Congrès International de Géographie Amsterdam 1938*, Leiden, Vol. 2, 1938, Sec. IIIc, pp. 397–404.

GONGGRIJP, L. Het erosie-onderzoek. *Tectona*, Buitenzorg, Vol. 34, 1941, pp. 200–220.

GOUROU, PIERRE. Les paysans du delta tonkinois: étude de géographie humaine. *Publications de l'Ecole Française d'Extrême-Orient, Vol. 27*. Paris, 1936

———. L'utilisation du sol en Indochine française. *Centre d'Etudes de Politique Etrangère, Travaux des Groupes d'Etudes Publ. 14*. Prepared in cooperation with the Institute of Pacific Relations. Paris, [1940].

GRIST, D. H. An Outline of Malayan Agriculture. *Malayan Planting Manual No. 2*. Straits Settlements and Federated Malay States Department of Agriculture, Kuala Lumpur, 1936.

HAGREIS, B. J. Ladangbouw. *Landbouw*, Buitenzorg, Vol. 6, 1930–31, pp. 43–76 and 629–642.

HAHN, EDUARD. Das Alter der wirtschaftlichen Kultur der Menschheit: Ein Rückblick und ein Ausblick. *Verlags-Archiv Nr. 13*. Heidelberg, 1905.

———. Die Entstehung der Pflugkultur. Heidelberg, 1909.

———. Von der Hacke zum Pflug. *Wissenschaft und Bildung, Einzeldarstellungen aus allen Gebieten des Wissens 127*. Leipzig, 1914.

HAILEY, W. M., BARON HAILEY. An African Survey: A Study of Problems Arising in Africa South of the Sahara. Issued by the Committee of the African Research Survey under the auspices of the Royal Institute of International Affairs. London, New York, and Toronto, 1938.

HAINES, W. B. The Uses and Control of Natural Undergrowth on Rubber Estates. *Planting Manual No. 6*. Rubber Research Institute of Malaya, Kuala Lumpur, 1934.

HARDJAWIRAGA and KOESRIN. Ajo, Menjang Kolonisasi. Batavia, 1940.

HARDJAWISASTRA. Bojo Menjang Tanah Sabrang. Batavia, 1938.

HART, G. H. C. Recent Development in the Netherlands-Indies. *Geogr. Journ.*, London, Vol. 99, 1942, pp. 81–102.

———. Towards Economic Democracy in the Netherlands Indies. Prepared for the Netherlands and Netherlands East Indies Council of the Institute of Pacific Relations. New York, 1942.

———. Volkstal en bestaansmiddelen in Nederlandsch-Indië. *Tectona*, Buitenzorg, Vol. 30, 1937, pp. 571–596.

HASSELMAN, C. J. Algemeen overzicht van de uitkomsten van het welvaartonderzoek, gehouden op Java en Madoera in 1904–05. The Hague, 1914.

HAYDEN, J. R. The Philippines: A Study in National Development. New York, 1942.

HAYNES, A. S. Memorandum on the Need of a Definite Policy for the Extension of Rice Cultivation in the Federated Malay States. *Proc. Federal Council of the Federated Malay States 1933*, Paper No. 28. Kuala Lumpur, 1933.

Hearings before the Committee on the Philippines of the U.S. Senate. *57th Congr., 1st Sess., S. D. No. 331*, Part 2, 1902.

HENRY, Y. M., and MAURICE DE VISME,. Documents de démographie et riziculture en Indochine. Government General of Indochina, Hanoi, 1928.

HESKE, FRANZ. Der tropische Wald als Rohstoffquelle. *Zeitschrift für Weltforstwirtschaft*, Neudamm and Berlin, Vol. 6, 1939, pp. 413–485.

Het B. B. [Binnenlandsch Bestuur] Congres 1938. Verslag van het Congres der Vereeniging van Ambtenaren bij het Binnenlandsch Bestuur in Nederlandsch-Indië, gehouden te Batavia van 17 tot en met 20 Mei 1938. Batavia, 1938.

HEYNE, K. De nuttige planten van Nederlandsch-Indië. Buitenzorg, 1927. 2nd ed. In 3 vols.

HOLLE, K. F. Eenige gegevens omtrent den inlandschen landbouw op tegalans en het nut van terrassen. *Tijdschrift voor Nijverheid en Landbouw in Nederlandsch-Indië*, Batavia, Vol. 48, 1894, pp. 1–35.

HOLT, R. L., and P. F. RUSSELL. Malaria and Anopheles Reconnaissance in the Philippines. *Philippine Journ. of Science*, Manila, Vol. 49, No. 3, 1932, pp. 305–369.

HOWE, F. C. Philippine Homestead Settlement Plans. *National Research Council of the Philippines Bull. 17*. Manila, 1938, pp. 113–116.

Imperial Bureau of Soil Science. Tropical Soils in Relation to Tropical Crops. *Technical Communication No. 34*. Harpenden, Eng., 1936.

Indisch Verslag: Vol. 1, Tekst van het verslag van bestuur en staat van Nederlandsch-Indië (Netherlands Indian Report); Vol. 2, Statistisch jaaroverzicht van Nederlandsch-Indië (Statistical Abstract). Departement van Economische Zaken, Centraal Kantoor voor de Statistiek. Landsdrukkerij, Batavia. Annual, 1931—. This publication is a continuation of the *Statistisch jaaroverzicht van Nederlandsch-Indië*, published by the Centraal Kantoor voor de Statistiek of the Departement van Landbouw, Nijverheid en Handel, Batavia, annually, 1922–23 to 1930.

International Institute of Agriculture. Use of Leguminous Plants in Tropical Countries as Green Manure, As Cover and As Shade. Rome, 1936.

International Labour Office. Labour Conditions in Indo-China. *Studies and Reports, Ser. B, No. 26*. Geneva, 1938.

IRELAND, ALLEYNE. The Far Eastern Tropics: Studies in the Administration of Tropical Dependencies, Hong Kong, British North Borneo, etc. Boston, 1905.

Jaarverslag van de Centrale Commissie voor Migratie en Kolonisatie van Inheemschen. Batavia, Annual, 1938–1941.

Jaarverslag van den Kolonisatie Raad. Batavia, Annual, 1938–1940.

Jaarverslag van het Departement van Verkeer en Waterstaat over 1939. Batavia, 1940.

JOCHEMS, S. C. J. De begroeiing der tabakslanden in Deli en hare beteekenis voor de tabakscultuur. *Mededeelingen van het Deli Proefstation te Medan Ser. 2, No. 59*. Medan, n.d.

JONKERS, A. Monographie van het landrentedistrict Kebonongan. Unpublished ms. dated Jogjakarta, 1940.

JOOSTEN, J. H. L. Ontwikkeling en problemen van de bemoeienis van den landbouwvoorlichtingsdienst met de bodembescherming in West Java. *Landbouw*, Buitenzorg, Vol. 17, 1941, pp. 1063–1078.

KASERER, HERMANN. Beobachtungen über die Bewurzelung der Kulturpflanzen bei Reinsaat und Mischsaat. *Zeitschrift für das Landwirtschaftliche Versuchswesen in Deutsch Österreich*, Vienna, Vol. 14, 1911, pp. 1022–1030.
KEESING, F. M., and MARIE KEESING. Taming Philippine Headhunters: A Study of Government and of Cultural Change in Northern Luzon. Stanford University, [1934].
KERN, FRITZ. Die Anfänge der Weltgeschichte. Leipzig, 1933.
KING, J. M. Mixed Farming in Northern Nigeria. *Empire Journ. of Experimental Agriculture*, Oxford, Vol. 7, 1939, pp. 271–298.
KOLFF, C. H. VAN DER. Hahn's hypothese der ploegcultuur voor Indië ontwikkeld. *Koloniale Studiën*, Weltevreden, Vol. 10, 1926, Part II, pp. 884–907.
Kolonisatie Bulletin, Nos. 1–10. Uitgave van de Centrale Commissie voor Emigratie en Kolonisatie van Inheemschen. Batavia, 1938–1940.
KOOLS, J. F. Hoema's, hoemablokken en boschreserves in de Residentie Bantam. Wageningen, 1935.
KRÄMER, AUGUSTIN. Der Taro und die Nasskultur. *Petermanns Mitteilungen*, Gotha, Vol. 74, 1928, pp. 165–172.
KRUYT, A. C. De To Rongkong in Midden-Celebes. *Bijdragen tot de Taal-, Land- en Volkenkunde van Nederlandsch-Indië*, The Hague, Vol. 76, 1920, pp. 366–397.
———. De To Seko in Midden-Celebes. *Bijdragen tot de Taal-, Land- en Volkenkunde van Nederlandsch-Indië*, The Hague, Vol. 76, 1920, pp. 398–430.
Land Usage and Soil Erosion in Africa. *Journ. Royal African Soc.*, London and New York, Vol. 37, No. 146, Supplement, 1938.
LATTIMORE, OWEN. Inner Asian Frontiers of China. *Amer. Geogr. Soc. Research Ser. No. 21*. New York, 1940.
LEKKERKERKER, C. Java's overbevolking. *Tijdschrift van het Koninklijk Nederlandsch Aardrijkskundig Genootschap Amsterdam*, Leiden, Ser. 2, Vol. 54, 1937, pp. 866–885.
MAASSEN, C. C. J. De Javaansche landbouwkolonisatie in de Buitengewesten. Batavia, 1937. 2nd ed.
———. De Javaansche landbouwkolonisatie in de Buitengewesten. *In Het B. B. Congres 1938*, Batavia, 1938, pp. 21-40.
———. Javanese Agricultural Migration to the Outer Provinces. *Bull. Colonial Institute of Amsterdam*, Amsterdam, Vol. 2, 1938–1939, pp. 183–192.
MENGHIN, OSWALD. Weltgeschichte der Steinzeit. Wien, 1931.
MEULEN, W. A. VAN DER. Irrigation in the Netherlands Indies. *The Netherlands Indies, Bull. Colonial Institute*, Amsterdam, Vol. 3, Nos. 3–4, 1940, pp. 142–159.
MIJERS, W. N. Herbebosching op Madoera. *Tectona*, Buitenzorg, Vol. 34, 1941, pp. 909–942.
MILLER, H. H. Principles of Economics Applied to the Philippines. Boston and New York, [1932].
MILLS, VINCENTE. Planned Surveys. *Technical Bull. No. 8*. Dept. of Agriculture and Commerce, Manila, 1937.
MILSUM, J. N. Fruit Culture in Malaya. *Dept. of Agriculture Bull. No. 29*. Federated Malay States, Kuala Lumpur, 1919.
MITCHELL, K. L. Industrialization of the Western Pacific: An Economic Survey of the Pacific Area, Part III. Institute of Pacific Relations International Secretariat, New York, 1942.

Mohr, E. C. J. The Relation between Soil and Population Density in the Netherlands East Indies. *Comptes Rendus du Congrès International de Géographie Amsterdam 1938*, Leiden, Vol. 2, 1938, Sec. IIIc, pp. 478–493.

Nano, J. F. Kaiñgin Laws and Penalties in the Philippines. *Philippine Journ. of Forestry*, Manila, Vol. 2, 1939, pp. 87–92.

Nederlandsch-Indische Maatschappij voor Nijverheid en Landbouw. Handleiding voor de gouvernements—koffiekultuur. Batavia, 1873.

Netherlands East Indies. Gouvernementsbesluit (Government Order) No. 23. January 12, 1937.

———. Gouvernementsbesluit (Government Order) No. 30. January 29, 1937.

———. Gouvernementsbesluit (Government Order) No. 18. May 31, 1939.

Netherlands East Indies, Algemeene Secretarie. Rondschrijven (Circular) No. 645/B, March 25, 1937.

Nguyen-cong-Tieu. L'azolle cultivée comme engrais vert. *Bull. économique de l'Indochine*, Hanoi-Haiphong, Vol. 33, 1930 Sec. B, pp. 335–350. Also in *Proc. Fourth Pac. Sci. Congr. Java 1939*, Batavia-Bandoeng, Vol. 4, 1930, pp. 355–370.

Nicol, Hugh. Mixed Cropping in Primitive Agriculture. *Empire Journ. of Experimental Agriculture*, Oxford, Vol. 3, 1935, pp. 189–195.

Nieuwe kolonisatiemogelijkheden. Unsigned article in *Kolonisatie Bulletin*, Batavia, No. 2, 1938, pp. 17–19.

Nippold, Walter. Rassen- und Kulturgeschichte der Negrito-Völker Südost-Asiens. Göttingen, 1936.

Ochse, J. J. *See* Departement van Economische Zaken

Pelzer, K. J. Die Arbeiterwanderungen in Südostasien: Eine wirtschafts- und bevölkerungsgeographische Untersuchung. Hamburg, 1935.

———. Population and Land Utilization: Economic Survey of the Pacific Area, Part I. American Council, Institute of Pacific Relations, New York, 1941.

Pendleton, R. L. Glimpses of Hinterlands in the Southern Islands: Zamboanga, Jolo and Cebu. *Philippine Agriculturist*, Manila, Vol. 23, 1935, pp. 823–832.

———. Land Utilization and Agriculture of Mindanao, Philippine Islands. *Geogr. Rev.*, New York, Vol. 32, 1942, pp. 180–210.

———. Soil Erosion in the Tropics. *Journ. of Forestry*, Washington, Vol. 38, 1940, pp. 753–762.

———. Some Interrelations between Agriculture and Forestry, Particularly in Thailand. *Journ. Thailand Research Soc., Natural History Suppl.*, Bangkok, Vol. 12, 1939, pp. 33–52.

Philippine Islands. Annual Report, Fourth [of the Philippine Commission, October 1, 1902–December 22, 1903] . . . U. S. Insular Affairs Bureau, Washington, 1904. In 3 vols.

———. [Annual] Report [of the Philippine Commission] for the Year ended June 20, 1910. . . . *In* War Department Reports 1910, Vol. 4. Washington, 1911.

———. [Annual] Report of the Governor General of the Philippine Islands to the Secretary of War 1917. *In* Annual Reports, U. S. War Department, Fiscal Year ending June 30, 1918. Washington, 1918.

———. Annual Report of the Governor General of the Philippine Islands 1928. *U. S. Congr. (71st), 2nd sess., H. D. 133.* Washington, 1930.

———. Annual Report of the Governor General of the Philippine Islands 1929. *U. S. Congr. (71st), 3rd Sess., H. D. 665.* Washington, 1931.

———. Annual Report of the Governor-General of the Philippine Islands 1930. *U. S. Congr. (72nd), 1st Sess., H. D. 160.* Washington, 1932.

———. Annual Report of the Governor General of the Philippine Islands 1932. *U. S. Congr. (73rd), 2nd Sess., H. D. 233.* Washington, 1934.

———. Annual Report of the Governor General of the Philippine Islands 1935 covering the period January 1 to November 14. *U. S. Congr. (75th), 1st Sess., H. D. 100.* Washington, 1937.

———. Public Laws enacted by the Philippine Legislature, September 1, 1900 to November 14, 1935. Manila, 1903–1937. In 31 vols.

Philippine Islands, Bureau of Labor. Activities of the Bureau of Labor. *Bulletin No. 27.* Manila, 1930.

Philippines (Commonwealth). Messages of the President. Manila, 1935 [i.e. 1936] –1941. Vols. 1–5.

———. Public Laws of the Commonwealth enacted by the National Assembly during the period December 21, 1935, to —. Manila, 1939. Vols. 1–5.

Philippines (Commonwealth), Labor Dept. Fact-Finding Survey Report. Unpublished ms. dated Manila, 1936.

Philippines (Commonwealth), Rural Progress Administration. Report of the General Manager on the Progress of its Activities. Unpublished ms. dated Manila, 1940.

Philippines, University of, College of Agriculture. A Handbook of Philippine Agriculture. Manila and Los Banos [1939].

Proceedings of the Convention of Bureau of Lands Officials Held on November 12–20, 1940. Manila, 1940.

Raffles, T. S. The History of Java. London, 1817. In 2 vols.

Ramos, Jose. Discrimination against the Landowners. *Herald Mid-Week Mag.*, Manila, April 20, 1941.

Richards, A. I. Land, Labour and Diet in Northern Rhodesia: An Economic Study of the Pemba Tribe. London, New York, and Toronto, 1939.

Rivers, W. H. R. "Irrigation and the Cultivation of Taro." *In his* Psychology and Ethnology, G. E. Smith, ed. London and New York, 1926.

Robequain, Charles. L'évolution économique de l'Indochine française. *Centre d'Etudes de Politique Etrangère, Travaux des Groupes d'Etudes Publ. 13.* Prepared in cooperation with the Institute of Pacific Relations. Paris, [1939].

Rookmaaker, H. R. De Javanenkolonisatie in de Lampongsche Districten. *Koloniale Studiën*, Weltevreden, Vol. 21, 1937, pp. 411–431.

Roosevelt, Theodore. Land Problems in Puerto Rico and the Philippine Islands. *Geogr. Rev.*, New York, Vol. 24, 1934, pp. 182–204.

Rosell, D. Z. The Koronadal Valley, Cotabato. *Philippine Mag.*, Manila, Vol. 36, 1939, pp. 493–508.

Sanvictores, José. The Agrarian Problem in Buenavista. *Labor Bull.*, Manila, Vol. 2, 1939, pp. 188–196.

Sauer, C. O. Destructive Exploitation in Modern Colonial Expansion. *Comptes Rendus du Congrès International de Géographie Amsterdam 1938*, Leiden, Vol. 2, 1938, Sect. III c, pp. 494–499.

Schalkwijk, W. C. De kolonisatie met van Java afkomstige gezinnen in de Residentie Lampongsche Districten over het jaar 1920. *Koloniale Studiën*, Weltevreden, Vol. 5, 1921, Part I, pp. 219–240.

———. De kolonisatieproeven in de Lampongsche Districten. *Koloniale Studiën*, Weltevreden, Vol. 2, 1918, Part II, pp. 415–440.

SCHELTEMA, A. M. P. A. Deelbouw in Nederlandsch-Indië. Wageningen, 1931.

———. The Food Consumption of the Native Inhabitants of Java and Madura. National Council for the Netherlands Indies and the Netherlands Indies Institute of Pacific Relations, Batavia, 1936.

SCHOKKENKAMP, J. De boschexploitatie in Nederlandsch-Indië. *Nederlandsch-Indische Geografische Mededeelingen*, Batavia, Vol. 1, 1941, pp. 92–115.

SCHOPHUYS, H. J. Het stroomgebied van de Barito. Landbouwkundige kenschets en landbouwvoorlichting. Wageningen, 1936.

SCHUITEMAKER, B. Korte schets van den landbouw in het oost-moesson-rijstgebied in de Zuider- en Oosterafdeeling van Borneo. *Landbouw*, Buitenzorg, Vol. 14, 1938, pp. 740–778.

Services du Protectorat. Activité colonisatrice du Tonkin: Colonisation dans la haute et moyenne region du Tonkin. *Bull. Economique de l'Indochine*, Hanoi, Vol. 41, 1938, pp. 757–779.

SHEPHERD, JACK. Industry in Southeast Asia. Institute of Pacific Relations, International Research Series. New York, 1941.

SILAYAN, H. S., Organized Land Settlement under Government Direction. *National Research Council of the Philippines Bulletin No. 17*. Manila, 1938, pp. 107–112.

SITCHON, N. G. Land Settlement Policies of the Philippine Government. Unpublished M.A. Thesis, University of the Philippines. Manila, 1932.

SITSEN, P. H. W. The Industrial Development of the Netherlands Indies. *Institute of Pacific Relations Netherlands and Netherlands Indies Council Bulletin 2*. New York, [1943].

SMITH, W. D. Geology and Mineral Resources of the Philippine Islands. *Bur. of Sci. Publ. No. 19*. Dept. of Agriculture and Natural Resources, Manila, 1924.

SOENARIO. Onderzoek naar de practische mogelijkheden van schuldbevrijding in het sawahgebied van het oude Regentschap Toeloengagoeng. *Volkscredietwezen*, Batavia, Vol. 28, 1940, pp. 85–160.

SOLLEWIJN GELPKE, J. H. F. Naar aanleiding van Staatsblad, 1878 No. 110. Batavia, 1901.

Statistisch jaaroverzicht van Nederlandsch-Indië. See Indisch Verslag . . .

STEBBING, E. P. The Forests of India. London, 1922–1926. In 3 vols.

SUCHTELEN, B. C. C. C. M. M. VAN. Parallelen (hongi tochten—cultuurstelsel—bijzonder uitvoerrecht op bevolkingsrubber). *Koloniaal Tijdschrift*, The Hague, Vol. 27, 1938, pp. 575–584.

TERRA, G. J. A. De voeding der bevolking en de erfcultuur. *Koloniale Studiën*, Weltevreden, Vol. 16, 1932, pp. 552–593.

———. *See also* Departement van Economische Zaken

The Cassava Industry in the Philippines. Dept. of Agriculture and Commerce, Manila, 1939.

THOMPSON, VIRGINIA. Postmortem on Malaya. Issued under the auspices of the Secretariat, Institute of Pacific Relations, International Research Series. New York, 1943.

———. Thailand, The New Siam. Issued under the auspices of the Secretariat, Institute of Pacific Relations, International Research Series. New York, 1941.

———. War and Further India's Rice. *Far Eastern Survey*, New York, Vol. 10, 1941, pp. 183–188.

THURNWALD, RICHARD. Economics in Primitive Communities. Published for the International Institute of African Languages and Cultures. London and New York, 1932.

Tran-Van-Huu. Note sur la culture du riz flottant en Cochinchine. *Bull. Agricole de l'Institut Scientifique de Saigon*, Saigon, Vol. 2, 1920, pp. 46–54.

U. S. Congress (57th). Statutes at Large . . . Vol. 32, Part 2. Washington, 1903.

U. S. War Department Army Map Service. Gazetteer to Maps of Java, Madura, and Bali, Map Ser. AMS T521, Scale 1:250,000. Compiled by U. S. Board on Geographical Names. Washington, June 1944. 2nd ed.

Vandenbosch, Amry. The Dutch East Indies: Its Government, Problems and Politics. Published in cooperation with the International Secretariat of the Institute of Pacific Relations. Berkeley and Los Angeles, 1941. 2nd ed.

Velmonte, J. E. Farm Tenancy Problems of Rice Production in the Philippines. *National Research Council of the Philippines Bull. No. 17*. Manila, 1938, pp. 128–132.

Verslag nopens de overwogen plannen en maatregelen betreffende de kolonisatie van Javaansche werklieden op de cultuurondernemingen ter Oostkust van Sumatra, in verband met de voorgenomen afschaffing der zoogenaamde poenale sanctie in de Koelie-Ordonnantie. Weltevreden, 1920.

Verslag van de commissie tot de bestudeering van staatsrechtelijke hervormingen: Part I, Indie's ontwikkeling tusschen den eersten en den tweeden wereldoorlog; Part II, Indie's wenschen. [Batavia], [1941]. Reprinted New York, 1944.

Villalobos, M. Memorandum to the Secretary of Justice [Philippine Commonwealth] on Hacienda Esperanza, dated April 26, 1939. Unpublished.

Vissers, P. M. De landbouwkolonisatie van Javanen te Belitang. *Kolonisatie Bull. No. 9*. Batavia, 1940, pp. 2–28.

Volkstelling 1930 (Census of 1930 in Netherlands India). Departement van Economische Zaken. Landsdrukkerij, Batavia, 1933–1936. In 8 vols.

Vollenhoven, C. van. De Indonesiër en zijn grond. Leiden, 1932.

Vonk, H. Systematisch beheer van het ladangareaal in Palembang. *Tectona*, Buitenzorg, Vol. 30, 1937, pp. 616–634.

Vries, E. de. Landbouw en welvaart in het Regentschap Pasoereoan: Bijdrage tot de kennis van de sociale economie van Java. Vol. 1, text; Vol. 2, maps. Wageningen, 1931.

Waibel, Leo. Probleme der Landwirtschaftsgeographie. *Wirtschaftsgeographische Abhandlungen Herausgegeben von Leo Waibel Nr. 1*. Breslau, 1933.

Wallace, Robert. India in 1887. Edinburgh, 1888.

Wellan, J. W. J. Zuid-Sumatra, Economisch overzicht van de Gewesten Djambi, Palembang, de Lampoengsche Districten en Benkoelen. [Authorized by the Zuid Sumatra Instituut, Amsterdam]. Wageningen, 1932.

Wickizer, V. D. Rice and Wheat in World Agriculture and Consumption. *Wheat Studies*, Stanford University, Vol. 17, 1940, pp. 261–314.

——— and M. K. Bennett. The Rice Economy of Monsoon Asia. *Food Research Institute, Grain Economics Ser. No. 3*. Published in cooperation with the International Secretariat Institute of Pacific Relations. Stanford University, 1941.

Willis, J. C. Agriculture in the Tropics: An Elementary Treatise. Cambridge, England, 1914. 2nd ed.

Worthington, E. B. Science in Africa: A Review of Scientific Research Relating to Tropical and Southern Africa. Issued by the Committee of the African Research Survey under the auspices of the Royal Institute of International Affairs. London, New York, and Toronto, 1938.

Zimmerman, C. C. Siam, Rural Economic Survey 1930–31. Issued by the Thailand Ministry of Commerce and Communications. Bangkok, 1931.

———. Some Phases of Land Utilization in Siam. *Geogr. Rev.*, New York, Vol. 27, 1937, pp. 378–393.

Zwaal, Jacobus van der. De Javanenkolonies Gedong Tataän en Wonosobo in de Lampongsche Districten. *Koloniaal Tijdschrift*, The Hague, Vol. 25, 1936, pp. 201–216.

———. Inlandsch gemeentewezen in Zuid-Sumatra en Javanen-transmigratie. Wageningen, 1936.

———. Nieuwe Javanenkolonies in de Lampongsche Districten, *Koloniaal Tijdschrift*, The Hague, Vol. 25, 1936, pp. 331–335.

SUPPLEMENTARY LIST OF REFERENCES

NOTE—The following works, although not cited in the footnotes, have proved of especial value in this study.

Bowman, Isaiah. The Pioneer Fringe. *Amer. Geogr. Soc. Spec. Publ. No. 13*. New York, 1931.

Bowman, Isaiah, ed. Limits of Land Settlement: A Report on Present-day Possibilities. New York, 1937.

Burger, E. J. Landverhuizing bij de inheemsche bevolking in Nederlandsch-Indië als koloniaal-economisch verschijnsel. Den Helder, 1927.

Camus, J. S. Rice in the Philippines. *Philippine Agricultural Rev.*, Manila, Vol. 14, 1920, pp. 7–86.

Coster, Ch. Begrooiing, grondsoorten en erosie. *Tectona*, Buitenzorg, Vol. 34, 1941, pp. 519–547.

———. Bovengrondsche afstrooming en erosie op Java. *Tectona*, Buitenzorg, Vol. 31, 1938, pp. 457–572. Also issued as *Korte Mededeeling van het Boschbouwproefstation, No. 64*. Buitenzorg.

Danhof, G. N. Tweede bijdrage tot oplossing van het alang-alang vraagstuk in de Lampongsche Districten. *Tectona*, Buitenzorg, Vol. 34, 1941, pp. 67–85.

Dobby, E. H. G. Settlement and Land Utilization, Malacca. *Geogr. Journ.*, London, Vol. 94, 1939, pp. 466–484.

Eidmann, F. E. Die Aufgaben der Forstwirtschaft bei der Lösung bevölkerungspolitischer Probleme in tropischen Ländern unter besonderer Berücksichtigung der Verhältnisse in Niederländisch-Indien. *Zeitschrift für Weltforstwirtschaft*, Berlin, Vol. 5, 1938, pp. 527–538.

Field, F. V., and E. B. Field. Philippine Inter-island Migration. *In* Filipino Immigration to Continental United States and to Hawaii, Bruno Lasker, ed. Chicago, 1931.

Fokkinga, J. Boschreserveering en inlandsche landbouw van overjarige gewassen op Java en Madoera. *Tectona*, Buitenzorg, Vol. 27, 1934, pp. 142–189.

Fuchs, F. W. Moderne Kolonisation in Niederländisch-Indien. *Koloniale Rundschau*, Leipzig, Vol. 29, 1938, pp. 316–342.

Groeneveldt, W. Kolonisatie van Javanen in het algemeen en in de Residentie Benkoelen in het bijzonder. *Landbouw*, Buitenzorg, Vol. 12, 1936, pp. 297–346.

———. Boschvernieling in Nederlandsch-Indië speciaal in de Buitengewesten. *Koloniale Studiën*, Batavia, Vol. 20, 1936, pp. 28–66.

———. Het transmigratievraagstuk voor de Buitengewesten. *Koloniale Studiën*, Batavia, Vol. 18, 1934, pp. 553–576, and Vol. 19, 1935, pp. 68–89.

Hall, A D. The Improvement of Native Agriculture in Relation to Population and Public Health. London, 1936.

HARGIS, O. D. *et al.* Report of the Mindanao Exploration Commission. Prepared by O. D. HARGIS, STANTON YOUNGBERG, R. L. PENDLETON, H. F. SMITH, and Captain H. J. CASEY, under the auspices of the President's Advisory Committee on Political Refugees and privately published by the Refugee Economic Corporation, New York City, for the President's Advisory Committee, 1939.

HART, G. H. C. Overheid en economisch leven in Nederlandsch-Indië. *Landbouw*, Buitenzorg, Vol. 13, Nos. 7 & 8, 1937, pp. 11–48.

HEIJTING (Heyting), H. G. De eerstnoodige kolonisatie. *De Indische Gids*, Amsterdam, Vol. 54, 1932, pp. 507–518.

———. De les van wortelvast Gedongtatakan (Javanen-kolonisatie). *De Indische Gids*, Amsterdam, Vol. 60, 1938, pp. 1106–1117.

HOLLE, K. F. Maatregelen tegen verspilling van den bouwgrond, tegen roofbouw en uitputting van den boden. *Tijdschrift voor het Binnenlandsch Bestuur*, Batavia, Vol. 7, 1892, pp. 74–90.

———. Roofbouw en boschvernieling. *Tijdschrift voor het Binnenlandsch Bestuur*, Batavia, Vol. 10, 1895, pp. 87–106.

HOOIJER, D. G. Emigratie naar de Residentie Benkoelen. *Indisch Genootschap*, The Hague, Feb. 13, 1912, pp. 149–167.

JACKS, G. V., and R. O. WHYTE. Erosion and Conservation. *Herbage Publication Series Bull. No. 25*. Imperial Bureau of Pasture and Forage Crops, Aberystwyth, 1938.

KROON, A. H. J. De bemoeienis van de landbouwvoorlichting met den ladangbouw. *Landbouw*, Buitenzorg, Vol. 13, Nos. 7 and 8, 1937, pp. 379–393.

LADEJINSKY, W. I. Agriculture in British Malaya. *Foreign Agriculture*, Washington, Vol. 5, 1941, pp. 103–125.

———. Agriculture of the Netherlands Indies. *Foreign Agriculture*, Washington, Vol. 4, 1940, pp. 511–574.

———. Thailand's Agricultural Economy. *Foreign Agriculture*, Washington, Vol. 6, 1942, pp. 165–184.

LASKER, BRUNO. Peoples of Southeast Asia. Prepared under the auspices of the American Council of the Institute of Pacific Relations. New York, 1944.

LEKKERKERKER, C. Land en Volk van Java: Part I, Inleiding en algemeene beschrijving met losse bijlagen over de administratieve indeeling en de sterkte der bevolkingsgroepen. Groningen, 1938.

NOTESTEIN, F. W. Problems of Policy in Relation to Areas of Heavy Population Pressure. *Milbank Memorial Fund Quarterly*, New York, Vol. 22, 1944, pp. 424–444.

PFLUEGER, O. W. The 'Kaingin' Problem in the Philippines and a Possible Method of Control. *Journ. of Forestry*, Washington, Vol. 28, 1930, pp. 66–71.

SAMONTE, V. J. The Commonwealth Policy of Small Landholding. *Philippine Social Science Rev.*, Manila, Vol. 10, 1938, pp. 231–250.

THORENAAR, A. Regeering, volk, bosch en woeste grond. *Tectona*, Buitenzorg, Vol. 30, 1937, pp. 817–856.

VAN VALKENBURG, S. Java, A Study of Population. *Papers Michigan Acad. Science, Arts and Letters*, Ann Arbor, Vol. 14, 1931, pp. 399–415.

VILLALOBOS, M. Memorandum to the Secretary of Justice [Philippine Commonwealth] on Hacienda Bahay Pare, dated April 12, 1939. Unpublished.

———. Memorandum to the Secretary of Justice [Philippine Commonwealth] on Usury in the Province of Nueva Ecija, dated July 29, 1939. Unpublished.

VONK, H., A. EVERS, and TENGKOE ABDOEL HAMID. De lebakrijstcultuur in Palembang. *Landbouw*, Buitenzorg, Vol. 13, 1937, pp. 437–455.

WHITTLESEY, DERWENT. Fixation of Shifting Cultivation. *Economic Geography*, Worcester, Mass., Vol. 13, 1937, pp. 139–154.

———. Shifting Cultivation. *Economic Geography*, Worcester, Mass., Vol. 13, 1937, pp. 35–52.

WULFF, A. De rijstcultuur in Malakka. *Landbouw*, Buitenzorg, Vol. 15, 1939, pp. 101–135.

ZWAAL, J. VAN DER. Transmigratie van Javanen naar de Buitengewesten. *Koloniaal Tijdschrift*, The Hague, Vol. 25, 1936, pp. 410–419.

INDEX

The use of "(ref.)" *after a page number indicates that a bibliographical reference to a work by the person named occurs on that page. For itemized lists of the maps and tables see pp.* ix–xi.

www.ingramcontent.com/pod-product-compliance
Ingram Content Group UK Ltd.
Pitfield, Milton Keynes, MK11 3LW, UK
UKHW041640031026
15587UKWH00001B/401

9 780313 238536